Mechanical Engineering Series

Frederick F. Ling
Series Editor

F.P.J. Rimrott

Introductory Attitude Dynamics

With 120 Figures

Springer-Verlag
New York Berlin Heidelberg
London Paris Tokyo

F.P.J. Rimrott
Department of Mechanical Engineering, University of Toronto, Toronto, Ontario, Canada M5S 1A4

Series Editor

Frederick F. Ling
Director, Columbia Engineering Productivity Center and Professor, Department of Mechanical Engineering, Columbia University, New York, New York 10027-6699, William Howard Hart Professor Emeritus, Department of Mechanical Engineering, Aeronautical Engineering and Mechanics, Rensselaer Polytechnic Institute, Troy, New York 12180-3590, USA

Library of Congress Cataloging-in-Publication Data
Rimrott, F. P. J. (Friedrich Paul Johannes).
Introductory attitude dynamics / F.P.J. Rimrott.
p. cm.—(Mechanical engineering series)
Includes bibliographies and indexes.
ISBN 0-387-96726-5 (U.S.)
1. Astrodynamics. 2. Rotational motion. I. Title. II. Series.
TL1050.R56 1988
629.4'11—dc19 88-4945

Printed on acid-free paper.

Printed and bound by R.R. Donnelley & Sons, Harrisonburg, Virginia.
Printed in the United States of America.

9 8 7 6 5 4 3 2 1

ISBN 0-387-96726-5 Springer-Verlag New York Berlin Heidelberg
ISBN 3-540-96726-5 Springer-Verlag Berlin Heidelberg New York

Series Preface

Mechanical engineering, an engineering discipline borne of the needs of the industrial revolution, is once again asked to do its substantial share in the call for industrial renewal. The general call is urgent as we face profound issues of productivity and competitiveness that require engineering solutions, among others. The Mechanical Engineering Series is a new series, featuring graduate texts and research monographs, intended to address the need for information in contemporary areas of mechanical engineering.

The series is conceived as a comprehensive one which will cover a broad range of concentrations important to mechanical engineering graduate education and research. We are fortunate to have a distinguished roster of consulting editors, each an expert in one of the areas of concentration. The names of the consulting editors are listed on the first page of the volume. The areas of concentration are: applied mechanics; biomechanics; computational mechanics; dynamic systems and control; energetics; mechanics of materials; processing; thermal science; and tribology.

Professor Marshek, the consulting editor for dynamic systems and control, and I are pleased to present the first volume of the series: *Introductory Attitude Dynamics* by Professor Rimrott. We note that this volume comes at the fortuitous juncture in history coinciding with the establishment of humanity's permanent presence in space.

Frederick F. Ling

Preface

Satellite dynamics has a unique fascination for student and lecturer alike. It is not only a logical subject that can be explained by the few basic principles of mechanics, but it has also contributed extensively to the formulation of mechanics.

With the launching of *Sputnik I* on 4 October 1957, engineers entered the field, and the study of the dynamics of satellites took its rightful place as a subject within engineering mechanics.

Although most satellite dynamics problems are solved today by computers, one still needs to be acquainted with the fundamentals in order to formulate such problems for efficient computation, to understand the theory behind the computations, to validate computer outputs, and to arrive quickly at approximate results.

The primary purpose of the present text is to acquaint engineering students with the fundamentals of satellite gyrodynamics. The text is intended for senior undergraduate or graduate students, as well as for engineers in the various branches of the aerospace industry.

Students using the text are expected to have an adequate command of elementary dynamics, differential and integral calculus, and vector and matrix algebra. Vectors and tensors appear in matrix form, in particular for the treatment of the gyrosatellite, since the matrix formulation is not only well suited for computer programming, but it also affords a quick and intelligible assessment of the problem situation, so essential for engineering practice.

The book avoids overly elaborate formulations. Since it is written for engineers, the text assumes that astronomical data are known quantities to be used in the design of spacecraft. In this respect the treatment of the subject matter differs fundamentally from that customary in the fields of astronomy or physics, where efforts are typically directed towards finding and establishing these data.

Satellite behavior can be described and interpreted by the assumption of a point-sized satellite only up to a certain level. Beyond that, the finite size and shape of the satellite body, and gyroscopic effects resulting from the rotation of such body, must be taken into account.

This text deals with attitude dynamics and is devoted to satellites of finite size. It begins with a discussion of the inertia moment tensor, Euler's law, Euler's angle, Euler's equations, and Euler's frequencies, followed by a thorough treatment of the concept of center of gravity versus center of mass. Next follows a discussion on libration, as a disturbance to ideal motion, and as a means of stabilization of satellites. We then look at gyrodynamics proper, and the attitude of the moment-free satellite, including the gyrostat. Particular attention is paid to the attitude behavior of torquefree single and dual spinners, and the new collinearity theorems are introduced and exploited to predict attitude stability and attitude drift.

The derivation of each significant formula is followed by a discussion of a practical sample problem, in order to acquaint the reader with typical situations, results, and numerical values. For the same purpose there are numerous problems following each chapter. The types of problems chosen cover the whole range from programmed learning-type problems to engineering applications.

The most important engineering and astronomical data, the nomenclature employed, and the answers to the problems are compiled in the Appendices.

Each chapter is essentially self-contained, allowing the teacher and the reader flexibility of instruction without loss of coherence.

The conventional practice of citing detailed references to the literature has not been followed in this text because of its introductory nature. Each chapter ends with a list of publications under the heading of "Suggested Reading" for those interested in tracing work back to its sources or for further study of the subject matter.

Toronto, Canada F. P. J. Rimrott

Contents

	Series Preface	v
	Preface	vii
1	The Gyroscope	1
1.1	Angular Momentum	1
1.2	Time Derivative of Angular Momentum	4
1.3	Inertia Tensor	5
1.4	The Huygens-Steiner Parallel Axes Theorem	8
1.5	Principal Inertia Moments	9
1.6	Kinetic Energy	11
1.7	Euler Equations	13
1.8	Euler Equations for Principal Axes	14
1.9	Euler Angles	15
1.10	Coordinate Transformation	17
1.11	Coordinate Transformation of Angular Velocities	18
1.12	Integrability	20
1.13	The Magnus Shape Triangle	22
	Suggested Reading	27
	Problems	28
2	Center of Gravity	40
2.1	Rod Satellite	41
2.2	Small Rod Satellite	45
2.3	Center of Gravity for a Small Satellite, General Case	49
2.4	Kepler Force for a Small Satellite	52
2.5	Torque about Mass Center, General Case	53
2.6	Potential	54
2.7	Locked Rotation	55
2.8	Orbital Period for Locked Rotation	55
2.9	Spacecraft with Booms	59
2.10	Microgravity	63
	Suggested Reading	65
	Problems	65

3 Libration 76

3.1 Libration about Position Vector and within Orbital Plane 76
3.2 Libration about Position Vector and Normal to the Orbital Plane 78
3.3 Libration about Tangent to Orbit and Normal to the Orbital Plane 79
3.4 Attitude Change about Tangent and within Orbital Plane 81
3.5 Attitude Stability 83
3.6 Libration of the Moon 83
Suggested Reading 85
Problems 86

4 Stability of Satellite Attitude in a Central Force Field 87

4.1 Cardan Angles 87
4.2 Spinning Axisymmetric Satellite 91
4.3 Arbitrary Satellite in Locked Rotation 95
4.4 Arbitrary Satellite on Elliptical Orbit in Quasi-Locked Rotation 98
Suggested Reading 100
Problems 101

5 Torquefree Gyros 105

5.1 Solution of Euler Equations for Torquefree Gyros 105
5.2 Phase Plane Representation 111
5.3 Kinetic Energy Ellipsoid and Poinsot Ellipsoid 115
5.4 Angular Momentum Ellipsoid 117
5.5 MacCullagh Ellipsoid 118
5.6 Velocity and Acceleration of an Arbitrary Point within a Torquefree Gyro 121
5.7 Stability of Rotation about a Principal Axis for a Rigid Gyro 122
5.8 Stability of Rotation about a Principal Axis for a Deformable Gyro 124
5.9 Collinearity Theorems 131
Suggested Reading 132
Problems 132

6 Torquefree Axisymmetric Gyros 139

6.1 Angular Velocity Components 139
6.2 Euler Frequencies 144
6.3 Velocity and Acceleration of a Point 148
6.4 The Inertial Force Field 153
6.5 Changing Inertia Moments 154
6.6 Capture 156
Suggested Reading 160
Problems 160

7 Deformable Axisymmetric Gyros 166

7.1 The Single Solid Body 166
7.2 The Axisymmetric Solid Body 167
7.3 Observations 168
7.4 Floating Coordinates 169
7.5 Auxiliary Inertia Products and Moments 170
7.6 Angular Momentum of a Deforming Body 173
7.7 Kinetic Energy of Rotation of a Deforming Body 178
7.8 Energy Rates 179
7.9 Application to a Constrained System with $\delta_z = 0$ 180
7.10 Application to a Constrained System with $\delta_u = \delta_v = 0$ 184
7.11 Conclusions 198
Suggested Reading 198
Problems 199

8 Secular Attitude Drift of a Torquefree Dissipative Axisymmetric Gyro 205

8.1 Attitude Drift Rate and Attitude Stability 205
8.2 The Attitude Drift Process 209
8.3 Circles in the vz-Plane 215
8.4 Arbitrary Initial and Stable Final State 219
8.5 Elastic Deformation and Attitude Stability 225
8.6 Effect of Deformation on Angular Velocities 230
8.7 Effect of Deformation on Kinetic Energy and Strain Energy 232
Suggested Reading 233
Problems 234

9 Despin 240

9.1 Despin by Yo-Yo Masses 240
9.2 Despin by Boom Extension 244
9.3 Eccentric Installation 246
9.4 Configuration Change Terms in Euler Equations 249
Suggested Reading 253
Problems 253

10 Torque about Body-Fixed Axes of an Axisymmetric Gyro 256

10.1 Torque Perpendicular to Axis of Symmetry 256
10.2 Torque about Axis of Symmetry 259
Suggested Reading 262
Problems 262

11 Rigid Gyrostats 264

11.1 Angular Momentum and Euler Equations 265
11.2 Attitude Stability in the Short Run of Torquefree Gyrostat 268
11.3 Torquefree Axisymmetric Gyrostats 270
11.4 Bearing Friction 273
11.5 Euler Angles for a Torquefree Axisymmetric Gyrostat 276
11.6 The Inertial Force Fields 278
11.7 Spinning Axisymmetric Gyrostat in a Central Force Field 281
Suggested Reading 287
Problems 287

12 Torquefree Dissipative Axisymmetric Gyrostats 290

12.1 Near-Rigid Rotors and Platforms 290
12.2 Secular Attitude Drift for a Torquefree Axisymmetric Gyrostat with Dissipative Platform 292
12.3 Secular Attitude Drift for a Torquefree Axisymmetric Gyrostat with Dissipative Rotor 297

12.4 Secular Attitude Drift for a Torquefree Axisymmetric Gyrostat with Dissipative Rotor and Dissipative Platform 299
Suggested Reading 305
Problems 306

Appendix A: Euler Angles 311
Appendix B: Euler Parameters 321
Appendix C: Cardan Angles 329
Appendix D: Engineering Data 342
Appendix E: Nomenclature 349
Appendix F: Answers to Selected Problems 352

Author Index 379
Subject Index 381

1

The Gyroscope

In order to describe the behavior of a *gyroscope* (or simply *gyro*), Euler's *angular momentum law* in the form

$$\mathbf{M} = \dot{\mathbf{H}} \tag{1.1}$$

is essential. The quantity $\mathbf{M}$ is the *torque* applied to the gyroscope from the outside. The quantity $\mathbf{H}$ is the *angular momentum* (or the *moment of momentum*) of the gyroscope, and $\dot{\mathbf{H}}$ is its first derivative with respect to time. The torque $\mathbf{M}$ and the angular momentum $\mathbf{H}$ are taken with respect to the same reference point.

Euler's axiom as basic law governing rotational motions represents the complement to Newton's second law, which governs translational motions (Wittenburg, 1977; Truesdell, 1964). For a rigid body Euler's law can be derived directly from Newton's second law. For other cases[1] additional assumptions have to be included.

A concise general treatment of gyrodynamic theory, including gyrosatellites, is available in Magnus (1971), Arnold and Maunder (1961), Ishlinskij (1963), and Liu (1986), while Wittenburg (1977) discusses multi-body satellites and Hughes (1985) the attidude dynamics of satellites. There are also sections on gyrodynamics in Kaplan (1976) and Thomson (1963). The gyrodynamics of an elastic satellite, including its attitude stability, is described by Robe and Kane (1967).

1.1 Angular Momentum

The angular momentum $\mathbf{H}$ of a body of finite size and of mass m is a vector defined by

$$\mathbf{H} = \int_m \mathbf{r} \times \dot{\mathbf{r}}\, dm \tag{1.2}$$

where $\mathbf{r}$ is the position vector of the mass element dm, and $\dot{\mathbf{r}}$ is the velocity of the same mass element (Figure 1.1).

For a rigid body with inertia tensor

$$[I] = \begin{bmatrix} I_{xx} & I_{xy} & I_{xz} \\ I_{xy} & I_{yy} & I_{yz} \\ I_{xz} & I_{yz} & I_{zz} \end{bmatrix} \tag{1.3}$$

[1] For an example involving an elastic body see, e.g., Tabarrok et al. (1974).

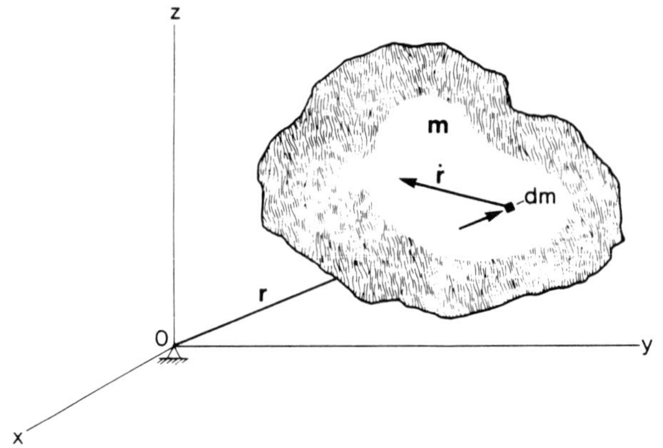

FIGURE 1.1. Position and velocity of a mass element.

and rotating at an angular velocity

$$\boldsymbol{\omega} = [\mathbf{e}_x \ \mathbf{e}_y \ \mathbf{e}_z] \begin{bmatrix} \omega_x \\ \omega_y \\ \omega_z \end{bmatrix} \tag{1.4}$$

such that $\dot{\mathbf{r}} = \boldsymbol{\omega} \times \mathbf{r}$, the angular momentum becomes

$$\mathbf{H} = [\mathbf{e}_x \ \mathbf{e}_y \ \mathbf{e}_z] \begin{bmatrix} I_{xx} & I_{xy} & I_{xz} \\ I_{xy} & I_{yy} & I_{yz} \\ I_{xz} & I_{yz} & I_{zz} \end{bmatrix} \begin{bmatrix} \omega_x \\ \omega_y \\ \omega_z \end{bmatrix} \tag{1.5}$$

or in shorthand notation

$$\mathbf{H} = \{\mathbf{e}_i\}^T [I_{ij}] \{\omega_j\}. \tag{1.6}$$

For the column matrix of the components of the angular momentum one obtains

$$\begin{bmatrix} H_x \\ H_y \\ H_z \end{bmatrix} = \begin{bmatrix} I_{xx} & I_{xy} & I_{xz} \\ I_{xy} & I_{yy} & I_{yz} \\ I_{xz} & I_{yz} & I_{zz} \end{bmatrix} \begin{bmatrix} \omega_x \\ \omega_y \\ \omega_z \end{bmatrix} \tag{1.7}$$

or

$$\{H_i\} = [I_{ij}] \{\omega_j\}. \tag{1.8}$$

The inertia tensor $[I_{ij}]$ maps the angular velocity $\{\omega_j\}$ into the angular momentum $\{H_i\}$, i.e., $I : \omega \rightarrow H$.

For a coordinate system that originates at the body's mass center C (Figure 1.2a), the angular momentum is, from equation (1.6),

$$\mathbf{H}_{\mathrm{C}} = \{\mathbf{e}\}^T [I] \{\omega\}.$$

In gyrodynamics there arise situations where the angular momentum with respect to a point other than the body's own mass center is required.

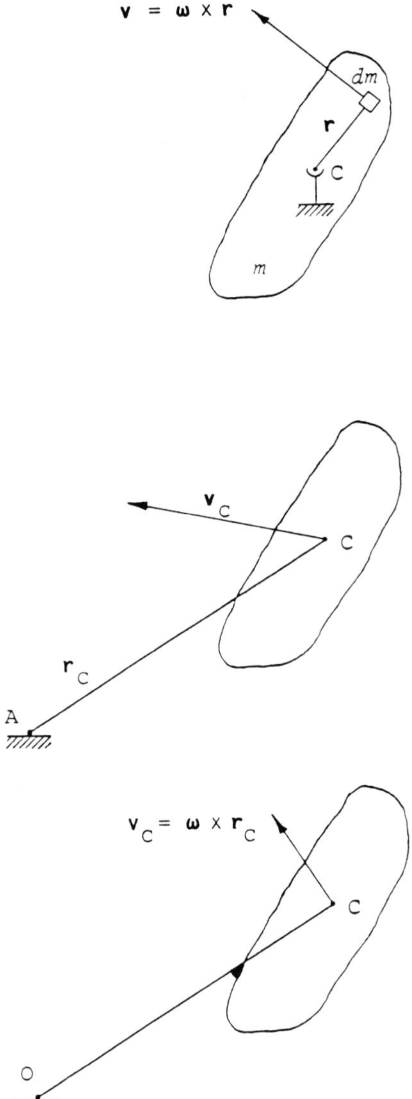

2a

2b

2c

FIGURE 1.2. The angular momentum with respect to points C, A, O.

Think, e.g., of the angular momentum of a spinning propeller on a yawing aircraft, which you may want to express with respect to the aircraft's mass center. In general, the angular momentum taken with respect to another point A (Figure 1.2b) is

$$\mathbf{H}_\mathrm{A} = \mathbf{H}_\mathrm{C} + m\mathbf{r}_\mathrm{C} \times \mathbf{v}_\mathrm{C}. \tag{1.9}$$

If $\mathbf{r}_\mathrm{C}$ is a *fixed* line on body m (i.e., $\mathring{\mathbf{r}}_\mathrm{C} = \mathbf{0}$), and O is a fixed point in space, then

$$\mathbf{r}_\mathrm{C} \times \mathbf{v}_\mathrm{C} = \mathbf{r}_\mathrm{C} \times (\boldsymbol{\omega} \times \mathbf{r}_\mathrm{C}) \tag{1.10}$$

and consequently

$$\mathbf{H}_\mathrm{O} = \{\mathbf{e}\}^T[I]_\mathrm{O}\{\omega\} \tag{1.11}$$

with $[I]_\mathrm{O}$ as defined by equation (1.37) later.

1.2 Time Derivative of Angular Momentum

If we use equation (1.6) and form the derivative with respect to time, we obtain, according to the chain rule of differentiation,

$$\dot{\mathbf{H}} = \{\mathbf{e}_i\}^T[I_{ij}]\{\dot{\omega}_j\} + \{\mathbf{e}_i\}^T[\dot{I}_{ij}]\{\omega_j\} + \{\dot{\mathbf{e}}_i\}^T[I_{ij}]\{\omega_j\}. \tag{1.12}$$

The first two terms on the right side of equation (1.2) represent the *relative rate of change* (with respect to the coordinate system used). We may combine the first two terms into

$$\mathring{\mathbf{H}} = \{\mathbf{e}_i\}^T\left[[I_{ij}]\{\dot{\omega}_j\} + [\dot{I}_{ij}]\{\omega_j\}\right] \tag{1.13}$$

which we read "*H* circle." For rigid bodies $[\dot{I}_{ij}] = [0]$.

The third term on the right side of equation (1.12) represents the *frame* (or *guide*) *rate* of change, i.e., the contribution made by the fact that the coordinate system used to describe the motion happens to rotate. It can be shown that

$$\boldsymbol{\Omega} \times \mathbf{H} = \{\dot{\mathbf{e}}_i\}^T[I_{ij}]\{\omega_j\} \tag{1.14}$$

where $\boldsymbol{\Omega}$ is the angular velocity of the *coordinate system* used to describe the motion. (The angular velocity $\boldsymbol{\Omega}$ *may* or *may not* be equal to the angular velocity $\boldsymbol{\omega}$ of the gyro, as given by equation (1.4).) Combining equations (1.12), (1.13), and (1.14), we have

$$\dot{\mathbf{H}} = \mathring{\mathbf{H}} + \boldsymbol{\Omega} \times \mathbf{H}. \tag{1.15}$$

We may, of course, also write the angular momentum in the following form:

$$\mathbf{H} = [\mathbf{e}_x\ \mathbf{e}_y\ \mathbf{e}_z]\begin{bmatrix} H_x \\ H_y \\ H_z \end{bmatrix} \tag{1.16}$$

and upon differentiation with respect to time obtain

$$\dot{\mathbf{H}} = [\mathbf{e}_x \ \mathbf{e}_y \ \mathbf{e}_z] \begin{bmatrix} \dot{H}_x \\ \dot{H}_y \\ \dot{H}_z \end{bmatrix} + [\dot{\mathbf{e}}_x \ \dot{\mathbf{e}}_y \ \dot{\mathbf{e}}_y] \begin{bmatrix} H_x \\ H_y \\ H_z \end{bmatrix}. \tag{1.17}$$

Upon introduction of the *relative rate* (i.e., with respect to the xyz coordinate system)

$$\overset{\circ}{\mathbf{H}} = [\mathbf{e}_x \ \mathbf{e}_y \ \mathbf{e}_z] \begin{bmatrix} \dot{H}_x \\ \dot{H}_y \\ \dot{H}_z \end{bmatrix} \tag{1.18}$$

and a cross product, which represents the *frame rate* (i.e., due to the rotation of the frame, i.e., of the xyz coordinate system),

$$\boldsymbol{\Omega} \times \mathbf{H} = [\dot{\mathbf{e}}_x \ \dot{\mathbf{e}}_y \ \dot{\mathbf{e}}_z] \begin{bmatrix} H_x \\ H_y \\ H_z \end{bmatrix} \tag{1.19}$$

the time derivative of the angular momentum becomes again

$$\dot{\mathbf{H}} = \overset{\circ}{\mathbf{H}} + \boldsymbol{\Omega} \times \mathbf{H} \tag{1.20}$$

as derived previously.

1.3 Inertia Tensor

The tensor

$$[I] = \begin{bmatrix} I_{xx} & I_{xy} & I_{xz} \\ I_{xy} & I_{yy} & I_{yz} \\ I_{xz} & I_{yz} & I_{zz} \end{bmatrix} \tag{1.21}$$

is called the *inertia tensor.*

The elements along the diagonal of matrix (1.21) are the *polar inertia moments,* the others are the *inertia products* (or *centrifugal moments,* or *deviation moments*).

The polar inertia moments are defined as

$$I_{xx} = \int_m (y^2 + z^2) dm \tag{1.22}$$

$$I_{yy} = \int_m (z^2 + x^2) dm \tag{1.23}$$

$$I_{zz} = \int_m (x^2 + y^2) dm. \tag{1.24}$$

The inertia products are defined by

$$I_{xy} = -\int_m xy\,dm = I_{yx} \tag{1.25}$$

$$I_{yz} = -\int_m yz\,dm = I_{zy} \tag{1.26}$$

$$I_{xz} = -\int_m xz\,dm = I_{zx}. \tag{1.27}$$

In shorthand notation, the individual entries (inertia moments as well as inertia products) of the inertia tensor may either be written

$$I_{ij} = \int_m ((x^2 + y^2 + z^2)\delta_{ij} - ij)dm \tag{1.28}$$

or

$$I_{ij} = \int_m (r^2\delta_{ij} - ij)dm \tag{1.29}$$

or

$$I_{ij} = \int_m (r_k r_k \delta_{ij} - r_i r_j)dm \tag{1.30}$$

with $i, j, k = x, y, z$, and $r_i, r_j = x, y, z$.

How an inertia tensor is obtained is most readily demonstrated by an example, for which we shall use the parallelepiped shown in Figure 1.3. It consists of homogeneous material and has a total mass m. The inertia tensor for the body *and* the reference coordinate system employed is

$$[I_{ij}] = m \begin{bmatrix} \frac{1}{3}(b^2 + c^2) & -\frac{1}{4}ab & -\frac{1}{4}ac \\ -\frac{1}{4}ab & \frac{1}{3}(a^2 + c^2) & -\frac{1}{4}bc \\ -\frac{1}{4}ac & -\frac{1}{4}bc & \frac{1}{3}(a^2 + b^2) \end{bmatrix}. \tag{1.31}$$

The definition for the inertia products is not uniform. In the literature, the following definition also appears:

$$P_{xy} = \int_m xy\,dm$$

$$P_{yz} = \int_m yz\,dm$$

$$P_{xz} = \int_m xz\,dm.$$

If this definition is employed, the inertia tensor assumes the following appearance

$$[I] = \begin{bmatrix} I_{xx} & -P_{xy} & -P_{xz} \\ -P_{xy} & I_{yy} & -P_{yz} \\ -P_{xz} & -P_{yz} & I_{zz} \end{bmatrix}. \tag{1.32}$$

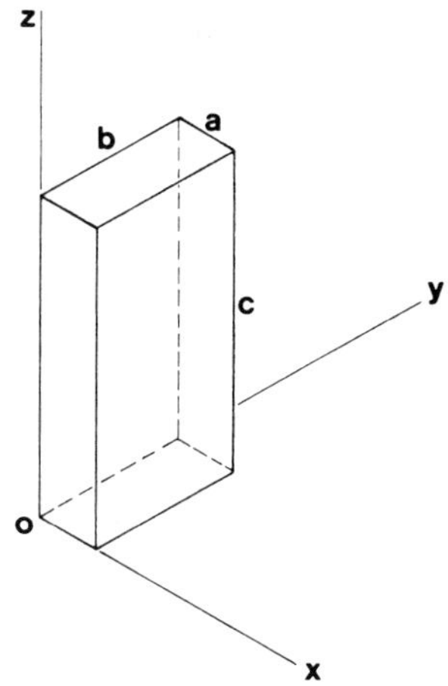

FIGURE 1.3. Parallelepiped.

If the $Oxyz$ coordinate system is located such that the inertia products vanish, one refers to the associated x, y, and z axes as *principal axes* and the associated inertia moments as *principal inertia moments.* For simplicity we introduce the symbols A, B, and C for the principal inertia moments. The inertia tensor in this case assumes the following form

$$[I] = \begin{bmatrix} A & 0 & 0 \\ 0 & B & 0 \\ 0 & 0 & C \end{bmatrix}. \tag{1.33}$$

Based on the original definitions (1.22), (1.23), and (1.24) it is seen that

$$\int_m (x^2 + y^2 + z^2) dm = \frac{1}{2}(I_{xx} + I_{yy} + I_{zz}). \tag{1.34}$$

Also, the individual inertia moments are interrelated by

$$I_{xx} + I_{yy} \geq I_{zz} \tag{1.35a}$$

$$I_{yy} + I_{zz} \geq I_{xx} \tag{1.35b}$$

$$I_{zz} + I_{xx} \geq I_{yy}. \tag{1.35c}$$

These conditions, which are rather restrictive, require that $I_{xx} = I_{yy}$ when $I_{zz} = 0$. They also require if I_{zz} is the largest of the three and $I_{xx} = I_{yy}$, that $I_{zz} > I_{xx} = I_{yy} > \frac{1}{2} I_{zz}$.

Since $(y - z)^2 \geq 0$, we find that $y^2 + z^2 \geq 2yz$, and consequently

$$I_{xx} \geq 2|I_{yz}| \tag{1.36a}$$

$$I_{yy} \geq 2|I_{xz}| \tag{1.36b}$$

$$I_{zz} \geq 2|I_{xy}|. \tag{1.36c}$$

Equations (1.35) also apply of course to the principal inertia moments

$$A + B \geq C$$

$$B + C \geq A$$

$$C + A \geq B.$$

1.4 The Huygens-Steiner Parallel Axes Theorem

The inertia tensor $[I]_O$ for an $Oxyz$ coordinate system parallel to a $C\xi\eta\zeta$ coordinate system with origin at the mass center C (Figure 1.4) can be obtained from

$$[I]_O = [I]_C + m \begin{bmatrix} (\overline{y}^2 + \overline{z}^2) & -\overline{xy} & -\overline{xz} \\ -\overline{xy} & (\overline{z}^2 + \overline{x}^2) & -\overline{yz} \\ -\overline{xz} & -\overline{yz} & (\overline{x}^2 + \overline{y}^2) \end{bmatrix}. \tag{1.37}$$

As an example we consider the parallelepiped of Figure 1.4. The inertia moment for a $C\xi\eta\zeta$ system is (see Appendix E3):

$$[I]_C = \begin{bmatrix} \frac{m}{12}(b^2 + c^2) & 0 & 0 \\ 0 & \frac{m}{12}(c^2 + a^2) & 0 \\ 0 & 0 & \frac{m}{12}(a^2 + b^2) \end{bmatrix}. \tag{1.38}$$

Since $\overline{x} = a/2$, $\overline{y} = b/2$ and $\overline{z} = c/2$, the second matrix of equation (1.37) becomes

$$\begin{bmatrix} \frac{m}{4}(b^2 + c^2) & -\frac{m}{4}ab & -\frac{m}{4}ac \\ -\frac{m}{4}ab & \frac{m}{4}(c^2 + a^2) & -\frac{m}{4}bc \\ -\frac{m}{4}ac & -\frac{m}{4}bc & \frac{m}{4}(a^2 + b^2) \end{bmatrix}. \tag{1.39}$$

Matrices (1.38) and (1.39) added give the inertia tensor with respect to the $Oxyz$ coordinate system, as previously obtained and given by equation (1.31).

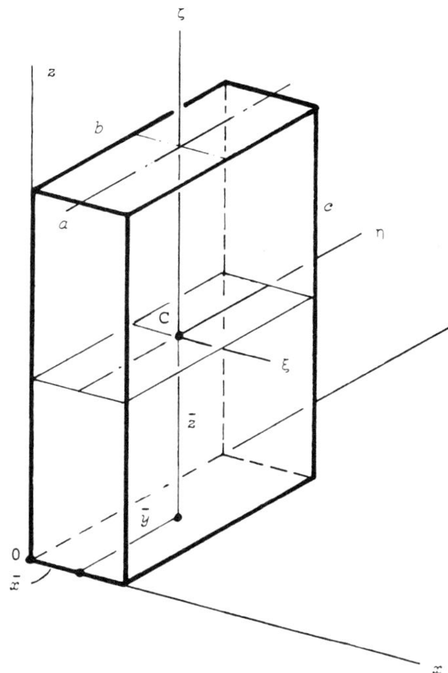

FIGURE 1.4. Parallel axes systems.

1.5 Principal Inertia Moments

By rotation of the reference coordinate system, the associated inertia tensor $[I_{ij}]$ can eventually be brought into the form

$$\begin{bmatrix} A & 0 & 0 \\ 0 & B & 0 \\ 0 & 0 & C \end{bmatrix}. \tag{1.40}$$

The *principal inertia moments* are obtained by solving the characteristic equation

$$\det([I_{ij}] - \lambda(\delta_{ij}]) = 0 \tag{1.41}$$

for the eigenvalues λ. Since equation (1.41) is a cubic equation in λ, we obtain three eigenvalues

$$\lambda_1 = A$$

$$\lambda_2 = B$$

$$\lambda_3 = C$$

which are the principal inertia moments.

As an example we consider an inertia tensor

$$[I_{ij}] = \begin{bmatrix} 20 & -2 & 0 \\ -2 & 30 & 0 \\ 0 & 0 & 40 \end{bmatrix}.$$

Equation (1.41) then represents the determinant

$$\begin{vmatrix} 20-\lambda & -2 & 0 \\ -2 & 30-\lambda & 0 \\ 0 & 0 & 40-\lambda \end{vmatrix} = 0$$

which leads to the characteristic equation (1.41)

$$(20-\lambda)(30-\lambda)(40-\lambda) - 4(40-\lambda) = 0.$$

Equation (1.41) is satisfied for $\lambda_3 = C = 40$.

Further, since, after canceling $(40-\lambda)$,

$$(20-\lambda)(30-\lambda) - 4 = 0$$

equation (1.41) is also satisfied for

$$\lambda_1 = A = 25 + \sqrt{25^2 + 596} = 30.385$$

$$\lambda_2 = B = 25 - \sqrt{25^2 - 596} = 19.615$$

eventually giving for the principal inertia tensor

$$\begin{bmatrix} 30.385 & 0 & 0 \\ 0 & 19.615 & 0 \\ 0 & 0 & 40 \end{bmatrix}.$$

The *direction* of a principal axis i is established by solving

$$\begin{bmatrix} I_x - \lambda_i & I_{xy} & I_{xz} \\ I_{xy} & I_y - \lambda_i & I_{yz} \\ I_{xz} & I_{yz} & I_z - \lambda_i \end{bmatrix} \begin{bmatrix} \cos\alpha_i \\ \cos\beta_i \\ \cos\gamma_i \end{bmatrix} = \begin{bmatrix} 0 \\ 0 \\ 0 \end{bmatrix} \tag{1.42}$$

for the direction cosines, which must also satisfy

$$\cos^2\alpha_i + \cos^2\beta_i + \cos^2\gamma_i = 1. \tag{1.43}$$

Continuing with the example, we write for the first principal axis

$$\begin{bmatrix} 20-30.385 & -2 & 0 \\ -2 & 30-30.385 & 0 \\ 0 & 0 & 40-30.385 \end{bmatrix} \begin{bmatrix} \cos\alpha_1 \\ \cos\beta_1 \\ \cos\gamma_1 \end{bmatrix} = \begin{bmatrix} 0 \\ 0 \\ 0 \end{bmatrix}$$

$$\begin{array}{rlrlrlcl} -10.385 & \cos\alpha_1 & -2 & \cos\beta_1 & + & 0 & & = 0 \\ -2 & \cos\alpha_1 & -0.385 & \cos\beta_1 & + & 0 & & = 0 \\ 0 & & +0 & & + & 9.615 & \cos\gamma_1 & = 0 \end{array}$$

and obtain

$$\alpha_1 = 79.1^\circ$$
$$\beta_1 = 169.1^\circ$$
$$\gamma_1 = 90^\circ.$$

For the second principal axis

$$\begin{bmatrix} 20 - 19.615 & -2 & 0 \\ -2 & 30 - 19.615 & 0 \\ 0 & 0 & 40 - 19.615 \end{bmatrix} \begin{bmatrix} \cos\alpha_2 \\ \cos\beta_2 \\ \cos\gamma_2 \end{bmatrix} = \begin{bmatrix} 0 \\ 0 \\ 0 \end{bmatrix}$$

we obtain

$$\alpha_2 = 10.9^\circ$$
$$\beta_2 = 79.1^\circ$$
$$\gamma_2 = 90^\circ.$$

For the third principal axis we have

$$\begin{bmatrix} 20 - 40 & -2 & 0 \\ -2 & 30 - 40 & 0 \\ 0 & 0 & 40 - 40 \end{bmatrix} \begin{bmatrix} \cos\alpha_3 \\ \cos\beta_3 \\ \cos\gamma_3 \end{bmatrix} = \begin{bmatrix} 0 \\ 0 \\ 0 \end{bmatrix}$$

and obtain

$$\alpha_3 = 90^\circ$$
$$\beta_3 = 90^\circ$$
$$\gamma_3 = 0^\circ.$$

1.6 Kinetic Energy

The kinetic energy of an arbitrary mass element dm of a gyroscope is

$$\frac{1}{2} dm\, |\dot{\mathbf{r}}|^2 \tag{1.44}$$

where, for the case that the velocity of the origin of the coordinate system is zero, the velocity $\dot{\mathbf{r}}$ of the mass element dm (Figure 1.5) is given by

$$\dot{\mathbf{r}} = \boldsymbol{\omega} \times \mathbf{r}. \tag{1.45}$$

The kinetic energy of the whole gyroscope is then

$$T = \frac{1}{2} \int_m |\boldsymbol{\omega} \times \mathbf{r}|^2 dm. \tag{1.46}$$

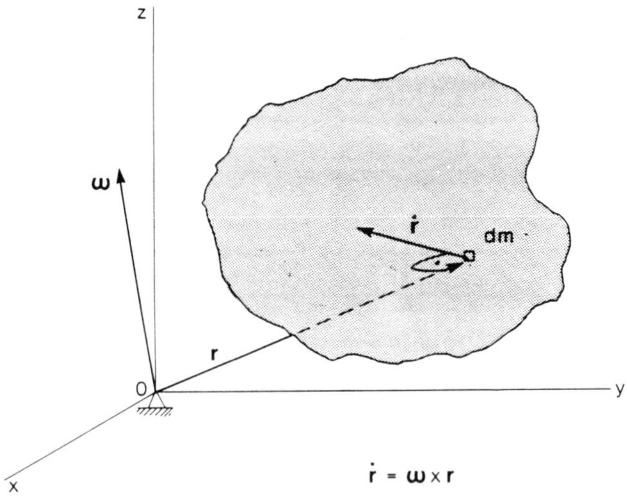

FIGURE 1.5. Position and velocity of a mass element in case of pure rotation.

Upon integration we obtain for the kinetic energy of the gyroscope rotating about 0,

$$T = \frac{1}{2}(I_{xx}\omega_x^2 + I_{yy}\omega_y^2 + I_{zz}\omega_z^2 + 2I_{xy}\omega_x\omega_y + 2I_{xz}\omega_x\omega_z + 2I_{yz}\omega_y\omega_z). \quad (1.47)$$

In case the $Oxyz$ coordinate system coincides with the principal axes, the centrifugal moments vanish and the kinetic energy becomes

$$T = \frac{1}{2}(A\omega_x^2 + B\omega_y^2 + C\omega_z^2). \quad (1.48)$$

The kinetic energy of a gyroscope of fixed size may also be expressed as a dot product of angular momentum $\mathbf{H}$ and angular velocity $\boldsymbol{\omega}$. We write equation (1.46) in the following manner:

$$T = \frac{1}{2}\int_m \dot{\mathbf{r}} \cdot (\boldsymbol{\omega} \times \mathbf{r})dm. \quad (1.49)$$

Making use of the identity $\dot{\mathbf{r}} \cdot (\boldsymbol{\omega} \times \mathbf{r}) = \boldsymbol{\omega} \cdot (\mathbf{r} \times \dot{\mathbf{r}})$ for a mixed triple product, and recalling that the angular velocity is the same for each mass element dm, we get

$$T = \frac{1}{2}\boldsymbol{\omega} \cdot \int_m \mathbf{r} \times \dot{\mathbf{r}}\, dm. \quad (1.50)$$

The integral in equation (1.50) represents the angular momentum $\mathbf{H}$ of the gyroscope (with respect to the origin O of the coordinate system). The kinetic energy can thus be written

$$T = \frac{1}{2}\boldsymbol{\omega} \cdot \mathbf{H}. \quad (1.51)$$

But remember that this equation presupposes that the gyroscope changes neither shape or size, nor can it consist of several components that move or can move with respect to each other (i.e., $\overset{\circ}{\mathbf{r}} = 0$ in the equation $\dot{\mathbf{r}} = \overset{\circ}{\mathbf{r}} + \boldsymbol{\omega} \times \mathbf{r}$).

Since $\boldsymbol{\omega} \cdot \mathbf{H} = \{\omega_i\}^T\{H_i\} - \{\omega_i\}^T[I_{ij}]\{\omega_j\}$, the kinetic energy can also be expressed by

$$T = \frac{1}{2}\{\omega_i\}^T[I_{ij}]\{\omega_j\}. \tag{1.52}$$

1.7 Euler Equations

We restrict the subsequent discussion to a case of practical importance, where the origin of the coordinate system coincides with mass center C of the gyroscope. The angular velocity of the $Cxyz$ coordinate system is $\boldsymbol{\Omega}$.

Euler's angular momentum law states that the applied torque equals the time derivative of the angular momentum

$$\mathbf{M} = \dot{\mathbf{H}}. \tag{1.53}$$

Or, using components

$$\begin{bmatrix} M_x \\ M_y \\ M_z \end{bmatrix} = \begin{bmatrix} (\dot{H})_x \\ (\dot{H})_y \\ (\dot{H})_z \end{bmatrix}. \tag{1.54}$$

The reader is reminded that it is general practice to write

$$\dot{H}_x = \frac{d}{dt}H_x = (H_x)^{\cdot} \tag{1.55}$$

i.e., $\dot{H}_x$ represents the time derivative of the x-component H_x of the angular momentum $\mathbf{H}$. In order to be able to distinguish between the x-component of $\dot{\mathbf{H}}$ on the one hand and the derivative of the x-component of $\mathbf{H}$ on the other, one must write for the x-component of the derivative $\dot{\mathbf{H}}$

$$\mathbf{e}_x(\dot{H})_x = \left(\frac{d}{dt}\mathbf{H}\right)_x \tag{1.56}$$

as has been done in equation (1.54).

For a rotating coordinate system

$$\dot{\mathbf{H}} = \overset{\circ}{\mathbf{H}} + \boldsymbol{\Omega} \times \mathbf{H} \tag{1.57}$$

with the *relative* change

$$\overset{\circ}{\mathbf{H}} = [\mathbf{e}_x \ \mathbf{e}_y \ \mathbf{e}_z] \begin{bmatrix} \dot{H}_x \\ \dot{H}_y \\ \dot{H}_z \end{bmatrix} \tag{1.58}$$

and the *frame* change

$$\boldsymbol{\Omega} \times \mathbf{H} = \begin{vmatrix} \mathbf{e}_x & \mathbf{e}_y & \mathbf{e}_z \\ \Omega_x & \Omega_y & \Omega_z \\ H_x & H_y & H_z \end{vmatrix}. \tag{1.59}$$

Equation (1.57) resolved into components is

$$\begin{bmatrix} (\dot{H})_x \\ (\dot{H})_y \\ (\dot{H})_z \end{bmatrix} = \begin{bmatrix} \dot{H}_x + \Omega_y H_z - \Omega_z H_y \\ \dot{H}_y + \Omega_z H_x - \Omega_x H_z \\ \dot{H}_z + \Omega_x H_y - \Omega_y H_x \end{bmatrix}. \tag{1.60}$$

Combining equations (1.54) and (1.60), one obtains the three *Euler equations:*

$$M_x = \dot{H}_x + \Omega_y H_z - \Omega_z H_y \tag{1.61}$$

$$M_y = \dot{H}_y + \Omega_z H_x - \Omega_x H_z \tag{1.62}$$

$$M_z = \dot{H}_z + \Omega_x H_y - \Omega_y H_x. \tag{1.63}$$

1.8 Euler Equations for Principal Axes

In the majority of practical cases, the coordinate system can be arranged such that it is not only fixed to the gyroscope but also directed along the principal axes. For a coordinate system fixed to the gyroscope we conclude that angular velocity of coordinate system and angular velocity of gyro become equal, i.e.,

$$\boldsymbol{\Omega} = \boldsymbol{\omega}. \tag{1.64}$$

For a coordinate system along the principal axes, the inertia tensor becomes simply

$$[I] = \begin{bmatrix} A & 0 & 0 \\ 0 & B & 0 \\ 0 & 0 & C \end{bmatrix}. \tag{1.65}$$

The angular momentum becomes

$$\mathbf{H} = [\mathbf{e}_x \ \mathbf{e}_y \ \mathbf{e}_z] = \begin{bmatrix} A\omega_x \\ B\omega_y \\ C\omega_z \end{bmatrix}. \tag{1.66}$$

The relative rate of change becomes

$$\overset{\circ}{\mathbf{H}} = [\mathbf{e}_x \ \mathbf{e}_y \ \mathbf{e}_z] \begin{bmatrix} A\dot{\omega}_x \\ B\dot{\omega}_y \\ C\dot{\omega}_z \end{bmatrix} \tag{1.67}$$

and the Euler equations assume the form

$$M_x = A\dot{\omega}_x - (B - C)\omega_y\omega_z \tag{1.68}$$

$$M_y = B\dot{\omega}_y - (C - A)\omega_z\omega_x \tag{1.69}$$

$$M_z = C\dot{\omega}_z - (A - B)\omega_x\omega_y. \tag{1.70}$$

1.9 Euler Angles

In order to establish the *attitude* of a gyroscope in absolute space, three angular coordinates, the so-called *Euler angles* may be used. (See also Appendices A and B.) The sequence of performing the rotation, and the symbols used for these angles are given in different ways by various authors. As a consequence there exists no uniformity in the literature. Expressed in mnemonic symbols, the Euler angles are

$$\psi = \textit{precession angle}$$

$$\nu = \textit{nutation angle}$$

$$\sigma = \textit{spin angle.}$$

In Figure 1.6, the gyroscope is shown at an attitude $\psi = 0$, $\nu = 0$, $\sigma = 0$. In Figure 1.7 its attitude is characterized by $\psi = \psi$, $\nu = \nu$, $\sigma = \sigma$. The $CXYZ$ coordinate system is fixed in space, while the $Cxyz$ coordinate system is body-fixed (i.e., fixed to the gyroscope). The x_2-line, representing the intersection of the xy-plane with the XY-plane, is called the *node line.*

The time derivatives of the Euler angles are the *Euler frequencies* (or *Euler rates*):

$$\dot{\psi} = \textit{precession}$$

$$\dot{\nu} = \textit{nutation}$$

$$\dot{\sigma} = \textit{spin.}$$

The terminology used here is that of classical mechanics. As far as the terms precession and nutation are concerned, we find that engineers often refer to the self-generated velocities of the torquefree gyroscope as *nutation*, while they refer to all velocities resulting from externally applied torques as *precession*. This practice is not adopted in the present book.

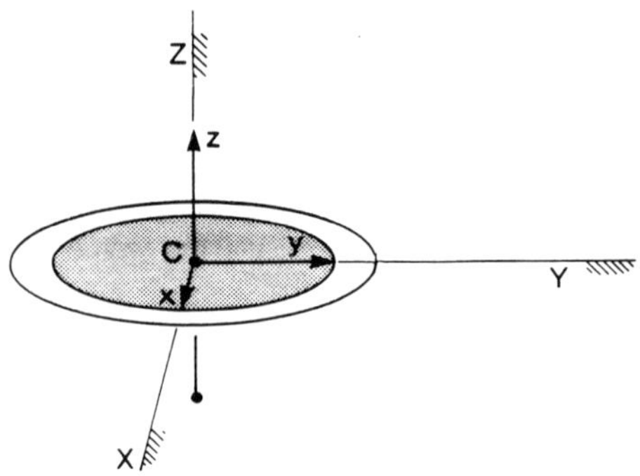

FIGURE 1.6. Gyroscope disk located within the reference plane.

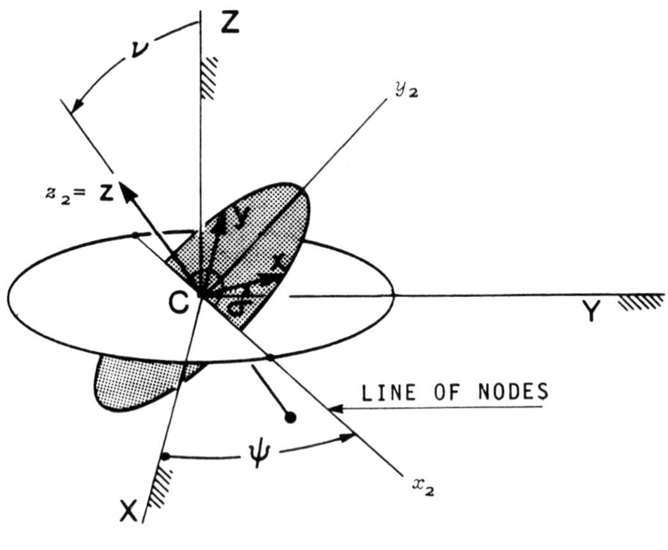

FIGURE 1.7. Gyroscope disk at nutation angle ν, precession angle ψ, and spin angle σ.

1.10 Coordinate Transformation

Any coordinate transformation, from a coordinate system 0 (i.e., $Cx_0y_0z_0$) to a coordinate system 1 (i.e., $Cx_1y_1z_1$), follows the relation

$$\{r\}_1 = [T]_{1\leftarrow 0}\{r\}_0. \tag{1.71}$$

For three successive transformations (Figure 1.8), one obtains

$$\{r\}_3 = [T]_{3\leftarrow 2}[T]_{2\leftarrow 1}[T]_{1\leftarrow 0}\{r\}_0 \tag{1.72}$$

with the transformation matrices

$$[T]_{1\leftarrow 0} = \begin{bmatrix} \cos\psi & \sin\psi & 0 \\ -\sin\psi & \cos\psi & 0 \\ 0 & 0 & 1 \end{bmatrix} \tag{1.73}$$

$$[T]_{2\leftarrow 1} = \begin{bmatrix} 1 & 0 & 0 \\ 0 & \cos\nu & \sin\nu \\ 0 & -\sin\nu & \cos\nu \end{bmatrix} \tag{1.74}$$

$$[T]_{3\leftarrow 2} = \begin{bmatrix} \cos\sigma & \sin\sigma & 0 \\ -\sin\sigma & \cos\sigma & 0 \\ 0 & 0 & 1 \end{bmatrix}. \tag{1.75}$$

If we write

$$\{r\}_3 = [T]_{3\leftarrow 0}\{r\}_0 \tag{1.76}$$

then

$$[T]_{3\leftarrow 0} = [T]_{3\leftarrow 2}[T]_{2\leftarrow 1}[T]_{1\leftarrow 0}. \tag{1.77}$$

Multiplication of the three matrices results in

$$[T]_{3\leftarrow 0} =$$

$$\begin{bmatrix} \cos\psi\cos\sigma - \sin\psi\cos\nu\sin\sigma & \sin\psi\cos\sigma + \cos\psi\cos\nu\sin\sigma & \sin\nu\sin\sigma \\ -\cos\psi\sin\sigma - \sin\psi\cos\nu\cos\sigma & -\sin\psi\sin\sigma + \cos\psi\cos\nu\cos\sigma & \sin\nu\cos\sigma \\ \sin\psi\sin\nu & -\cos\psi\sin\nu & \cos\nu \end{bmatrix}. \tag{1.78}$$

The inverse relationship is

$$\{r\}_0 = [T]_{0\leftarrow 3}\{r\}_3$$

with $[T]_{0\leftarrow 3} = [T]^T_{3\leftarrow 0}$ or

$$[T]_{0\leftarrow 3} =$$

$$\begin{bmatrix} \cos\psi\cos\sigma - \sin\psi\cos\nu\sin\sigma & -\cos\psi\sin\sigma - \sin\psi\cos\nu\cos\sigma & \sin\psi\sin\nu \\ \sin\psi\cos\sigma + \cos\psi\cos\nu\sin\sigma & -\sin\psi\sin\sigma + \cos\psi\cos\nu\cos\sigma & -\cos\psi\sin\nu \\ \sin\nu\sin\sigma & \sin\nu\cos\sigma & \cos\nu \end{bmatrix}.$$

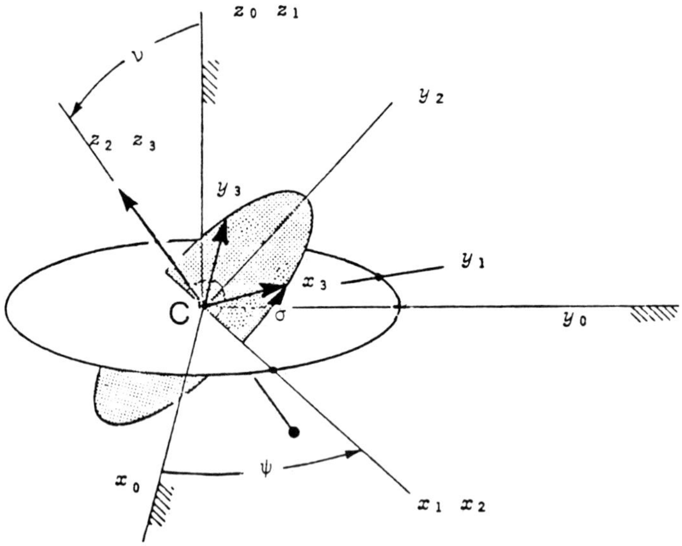

FIGURE 1.8. Successive coordinate transformations: $0 \rightarrow 1$ through angle ψ; $1 \rightarrow 2$ through angle ν; and $2 \rightarrow 3$ through angle σ.

1.11 Coordinate Transformation of Angular Velocities

Frequently one needs to transform from the components ω_x, ω_y, ω_z of the angular velocity $\boldsymbol{\omega}$ of a gyroscope to the Euler frequences $\dot{\psi}$, $\dot{\nu}$, $\dot{\sigma}$ or vice versa. We select as an intermediate system an orthogonal coordinate system consisting of the line x_2 of nodes, a perpendicular direction y_2, and the z-axis (Figure 1.9), such that $Cx_2y_2z_2$ forms an orthogonal right-hand system. The projected Euler frequencies are:
onto the line x_2 of nodes

$$\dot{\nu}$$

onto the z-axis

$$\dot{\psi}\cos\nu + \dot{\sigma}$$

and onto the y_2-coordinate

$$\dot{\psi}\sin\nu.$$

Another projection, this time onto the body-fixed x- and y-coordinates (Figure 1.10), results in the desired transformation:

$$\begin{bmatrix} \omega_x \\ \omega_y \\ \omega_z \end{bmatrix} = \begin{bmatrix} \sin\nu\sin\sigma & \cos\sigma & 0 \\ \sin\nu\cos\sigma & -\sin\sigma & 0 \\ \cos\nu & 0 & 1 \end{bmatrix} \begin{bmatrix} \dot{\psi} \\ \dot{\nu} \\ \dot{\sigma} \end{bmatrix}. \qquad (1.79)$$

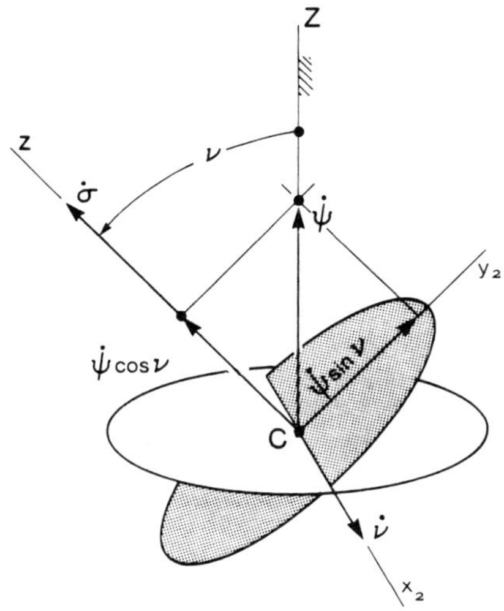

FIGURE 1.9. Components of the Euler frequencies along node line x_2, coordinate y_2, and spin axis z.

The inverse transformation is

$$\begin{bmatrix} \dot{\psi} \\ \dot{\nu} \\ \dot{\sigma} \end{bmatrix} = \begin{bmatrix} \frac{\sin\sigma}{\sin\nu} & \frac{\cos\sigma}{\sin\nu} & 0 \\ \cos\sigma & -\sin\sigma & 0 \\ -\frac{\sin\sigma}{\tan\nu} & -\frac{\cos\sigma}{\tan\nu} & 1 \end{bmatrix} \begin{bmatrix} \omega_x \\ \omega_y \\ \omega_z \end{bmatrix}. \tag{1.80}$$

Note that the transformation matrix of equation (1.80) is the inverse, not the transpose, of the transformation matrix of equation (1.79), because the transformation matrices are not orthogonal.

The transformation with respect to the space-fixed $OXYZ$ coordinate system is

$$\begin{bmatrix} \omega_X \\ \omega_Y \\ \omega_Z \end{bmatrix} = \begin{bmatrix} 0 & \cos\psi & \sin\psi\sin\nu \\ 0 & \sin\psi & -\cos\psi\sin\nu \\ 1 & 0 & \cos\nu \end{bmatrix} \begin{bmatrix} \dot{\psi} \\ \dot{\nu} \\ \dot{\sigma} \end{bmatrix}. \tag{1.81}$$

Note that the determinants of the transformation matrices in equations (1.79) and (1.81) vanish when $\nu = 0$, pointing to a *degeneracy* (or *singularity*) of the Euler angle system. Or, in other words, when $\nu = 0$ the other two angles, ψ and σ, lie in one and the same plane and are no longer distinguishable, as an inspection of Figure 1.7 quickly indicates.

The inverse of equation (1.81) is

$$\begin{bmatrix} \dot{\psi} \\ \dot{\nu} \\ \dot{\sigma} \end{bmatrix} = \begin{bmatrix} -\frac{\sin\psi}{\tan\nu} & \frac{\sin\psi}{\tan\nu} & 1 \\ \cos\psi & \sin\psi & 0 \\ \frac{\sin\psi}{\sin\nu} & -\frac{\cos\psi}{\sin\nu} & 0 \end{bmatrix} \begin{bmatrix} \omega_X \\ \omega_Y \\ \omega_Z \end{bmatrix}.$$

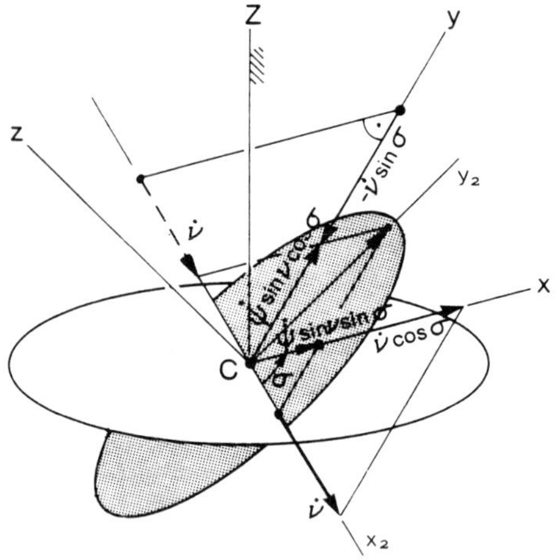

FIGURE 1.10. Components of the Euler frequencies along the body-fixed axes x and y.

The angular velocity vector can thus be expressed as (Figure 1.7)

$$\boldsymbol{\omega} = [\mathbf{e}_x\ \mathbf{e}_y\ \mathbf{e}_z] \begin{bmatrix} \omega_x \\ \omega_y \\ \omega_z \end{bmatrix} = [\mathbf{e}_X\ \mathbf{e}_Y\ \mathbf{e}_Z] \begin{bmatrix} \omega_X \\ \omega_Y \\ \omega_Z \end{bmatrix}$$

$$= \quad [\mathbf{e}_Z\ \mathbf{e}_{x_2}\ \mathbf{e}_z] \begin{bmatrix} \dot{\psi} \\ \dot{\nu} \\ \dot{\sigma} \end{bmatrix} .$$

1.12 Integrability

With the angular velocity components known, the obvious question arises, as to whether an associated angular position can be obtained by integration. It will be shown that the Euler frequencies can readily be integrated to yield the Euler angles

$$\psi = \psi_0 + \int \dot{\psi}\, dt \tag{1.82}$$

$$\nu = \nu_0 + \int \dot{\nu}\, dt \tag{1.83}$$

$$\sigma = \sigma_0 + \int \dot{\sigma}\, dt \tag{1.84}$$

and thus give the attitude of the gyro as a function of time.

The angular velocity components ω_x, ω_y, and ω_z along the body-fixed axes x, y, and z, on the other hand can unfortunately not be integrated to obtain associated angles by means of which the attitude to the gyro might be described.

The angular velocity components ω_X, ω_Y, and ω_Z, along the spaced-fixed coordinate axes X, Y, and Z can also not be integrated to obtain the attitude.

As proof we use the integrability condition

$$\frac{\partial P}{\partial y} = \frac{\partial Q}{\partial x} \tag{1.85}$$

for an arbitrary total differential

$$dz = Pdx + Qdy. \tag{1.86}$$

From (1.79), e.g.,

$$\omega_x dt = \sin\nu \sin\sigma \, d\psi + \cos\sigma \, d\nu$$

with $P = \sin\nu \sin\sigma$, $Q = \cos\nu$, and

$$\frac{\partial P}{\partial \nu} = \cos\nu \sin\sigma$$

and

$$\frac{\partial Q}{\partial \psi} = 0$$

which obviously do not satisfy integrability condition (1.85).

For the Euler frequencies, on the other hand, we find that integrability condition (1.85) is satisfied. From equation (1.80)

$$d\nu = \cos\sigma(\omega_x dt) - \sin\sigma(\omega_y dt)$$

with $P = \cos\sigma$ and $Q = -\sin\sigma$, we obtain

$$\frac{\partial P}{\omega_y \partial t} = \frac{\partial \cos\sigma}{\omega_y \partial t} = -\sin\sigma \frac{\partial \sigma}{\omega_y \partial t} = -\sin\sigma \frac{-\cos\sigma}{\tan\nu}$$

$$\frac{\partial Q}{\omega_x \partial t} = \frac{\partial(-\sin\sigma)}{\omega_x \partial t} = -\cos\sigma \frac{\partial \sigma}{\omega_x \partial t} = -\cos\sigma \frac{-\sin\sigma}{\tan\nu}.$$

The reader should note that $\partial\sigma/\omega_y \partial t$ etc. stands for the *partial* derivative, i.e., $\partial\sigma/\omega_y \partial t = -\cos\sigma/\tan\nu$ from equation (1.80).

1.13 The Magnus Shape Triangle

The occasion sometimes arises that one wants to delimitate ***stable*** and ***unstable*** gyro configurations, as expressed by the gyro's three principal inertia moments. Magnus (1971) introduced a ***shape triangle*** for this purpose. An obvious difficulty when attempting to represent a gyro configuration on a graph is caused by the fact that one wants to represent *three* properties (i.e., the principal inertia moments A, B, and C) in a *two*-dimensional graph. Among several other possibilities (e.g., using A/C and B/C as coordinates), the Magnus shape triangle has the property of treating all three inertia moments equally. In order to represent the ***shape*** of a gyroscope, its actual size is of no significance and one can thus use

$$A^* + B^* + C^* = 1 \tag{1.87}$$

with

$$A^* = \frac{A}{A+B+C} \tag{1.88a}$$

$$B^* = \frac{B}{A+B+C} \tag{1.88b}$$

$$C^* = \frac{C}{A+B+C}. \tag{1.88c}$$

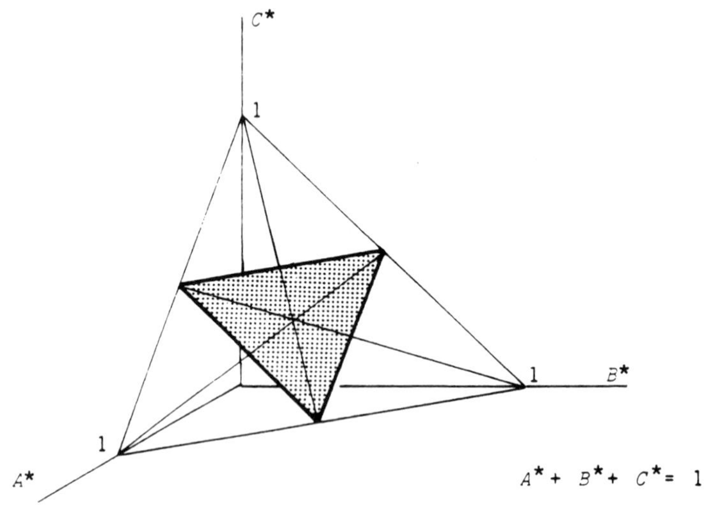

FIGURE 1.11. Location of the Magnus shape triangle in a three-dimensional $A^*B^*C^*$ coordinate system.

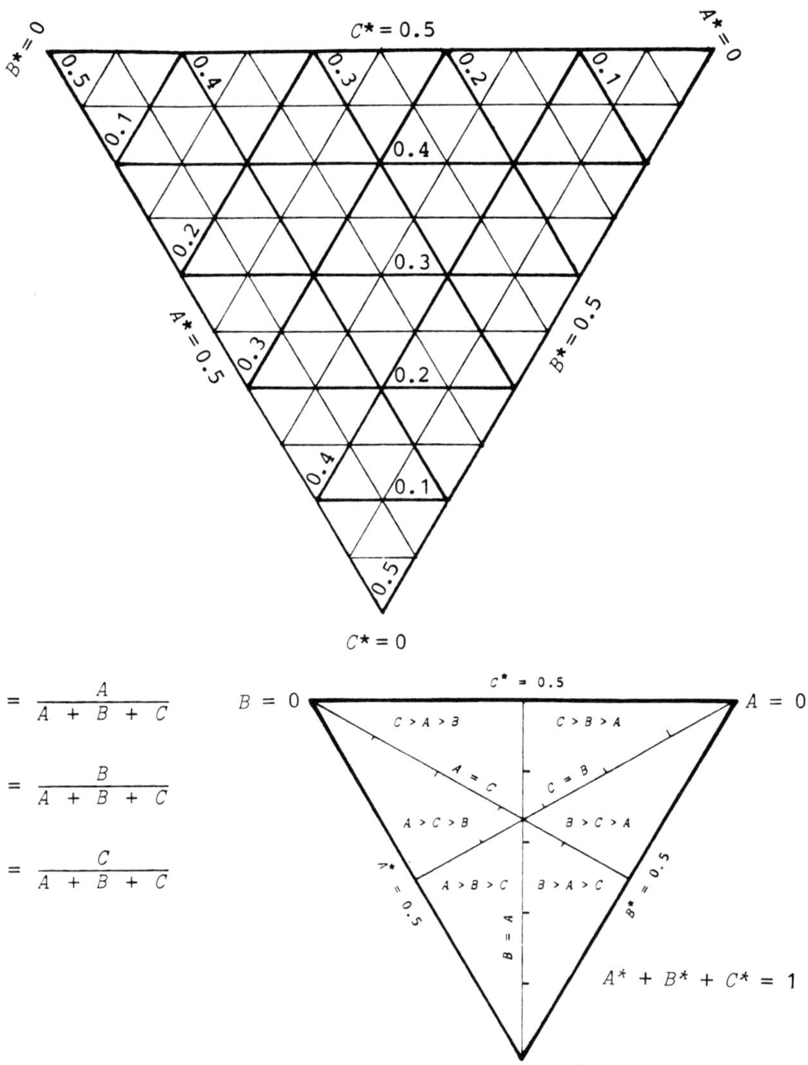

$$A^* = \frac{A}{A + B + C}$$

$$B^* = \frac{B}{A + B + C}$$

$$C^* = \frac{C}{A + B + C}$$

FIGURE 1.12. Magnus shape triangle.

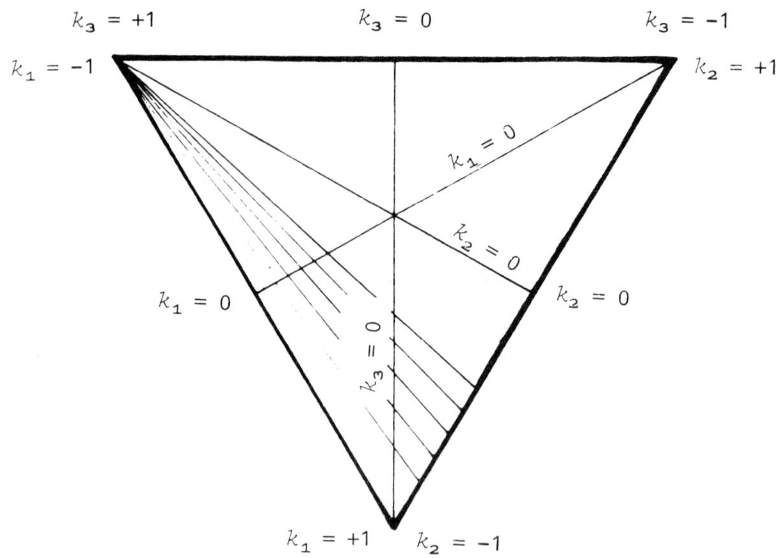

FIGURE 1.13. Magnus shape triangle in modified form.

If one then introduces temporarily an $A^*B^*C^*$ coordinate system as shown in Figure 1.11, one arrives at the conclusion that any possible configuration can only have inertia moments that lie within the shaded triangle. The top line of the triangle, for instance, represents the case $C^* = \frac{1}{2}$; the left corner $B^* = 0$, $C^* = \frac{1}{2}$; and the right corner $A^* = 0$, $C^* = \frac{1}{2}$. The line connecting the point ($A^* = 0$, $C^* = \frac{1}{2}$, $B^* = \frac{1}{2}$) with the point ($A^* = 1$, $B^* = 0$, $C^* = 0$) represents the case $B^* = C^*$. In Figure 1.12 the Magnus shape triangle is shown in the manner in which it usually appears in the literature. Recognizing that the point (A^*, B^*, C^*) of a given gyroscope is not too easy to locate in the Magnus shape triangle, experienced engineers draw the triangle such that the bisectors are a round five length units long each.

Another approach is that of introducing the *Smelt* parameters (DeBra and Delp, 1961)

$$k_1 = \frac{B - C}{A} \tag{1.89a}$$

$$k_2 = \frac{C - A}{B} \tag{1.89b}$$

$$k_3 = \frac{A - B}{C} \tag{1.89c}$$

as a consequence of which the Magnus shape triangle appears as shown in Figure 1.13. In this case it is advisable to make the sides of the triangle a round twenty length units long each.

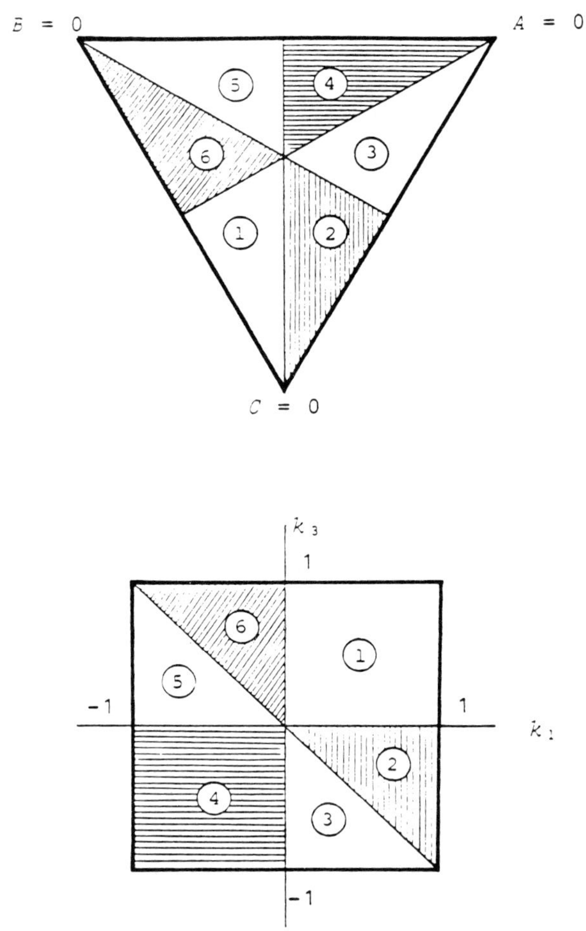

FIGURE 1.14. Magnus triangle and Cartesian plot of inertia moments.

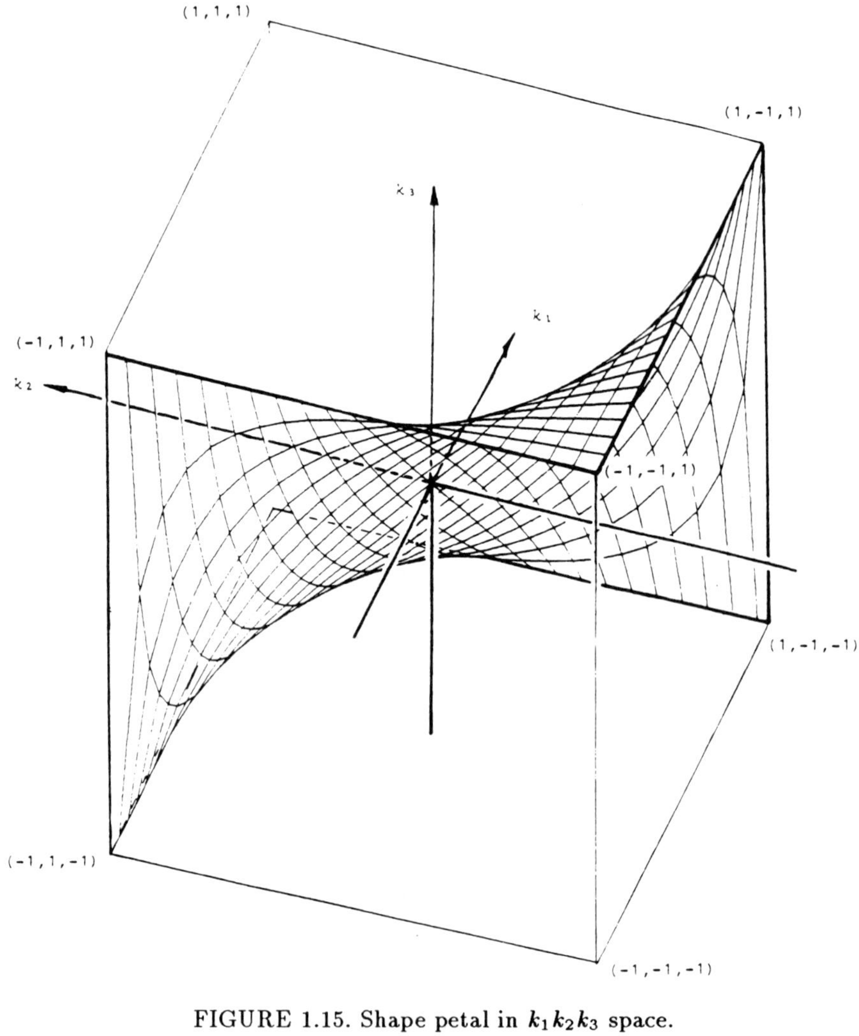

FIGURE 1.15. Shape petal in $k_1 k_2 k_3$ space.

In use are also diagrams of the type shown in Figures 1.14 and 1.15, introduced by DeBra and Delp (1961). Note that the symmetry and equality of treatment for each inertia moment characteristic of the Magnus shape triangle have been sacrificed in favor of a Cartesian representation. The reader is warned that the definitions of k_1, k_2, and k_3 are not uniform in the literature and care must be taken to ascertain exactly what each symbol stands for.

The Smelt parameters (1.89) are not independent of each other but satisfy

$$k_1 + k_2 + k_3 + k_1 k_2 k_3 = 0. \tag{1.90}$$

Equation (1.90) when plotted in $k_1 k_2 k_3$ space, results in the *shape petal* (Chesser and Rimrott, 1985) of Figure 1.15,

Suggested Reading

1. Arnold, R.N., L. Maunder. *Gyrodynamics*, Academic Press, 1961, 484 pp.

2. Chesser, H.; F.P.J. Rimrott. "Magnus Triangle and Smelt Petal," *Proceedings* (Ed. H. Rasmussen) CANCAM '85, 1985, CSME Montreal, A137–A138.

3. DeBra, D.B.; R.H. Delp. "Rigid Body Attitude Stability and Natural Frequencies in a Circular Orbit," *The Journal of the Astronautical Sciences*, 8, 1961, 14–17.

4. Greenwood, D.T. *Principles of Dynamics,* Second Edition, Prentice-Hall, 1988, 552 pp.

5. Hughes, P.C. *Spacecraft Attitude Dynamics*, Wiley and Sons, 1985, 576 pp.

6. Ishlinskij, A.Y. *Mechanics of Gyroscopic Systems* (in Russian), Academy Press Moskva, 1963, 313 pp.

7. Kane, T.R.; P.W. Likins; D.A. Levinson. *Spacecraft Dynamics*, McGraw-Hill, 1983, 436 pp.

8. Kaplan, M.H. *Modern Spacecraft Dynamics and Control*, John Wiley and Sons, 1976, 415 pp.

9. Liu, Y.-Z. *Gyrodynamics* (in Chinese), Science Press, Beijing, 1986, 310 pp.

10. Magnus, K. *Kreisel*, Springer-Verlag, 1971, 353 pp.

11. Robe, T.R.; T.R. Kane. "Dynamics of an Elastic Satellite," *International Journal of Solids and Structures*, 3, 1967, 333–352, 691–703, 1031–1051.

12. Tabarrok, B.; C.M. Leech; Y.I. Kim. "On the Dynamics of an Axially Moving Beam," *Journal of the Franklin Institute*, 297, 3, 1974, 201–220.

13. Thomson, W.T. *Introduction to Space Dynamics*, John Wiley and Sons, 1963, 317 pp.

14. Truesdell, C. "Die Entwicklung des Drallsatzes," *ZAMM*, 44, 4/5, 1964, 149–158.

15. Wittenburg, J. *Dynamics of Systems of Rigid Bodies*, B.G. Teubner, 1977, 224 pp.

Problems

1.1. A flat disk gyro, located on the surface of the Earth, has a radius $r = 1$ m, a mass $m = 50$ kg, and is suspended as shown. It spins at a constant angular frequency $\dot{\sigma} = 3000$ rad/s about the horizontal body-fixed z-axis. The disk also precesses about the vertical space-fixed Z-axis at a constant angular frequency $\dot{\psi} = 0.01$ rad/s. Use the shaft-fixed $Cuvz$ coordinate system (the shaft does not participate in the disk's spin) and determine (a) the gyro's inertia tensor $[I_{ij}]_C$; (b) the gyro's inertia tensor $[I_{ij}]_O$; (c) the angular velocity vector $\boldsymbol{\omega}$ of the gyro; (d) the angular velocity vector $\boldsymbol{\Omega}$ of the $Cuvz$ coordinate system; (e) the angular momentum $\mathbf{H}$; (f) the time derivative $\dot{\mathbf{H}}$ of the angular momentum; (g) the external moment $\mathbf{M}$ acting on the gyro; (h) the length l of the shaft; and (i) sketch $\boldsymbol{\omega}$, $\boldsymbol{\Omega}$, and $\mathbf{M}$ into the figure.

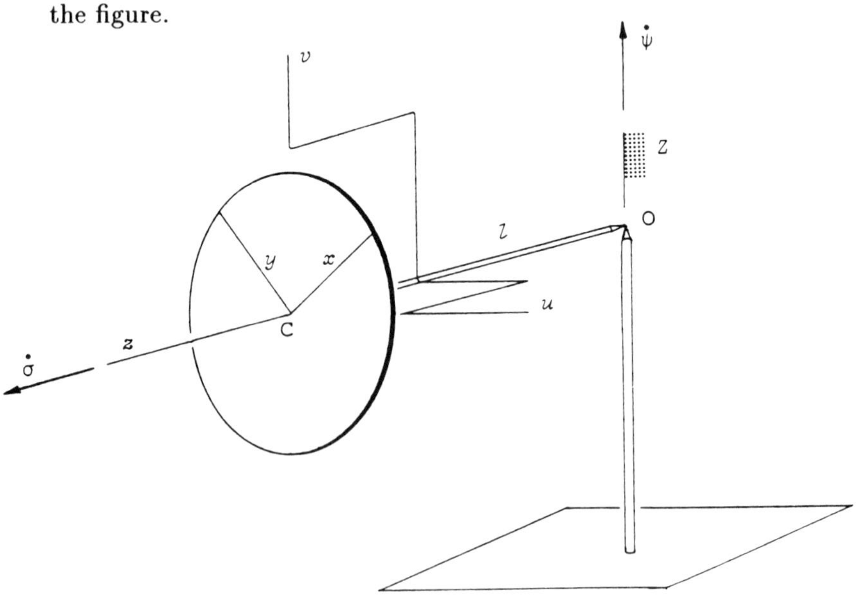

1.2. Show that

$$\{\mathbf{e}_i\}^T = \{\mathbf{e}_J\}^T [T_{Ji}].$$

1.3. Determine the transformation matrix $[T_{iJ}]$ between a (fixed) Cartesian coordinate system with the unit vectors $\mathbf{e}_x$, $\mathbf{e}_y$, and $\mathbf{e}_z$ ($J = x, y, z$) and a (rotating) cylindrical coordinate system with the unit vectors $\mathbf{e}_r$, $\mathbf{e}_\theta$, and $\mathbf{e}_z$ ($i = r, \theta, z$). The z-axes coincide. The angle between the unit vectors $\mathbf{e}_r$ and $\mathbf{e}_x$ is θ.

1.4. Given is the position vector

$$\mathbf{r} = \{\mathbf{e}_i\}^T \{r_i\}$$

with $i = x, y, z$, and the angular velocity of the $Oxyz$ coordinate system

$$\boldsymbol{\Omega} = \{\mathbf{e}_i\}\{\Omega_i\}.$$

The first derivative of the position vector is

$$\dot{\mathbf{r}} = \{\mathbf{e}_i\}^T \{\dot{r}_i\} + \{\dot{\mathbf{e}}_i\}^T \{r_i\}$$

which may also be written

$$\dot{\mathbf{r}} = \overset{\circ}{\mathbf{r}} + \boldsymbol{\Omega} \times \mathbf{r}.$$

Show that

$$\{\dot{\mathbf{e}}_i\}^T \{r_i\} = \boldsymbol{\Omega} \times \mathbf{r}.$$

1.5. Given the standard basis vectors

$$\mathbf{e}_x = [\mathbf{e}_x \ \mathbf{e}_y \ \mathbf{e}_z] \begin{bmatrix} 1 \\ 0 \\ 0 \end{bmatrix}$$

etc. Show that

$$\dot{\mathbf{e}}_x = \boldsymbol{\Omega} \times \mathbf{e}_x$$
$$\dot{\mathbf{e}}_y = \boldsymbol{\Omega} \times \mathbf{e}_y$$
$$\dot{\mathbf{e}}_z = \boldsymbol{\Omega} \times \mathbf{e}_z$$

and subsequently find

$$[\dot{\mathbf{e}}_i]^T = [\dot{\mathbf{e}}_x \ \dot{\mathbf{e}}_y \ \dot{\mathbf{e}}_z] =$$

in terms of the basis.

1.6. By how much do

$$(\dot{H})_x \quad \text{and} \quad (\dot{H_x})$$

differ, if the $Cxyz$ coordinate system used rotates at an angular velocity $\boldsymbol{\Omega}$? Which of the two quantities is usually simplified to $\dot{H}_x$?

1.7. Based on the definition for the inertia moments, (a) show that

$$I_{xx} + I_{yy} \geq I_{zz}$$

$$I_{yy} + I_{zz} \geq I_{xx}$$

$$I_{zz} + I_{xx} \geq I_{yy} .$$

(b) If $I_{zz} = 0$, what is the relationship between I_{xx} and I_{yy}? (c) If $I_{zz} = I_{yy} = 0$, what is I_{xx}? (d) If $I_{zz} \geq I_{xx}$ and $I_{zz} \geq I_{yy}$, establish the upper and lower limits for I_{zz} in terms of $I_{xx} + I_{yy}$.

1.8. Given is a plate of ISO-A size, i.e., $c \simeq 0$, $b = \sqrt{2}a$. Determine the principal inertia tensor

$$[I] = \frac{ma^2}{12} \begin{bmatrix} & 0 & 0 \\ 0 & & 0 \\ 0 & 0 & \end{bmatrix} .$$

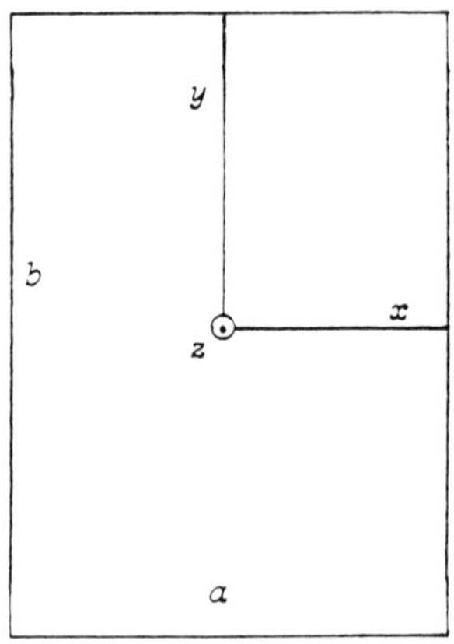

1.9. Using a spatial dumbbell configuration, determine c for a gyroscope where $A = B$ and $C/A = 2/3$.

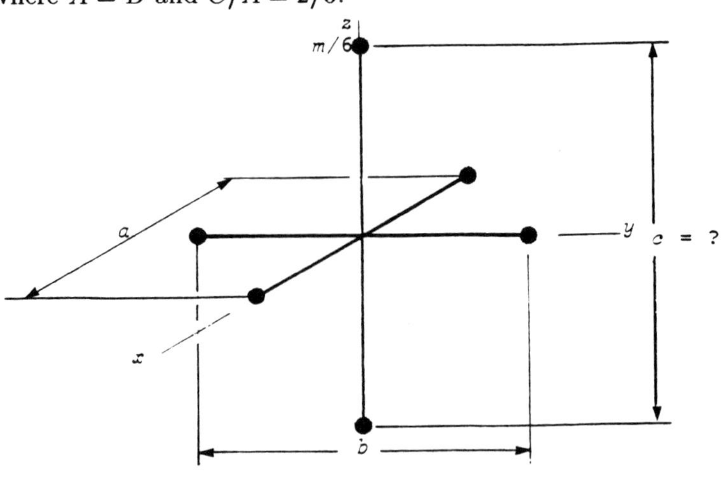

1.10. A circular cylindrical gyro of uniform mass distribution has a height l and a radius a. What is the ratio l/a for the inertia tensor if it is to assume the spherical form

$$[I] = \begin{bmatrix} A & 0 & 0 \\ 0 & A & 0 \\ 0 & 0 & A \end{bmatrix}.$$

1.11. Given is a sphere of uniform mass distribution and radius R. Determine the inertia moment tensor for the $Oxyz$ coordinate system shown.

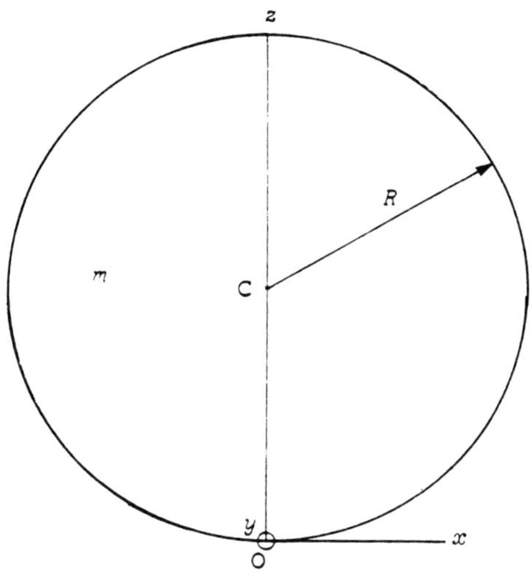

1.12. A flat plate has a ratio of sides 1 to 2. Locate the principal axes of a coordinate system whose origin is located at 0. The mass distribution is uniform.

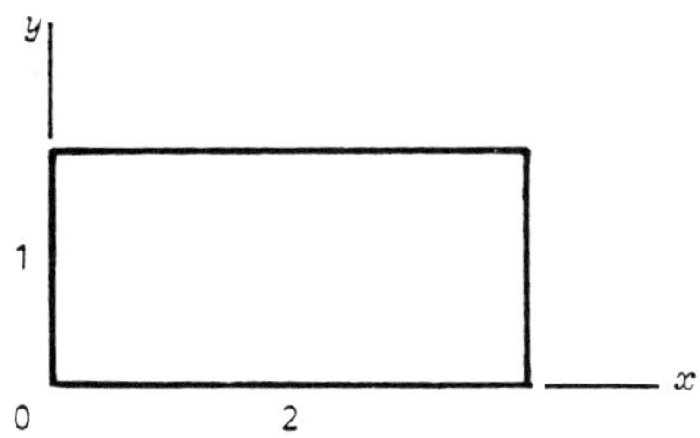

1.13. Given is the matrix

$$[I_{ij}] = \begin{bmatrix} 3 & 1 & 0 \\ 1 & 3 & 0 \\ 0 & 0 & 4 \end{bmatrix}.$$

Find the three eigenvalues.

1.14. Given is a cube of uniform mass distribution and side length a. The origin O of the xyz coordinate system is in one corner of the cube. (a) Determine the inertia tensor. (b) Determine the principal direction cosines of the principal axis associated with the smallest principal inertia moment.

1.15. Given ia a parallelepiped body of uniform density and sides

$$a = 1\,\text{m}$$

$$b = 2\,\text{m}$$

$$c = 3\,\text{m}.$$

(a) Determine the inertia tensor (in matrix form) for an $Oxyz$ coordinate system along the edges as shown.

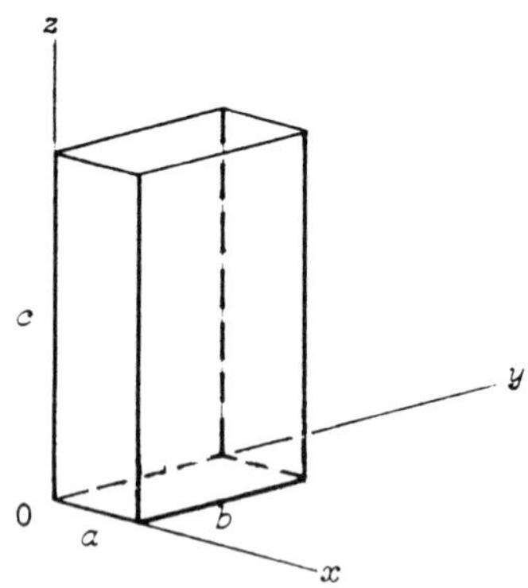

(b) Determine the principal inertia tensor (in matrix form).

(c) Compute and sketch the principal axes.

1.16. Expand the expression for the rotational kinetic energy

$$T = \frac{1}{2}\{\omega_i\}^T[I_{ij}]\{\omega_j\}$$

by using a matrix multiplication tree.

	$\{\omega_j\}$
$[I_{ij}]$	
$\{\omega_i\}^T$	

1.17. Given is a rectangular parallelepiped made of steel. Its sides are $a =$ 10 mm, b = 30 mm, c = 80 mm. It is fixed at corner point O and momentarily rotating at

$$\boldsymbol{\omega} = [\mathbf{e}_x\ \mathbf{e}_y\ \mathbf{e}_z] \begin{bmatrix} 60 \\ 80 \\ 240 \end{bmatrix} \quad \text{rev/min.}$$

Make (a) a sketch, determine (b) the angular momentum, and (c) the kinetic energy.

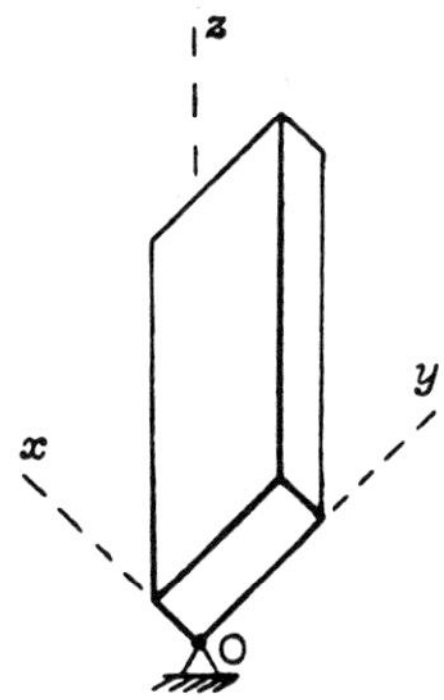

1.18. Given are the Euler frequencies. Find the angular velocity components along the space-fixed coordinates $CXYZ$.

$$\begin{bmatrix} \omega_X \\ \omega_Y \\ \omega_Z \end{bmatrix} = \begin{bmatrix} & & \\ & & \\ & & \end{bmatrix} \begin{bmatrix} \dot{\psi} \\ \dot{\nu} \\ \dot{\sigma} \end{bmatrix}.$$

1.19. Given is the matrix

$$[T] = \begin{bmatrix} \sin\nu\sin\sigma & \cos\sigma & 0 \\ \sin\nu\cos\sigma & -\sin\sigma & 0 \\ \cos\nu & 0 & 1 \end{bmatrix}.$$

Find the inverse.

1.20. Multiply the transformation matrices

$$[T]_1 = \begin{bmatrix} \sin\nu\sin\sigma & \cos\sigma & 0 \\ \sin\nu\cos\sigma & -\sin\sigma & 0 \\ \cos\nu & 0 & 1 \end{bmatrix}$$

and

$$[T]_2 = \begin{bmatrix} \frac{\sin\sigma}{\sin\nu} & \frac{\cos\sigma}{\sin\nu} & 0 \\ \cos\sigma & -\sin\sigma & 0 \\ -\frac{\sin\sigma}{\tan\nu} & -\frac{\cos\sigma}{\tan\nu} & 1 \end{bmatrix}.$$

1.21. A spatial dumbbell configuration with concentrated mass elements has $a = 1$ m, $b = 2$ m, and $c = 2 + 2t$ m, where t is measured in seconds. (a) Find the principal inertia tensor for $t = 0$. (b) Find the principal inertia tensor for $t = 2s$. (c) Determine the time derivative of the inertia tensor.

1.22. Derive the Euler equations for principal inertia moments for the case that $\dot{A} \neq 0$, $\dot{B} \neq 0$, and $\dot{C} \neq 0$.

1.23. Design a spatial dumbbell satellite with concentrated tip masses such that $\dot{A} = \dot{B} =$ constant and $\dot{C} = 0$.

1.24. For axisymmetric gyros with

$$[I] = \begin{bmatrix} A & 0 & 0 \\ 0 & A & 0 \\ 0 & 0 & C \end{bmatrix}$$

it is sometimes advantageous to use an $Ouvz$ coordinate system, which rotates at

$$\boldsymbol{\Omega} = [\mathbf{e}_u \ \mathbf{e}_v \ \mathbf{e}_z] \begin{bmatrix} \dot{\nu} \\ \dot{\psi} \sin \nu \\ \dot{\psi} \cos \nu \end{bmatrix}$$

while the gyro itself rotates at

$$\boldsymbol{\omega} = [\mathbf{e}_u \ \mathbf{e}_v \ \mathbf{e}_z] \begin{bmatrix} \dot{\nu} \\ \dot{\psi} \sin \nu \\ \dot{\psi} \cos \nu + \dot{\sigma} \end{bmatrix}.$$

Show that the gyro's angular momentum becomes

$$\mathbf{H} = [\mathbf{e}_u \ \mathbf{e}_v \ \mathbf{e}_z] \begin{bmatrix} A\dot{\nu} \\ A\dot{\psi} \sin \nu \\ C(\dot{\psi} \cos \nu + \dot{\sigma}) \end{bmatrix}$$

and the Euler equations become

$$M_u = A(\ddot{\nu} - \dot{\psi}^2 \sin \nu \cos \nu) + C\dot{\psi}(\dot{\psi} \cos \nu + \dot{\sigma}) \sin \nu$$

$$M_v = A(\ddot{\psi} \sin \nu + 2\dot{\nu}\dot{\psi} \cos \nu) - C\dot{\nu}(\dot{\psi} \cos \nu + \dot{\sigma})$$

$$M_z = C\frac{d}{dt}(\dot{\psi} \cos \nu + \dot{\sigma}).$$

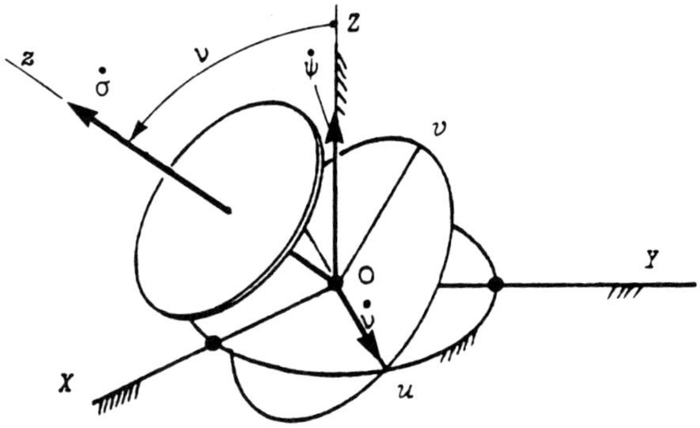

1.25. Sketch a Magnus shape triangle and locate the following gyros in it.

$$\text{(a)} \quad [I] = \begin{bmatrix} 0.5 & 0 & 0 \\ 0 & 0.5 & 0 \\ 0 & 0 & 0 \end{bmatrix}$$

$$\text{(b)} \quad [I] = \begin{bmatrix} 0.3 & 0 & 0 \\ 0 & 0.2 & 0 \\ 0 & 0 & 0.5 \end{bmatrix}$$

$$\text{(c)} \quad [I] = \begin{bmatrix} 0.35 & 0 & 0 \\ 0 & 0.25 & 0 \\ 0 & 0 & 0.4 \end{bmatrix}$$

$$\text{(d)} \quad [I] = \begin{bmatrix} 0.45 & 0 & 0 \\ 0 & 0.15 & 0 \\ 0 & 0 & 0.4 \end{bmatrix}$$

$$\text{(e)} \quad [I] = \begin{bmatrix} 0.35 & 0 & 0 \\ 0 & 0.45 & 0 \\ 0 & 0 & 0.2 \end{bmatrix}.$$

1.26. Locate a gyro, whose inertia tensor is

$$[I] = \begin{bmatrix} 0.45 & 0 & 0 \\ 0 & 0.3 & 0 \\ 0 & 0 & 0.25 \end{bmatrix}$$

(a) in the Magnus shape triangle, and make a sketch.

(b) Compute k_1, k_2, and k_3, sketch a modified Magnus shape triangle and locate the gyro in it.

1.27. A gyro has reached

$$\psi = 750°$$
$$\nu = 30°$$
$$\sigma = 3630°$$

and at that moment in time, its angular velocity is

$$\boldsymbol{\omega} = [\mathbf{e}_x\ \mathbf{e}_y\ \mathbf{e}_z]\begin{bmatrix} 3 \\ 4 \\ 5 \end{bmatrix} \text{ rad/s.}$$

Determine the Euler frequencies.

1.28. A rigid rod satellite of length $l = 20$ m, and mass $m = 120$ kg, rotates about its x-axis at a rate of 0.2 rad/s. Determine (a) the inertia matrix of the satellite; (b) its angular momentum vector (about point 0); (c) its kinetic energy; and (d) the velocity of point C_1.

The link connecting portions 1 and 2 of the satellite is then severed to allow each half to move freely. The separation process (e.g., by pyrotechnics) is quick and there is no interference between the two portions during or after separation. Each portion can be modeled as a rod satellite of 10-m length and 60-kg mass. Determine (e) the inertia matrix of portion 1 (with respect to C_1); (f) its angular momentum (with respect to C_1) assuming that portion 1 continues to rotate at 0.2 rad/s; (g) its kinetic energy of translation; (h) its kinetic energy of rotation; (i) the energy loss during the separation process; (j) the angular momentum of portion 1 with respect to point O; and (k) the angular momentum loss during the separation process.

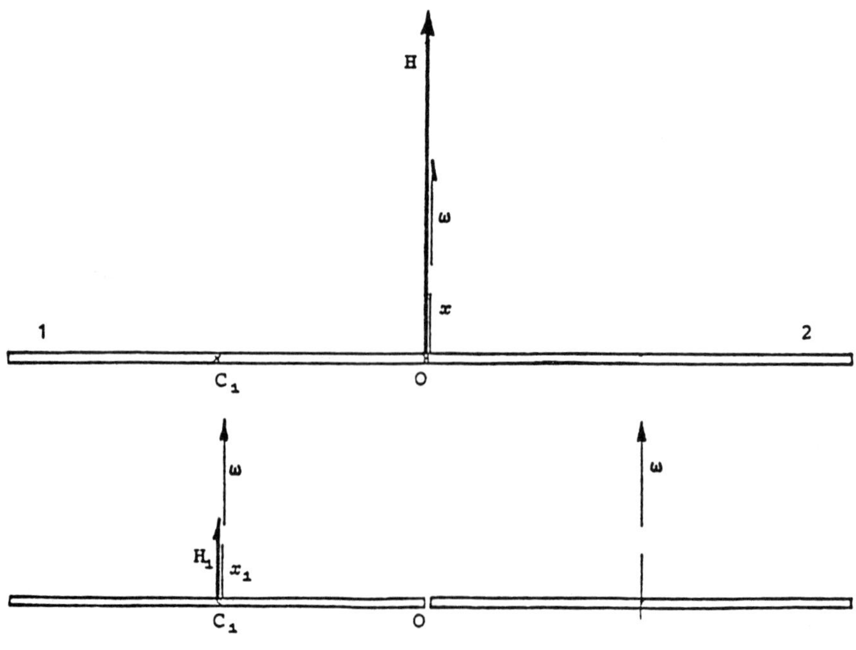

1.29. A torquefree gyro has an inertia tensor of

$$[I_{ij}] = \begin{bmatrix} 150 & 0 & 0 \\ 0 & 150 & 0 \\ 0 & 0 & 50 \end{bmatrix} \ \mathrm{m^2kg} \tag{P1.1}$$

with $i, j = x, y, z$. The gyro rotates at

$$\boldsymbol{\omega}_1 = [\mathbf{e}_x \ \mathbf{e}_y \ \mathbf{e}_z] \begin{bmatrix} 0 \\ 0 \\ 5 \end{bmatrix} \ \mathrm{rad/s}. \tag{P1.2}$$

(a) Determine the angular momentum $\mathbf{H}$. (b) Determine the kinetic energy T_1.

There is no external torque acting, while two antenna booms are extended. The new inertia tensor is

$$[I_{ij}] = \begin{bmatrix} 150 & 0 & 0 \\ 0 & 200 & 0 \\ 0 & 0 & 100 \end{bmatrix} \ \mathrm{m^2kg}. \tag{P1.3}$$

(c) Determine the new angular velocity $\boldsymbol{\omega}_2$. (d) Determine the new kinetic energy T_2.

Subsequently the antenna booms are pulled in again until the inertia tensor reassumes the value given by equation (P1.1). (e) Determine the new angular velocity $\boldsymbol{\omega}_3$. (f) Determine the new kinetic energy T_3. (g) If you find that T_2 is smaller than T_1, can you explain where the difference may have gone? (h) If you find that T_3 is larger than T_2, can you explain where the additional energy may have come from?

1.30. A large circular spacecraft is rotating slowly at an angular velocity $\Omega = 0.4$ rad/s to produce an artificial weight field. An astronaut, on his regulation daily jogging program, is instructed to run in the direction of rotation of the spacecraft. (a) How fast must he run (in m/s) to experience his regular weight? (b) In his leisure time, he prefers to run in the opposite direction, because it is more "fun." Explain why. [Assume that the spacecraft's angular velocity change due to the jogger's motion is negligible.]

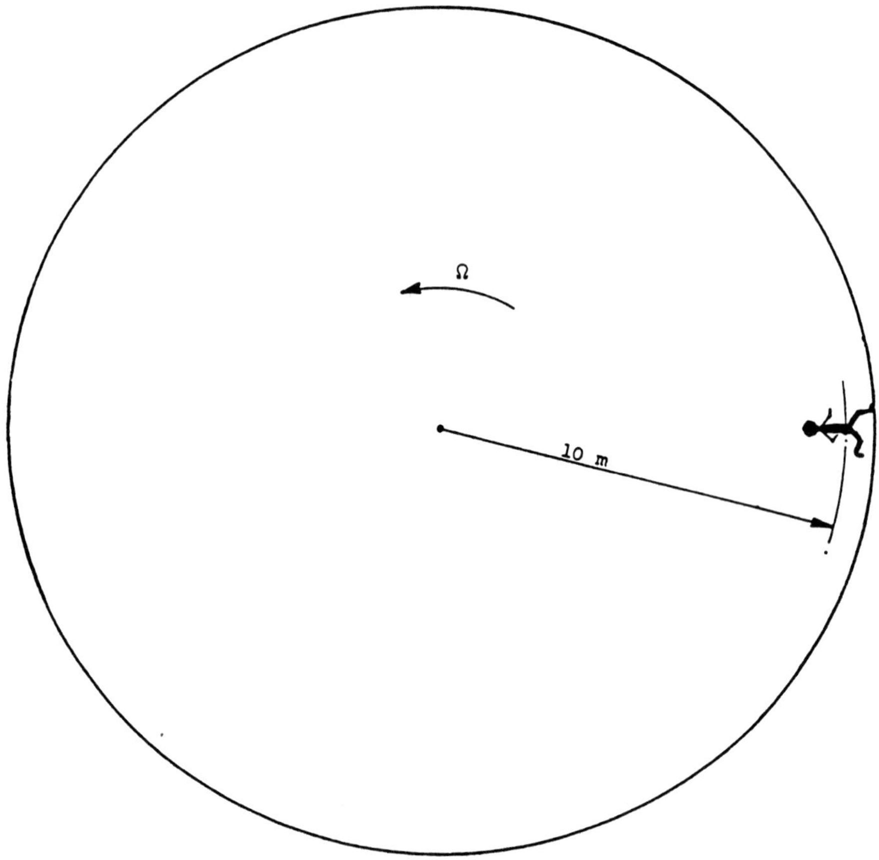

1.31. Show that the Smelt parameters k_1, k_2, and k_3 are functions of two variables only, by introducing A/C and B/C. Also show that $k_1 + k_2 + k_3 + k_1 k_2 k_3 = 0$.

1.32. Given is a thin rod of length l and mass m, mounted in a ball-and-socket joint as shown and rotating at an angular velocity

$$\boldsymbol{\omega} = [\mathbf{e}_x \ \mathbf{e}_y \ \mathbf{e}_z] \begin{bmatrix} 0 \\ \omega \\ 0 \end{bmatrix}$$

with ω = constant. The bars a and b are massless. Find (a) the inertia tensor; (b) the angular momentum; (c) the kinetic energy; (d) the angular momentum time derivative; (e) the external torque applied; (f) the angular momentum time derivative if $\omega = \alpha t$; and (g) the external torque applied if $\omega = \alpha t$.

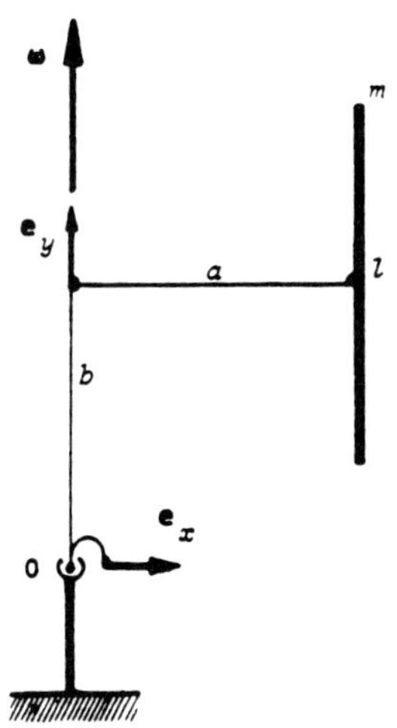

2

Center of Gravity

The position vector $\boldsymbol{\rho}_C$ linking the origin of an arbitrary coordinate system and the *mass center* C of a massive body is defined by

$$\boldsymbol{\rho}_C = \frac{1}{m}\int_m \boldsymbol{\rho}\, dm. \tag{2.1}$$

For the subsequent derivation of the location of the center of gravity of a satellite of finite size we assume that the potential field surrounding the master mass is spherical, as in the case of a master mass of point size. We shall show that the center of gravity is *not* a fixed point within the body but that its location is determined by the *attitude* (i.e., angular position with respect to the gradient of the potential field) of the satellite body and by the *distance* between satellite and master.

As a consequence we shall find that satellites of finite size are subject to a *torque* in addition to the attraction force. Another consequence is that the *orbital period* of satellites of *finite* size differs from the Kepler period (i.e., the orbital period of a *point* satellite).

For the general case, we can expect that a satellite changes its attitude continually and thus the location of its center of gravity.

The center of gravity of a body is that point at which the concentrated mass m of the body would have to be located in order to be attracted by the same gravitational force as the distributed mass of the body:

$$\mathbf{K} = -\frac{\mu m}{\rho_G^3}\boldsymbol{\rho}_G = -\mu\int_m \frac{\boldsymbol{\rho}}{\rho^3} dm. \tag{2.2}$$

where μ is the gravitational attraction constant of the master.

From equation (2.2) we obtain for the position vector $\boldsymbol{\rho}_G$ of the *center of gravity* G of a satellite body of finite size

$$\frac{\boldsymbol{\rho}_G}{\rho_G^3} = \frac{1}{m}\int_m \frac{\boldsymbol{\rho}}{\rho^3} dm. \tag{2.3}$$

Integral (2.3) can be solved explicitly only for a few special cases. We will at first consider such a special case and determine the location of the center of gravity for a one-dimensional *rod satellite* of mass m, length l, and inertia tensor

$$[I] = \begin{bmatrix} 0 & 0 & 0 \\ 0 & \frac{ml^2}{12} & 0 \\ 0 & 0 & \frac{ml^2}{12} \end{bmatrix}.$$

Subsequently, we will consider *small* satellites, i.e., those whose dimensions are small compared to the distance between master and satellite.

Then we will determine the torque of the Kepler force about the mass center of the satellite and devise an expression that is valid again for small satellites.

Eventually, the case of *locked rotation* is introduced, for which expressions for the orbital period for a given attitude of the satellite can be found.

2.1 Rod Satellite

Let us consider a rather simple case and determine the location of the center G of gravity of a rod satellite of uniform mass distribution at a *radial* attitude as shown in Figure 2.1.

Since we deal with a one-dimensional problem, equation (2.3) can be simplified to read

$$\frac{1}{\rho_G^2} = \frac{1}{m}\int_m \frac{dm}{\rho^2}. \tag{2.4}$$

Considering that the mass distribution is uniform, we introduce $dm = \frac{m}{l}d\rho$ and obtain

$$\frac{1}{\rho_G^2} = \frac{1}{l}\int_{\rho_C - l/2}^{\rho_C + l/2} \frac{d\rho}{\rho^2}. \tag{2.5}$$

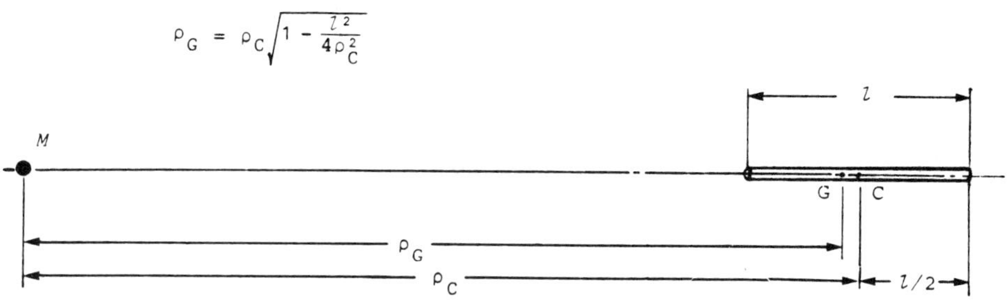

FIGURE 2.1. Center G of gravity and mass center C in case of radial attitude.

Integral (2.5) is easily solved. The location of the center of gravity is

$$\rho_G = \rho_C\sqrt{1 - \frac{l^2}{4\rho_C^2}} \tag{2.6}$$

where ρ_C gives the location of the mass center C of the satellite, which is assumed to be known. Judging by equation (2.6), the center of gravity is located *more closely* to the master than the mass center.

For the case in which the rod satellite assumes an attitude *perpendicular* to the connecting line between satellite mass center C and master (Figure

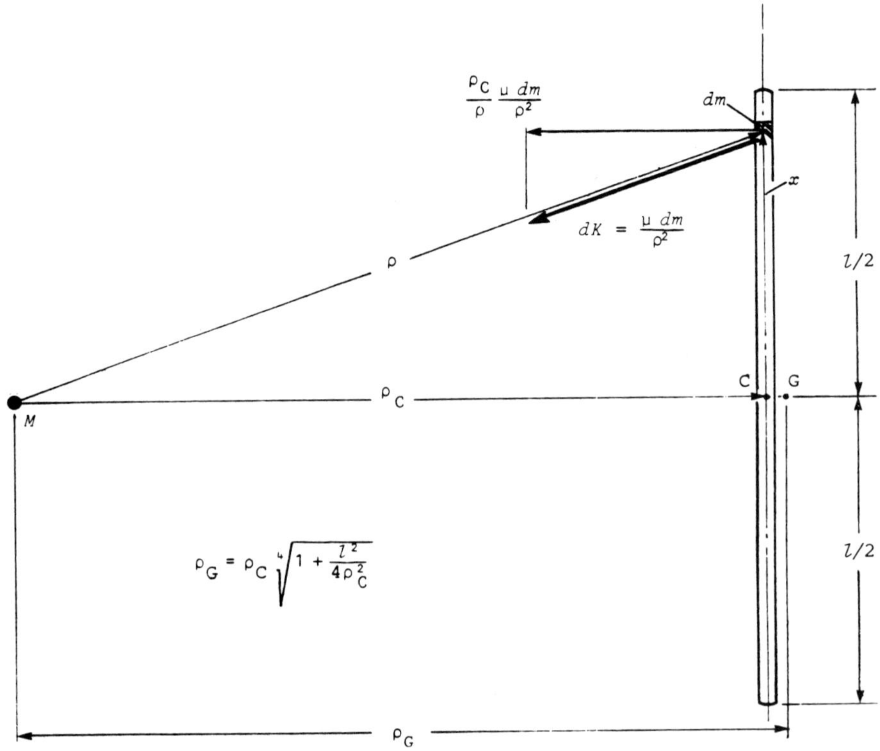

FIGURE 2.2. Center of gravity in case of tangential attitude.

2.2), we conclude for reasons of symmetry that the sum of all attraction force components perpendicular to the connecting line ρ_C must vanish. For the attraction force components parallel to ρ_C, we have

$$\frac{\mu m}{\rho_G^2} = \mu \int \frac{\rho_C}{\rho} \frac{dm}{\rho^2}. \tag{2.7}$$

Introducing $dm = \frac{m}{l} dx$ because of the uniform mass distribution, and $\rho = \sqrt{\rho_C^2 + x^2}$, and after some cancellations

$$\frac{1}{\rho_G^2} = \frac{\rho_C}{l} \int_{-l/2}^{+l/2} \frac{dx}{\left(\sqrt{\rho_C^2 + x^2}\right)^3} \tag{2.8}$$

which, upon integration, results in

$$\rho_G = \rho_C \sqrt[4]{1 + \frac{l^2}{4\rho_C^2}}. \tag{2.9}$$

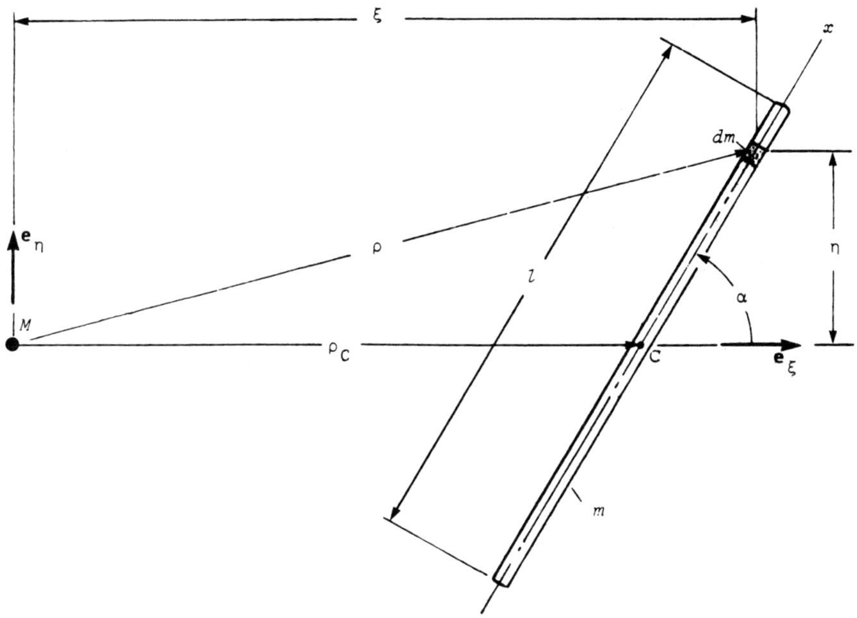

FIGURE 2.3. Rod satellite, inclined by an angle α with respect to the radial direction.

We conclude that the center of gravity G of the satellite for this case is located further *away* from the master mass than the mass center C of the satellite (Figure 2.2).

In Figure 2.3 the general case of a one-dimensional rod satellite is presented whose center line is inclined at an arbitrary angle α with respect to the position vector ρ_C of the satellite's mass center. Equation (2.3) resolved into components reads

$$\frac{\xi_G \mathbf{e}_\xi + \eta_G \mathbf{e}_\eta}{\rho_G^2} = \frac{1}{m} \int_m \frac{\xi \mathbf{e}_\xi + \eta \mathbf{e}_\eta}{\rho^3} dm. \tag{2.10}$$

We have made use of

$$\boldsymbol{\rho}_G = [\mathbf{e}_\xi \ \mathbf{e}_\eta] \begin{bmatrix} \xi_G \\ \eta_G \end{bmatrix}$$

and

$$\boldsymbol{\rho} = [\mathbf{e}_\xi \ \mathbf{e}_\eta] \begin{bmatrix} \xi \\ \eta \end{bmatrix}$$

and an $M\xi\eta$ coordinate system, whose ξ-axis coincides with the position vector ρ_C to the mass center C of the satellite. The ξ-coordinate of the

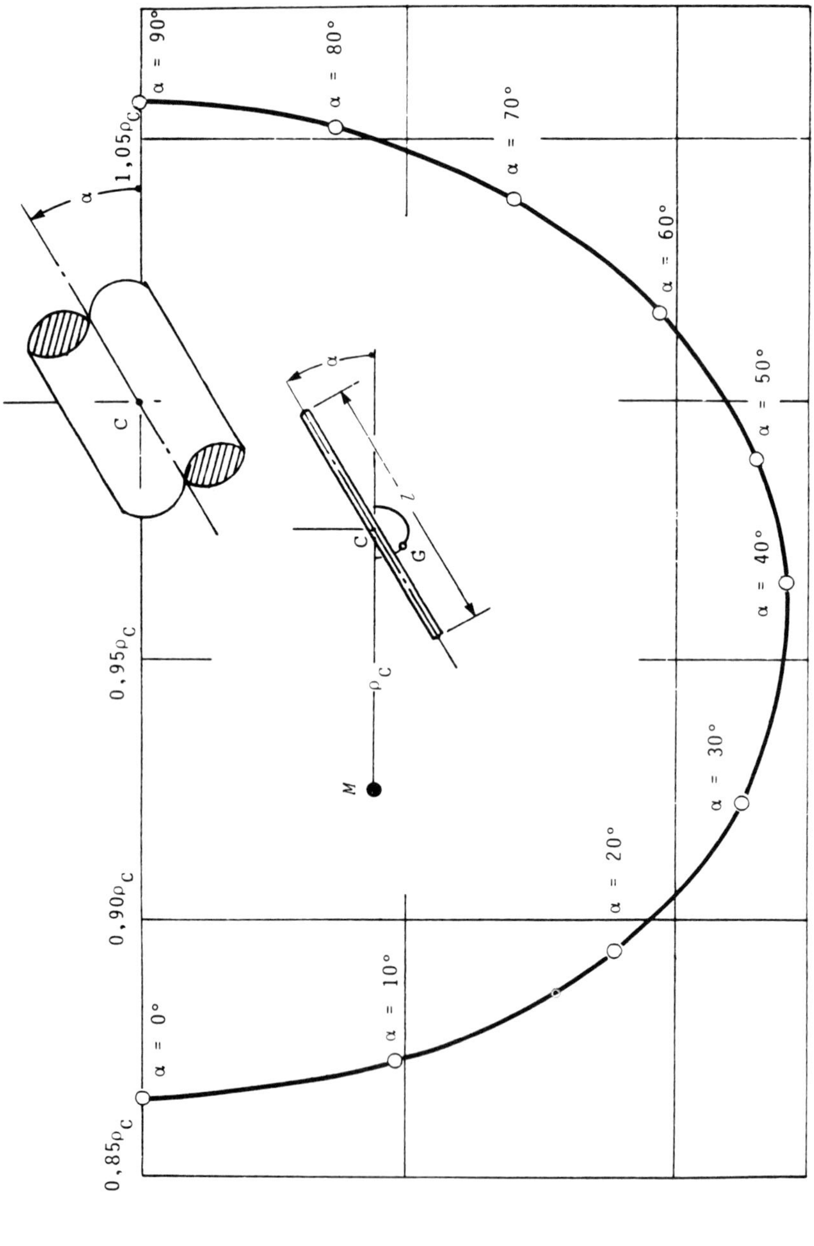

FIGURE 2.4. Location of the center G of gravity of a rod satellite for the case $\rho_C = l$.

center of gravity is thus

$$\frac{\xi_G}{\rho_G^3} = \frac{1}{m} \int_m \frac{\xi}{\rho^3} dm. \tag{2.11}$$

With $dm = \frac{m}{l} dx$ and $\xi = \rho_C + x \cos\alpha$ and

$$\rho = \sqrt{\rho_C^2 + 2\rho_C x \cos\alpha + x^2}$$

we obtain

$$\frac{\xi_G}{\rho_G^3} = \frac{1}{l\rho_C} \int_{-l/2}^{+l/2} \frac{1 + \frac{x}{\rho_C} \cos\alpha}{\sqrt{1 + 2\frac{x}{\rho_C} \cos\alpha + \frac{x^2}{\rho_C^2}}^{\,3}} dx \tag{2.12}$$

and upon integration

$$\frac{\xi_G}{\rho_G^3} = \frac{1}{2\rho_C^2} \left\{ \frac{1}{\sqrt{1 + \frac{l}{\rho_C} \cos\alpha + \frac{l^2}{4\rho_C^2}}} + \frac{1}{\sqrt{1 - \frac{l}{\rho_C} \cos\alpha + \frac{l^2}{4\rho_C^2}}} \right\}. \tag{2.13}$$

For the η-coordinate we have

$$\frac{\eta_G}{\rho_G^3} = \frac{1}{l\rho_C^2} \int_{-l/2}^{+l/2} \frac{\frac{x}{\rho_C} \sin\alpha}{\sqrt{1 + 2\frac{x}{\rho_C} \cos\alpha + \frac{x^2}{\rho_C^2}}^{\,3}} dx \tag{2.14}$$

and upon integration

$$\frac{\eta_G}{\rho_G^3} = -\frac{1}{l\rho_C \sin\alpha} \left\{ \frac{1 + \frac{l}{2\rho_C} \cos\alpha}{\sqrt{1 + \frac{l}{\rho_C} \cos\alpha + \frac{l^2}{4\rho_C^2}}} - \frac{1 - \frac{l}{2\rho_C} \cos\alpha}{\sqrt{1 - \frac{l}{\rho_C} \cos\alpha + \frac{l^2}{4\rho_C^2}}} \right\}. \tag{2.15}$$

For the theoretical case of a very long, rigid rod satellite, with $l = \rho_C$, the location of the gravity center is depicted in Figure 2.4.

2.2 Small Rod Satellite

In all practical cases a rod satellite can be considered *small*, i.e., $l << \rho_C$ (Figure 2.1). By expansion into series, which are broken off after the term with l^2/ρ_C^2, one gets for the abscissa of the center of gravity, from equation (2.13),

$$\frac{\xi_G}{\rho_G^3} = \frac{1}{\rho_C^2} \left\{ 1 + \frac{1}{8} \frac{l^2}{\rho_C^2} (3\cos^2\alpha - 1) \right\} \tag{2.16}$$

and for the ordinate, from equation (2.15)

$$\frac{\eta_G}{\rho_G^3} = -\frac{l^2}{8\rho_C^4}\sin 2\alpha. \tag{2.17}$$

Further approximation is possible by realizing that $\xi_G \simeq \rho_G$, i.e.,

$$\xi_G = \rho_C\left(1 - \frac{1}{16}\frac{l^2}{\rho_C^2}(3\cos^2\alpha - 1)\right). \tag{2.18}$$

Upon introduction of a $C\xi'\eta'$ coordinate system originating in the mass center of the rod satellite (Figure 2.5) with axes parallel to the $M\xi\eta$ system and

$$\xi_G' = \xi_G - \rho_C \tag{2.19}$$

the abscissa of the center of gravity becomes

$$\xi_G' = -\frac{1}{16}\frac{l^2}{\rho_C}(3\cos^2\alpha - 1) \tag{2.20}$$

and its ordinate

$$\eta_G' = -\frac{1}{8}\frac{l^2}{\rho_C}\sin 2\alpha. \tag{2.21}$$

The resultant Kepler force is defined by

$$\mathbf{K} = -\mu m\frac{\boldsymbol{\rho}_G}{\rho_G^3}. \tag{2.22}$$

From equations (2.16) and (2.17) for a small rod satellite

$$\frac{\boldsymbol{\rho}_G}{\rho_G^3} = \frac{1}{\rho_C^2}\left\{1 + \frac{1}{8}\frac{l^2}{\rho_C^2}(3cos^2\alpha - 1)\right\}\mathbf{e}_\xi - \frac{l^2}{8\rho_C^4}\sin 2\alpha\mathbf{e}_\eta \tag{2.23}$$

and thus the Kepler force becomes

$$\mathbf{K} = -\frac{\mu m}{\rho_C^2}\left[\left\{1 + \frac{1}{8}\frac{l^2}{\rho_C^2}(3\cos^2\alpha - 1)\right\}\mathbf{e}_\xi - \frac{l^2}{8\rho_C^2}\sin 2\alpha\mathbf{e}_\eta\right]. \tag{2.24}$$

Observing that the inertia moment $B = ml^2/12$, we can write for the Kepler force

$$\mathbf{K} = -\frac{\mu m}{\rho_C^2}\left[\left\{1 + \frac{3}{2}\frac{B}{m\rho_C^2}(3\cos^2\alpha - 1)\right\}\mathbf{e}_\xi - \frac{3}{2}\frac{B}{m\rho_C^2}\sin 2\alpha\mathbf{e}_\eta\right]. \tag{2.25}$$

The small rod satellite, which represents an important case in practical applications, has a resultent Kepler force of magnitude

$$K = \frac{\mu m}{\rho_G^2} \tag{2.26}$$

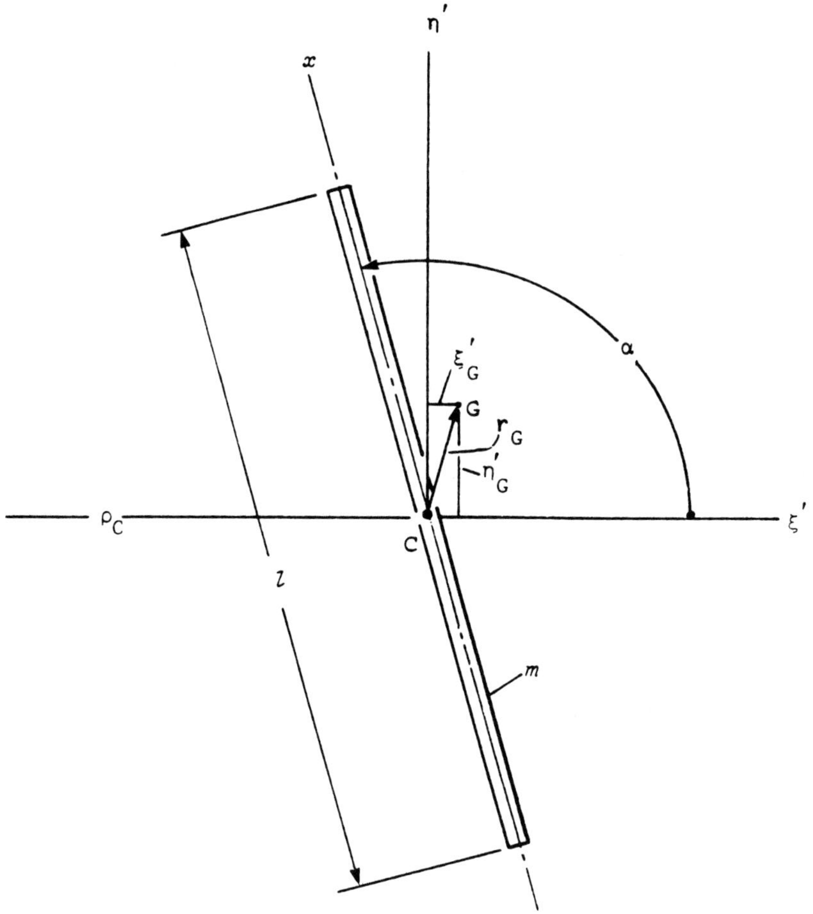

FIGURE 2.5. Location of center of gravity.

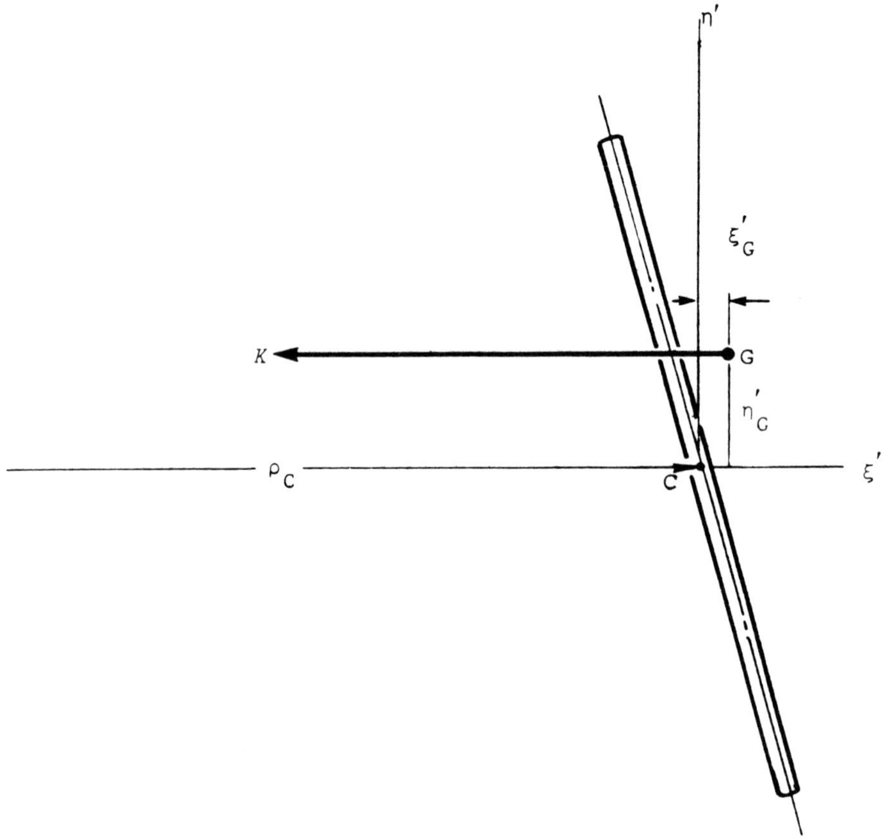

FIGURE 2.6. Torque of Kepler force about mass center.

acting at a moment arm of

$$\eta'_G = -\frac{1}{8}\frac{l^2}{\rho_C}\sin 2\alpha. \tag{2.27}$$

By multiplying the two and introducing a direction, we obtain for the moment of the Kepler force about the mass center of the satellite (Figure 2.6)

$$\mathbf{M} = \frac{\mu m}{8\rho_C\rho_G^2} l^2 \sin 2\alpha \frac{\mathbf{x} \times \boldsymbol{\rho}_C}{|\mathbf{x} \times \boldsymbol{\rho}_C|}. \tag{2.28}$$

This moment always tends to turn the rod satellite into a radial attitude (i.e., towards $\alpha = 0°$ or $\alpha = 180°$).

2.3 Center of Gravity for a Small Satellite, General Case

In Figure 2.7 a satellite is shown, whose mass center C is located at a position $\boldsymbol{\rho}_C$ from the point master M. We introduce a $Cxyz$ coordinate system fixed to the satellite with the coordinates representing the principal axes. The inertia tensor of the satellite is thus

$$[I] = \begin{bmatrix} A & 0 & 0 \\ 0 & B & 0 \\ 0 & 0 & C \end{bmatrix}. \tag{2.29}$$

The origin of the $M\xi\eta\zeta$ coordinate system is the point master M. The coordinates are parallel to the corresponding coordinates of the $Cxyz$ system.

Equation (2.3) can be resolved into components

$$\frac{\xi_G \mathbf{e}_\xi + \eta_G \mathbf{e}_\eta + \zeta_G \mathbf{e}_\zeta}{\rho_G^3} = \frac{1}{m}\int_m \frac{\xi \mathbf{e}_\xi + \eta \mathbf{e}_\eta + \zeta \mathbf{e}_\zeta}{\rho^3} dm. \tag{2.30}$$

Rather than treating the problem as a whole, we shall just look at one component. Because of the nature of the problem it does not matter which component we choose. Proceeding with ξ, we have

$$\frac{\xi_G}{\rho_G^3} = \frac{1}{m}\int_m \frac{\xi}{\rho^3} dm. \tag{2.31}$$

Since $\xi = \xi_C + x$,

$$\frac{\xi_C + x_G}{\rho_G^3} = \frac{1}{m}\int_m \frac{\xi_C + x}{\rho^3} dm. \tag{2.32}$$

The law of cosines applied to the triangle with the sides $\boldsymbol{\rho}_C$, $\boldsymbol{\rho}$, and $\mathbf{r}$ (Figure 2.7) gives

$$\rho = \rho_C \sqrt{1 + 2\frac{r}{\rho_C}\cos\phi + \frac{r^2}{\rho_C^2}}. \tag{2.33}$$

For *small* satellites, we let $r << \rho_C$. Upon an expansion into a series, which we break off after the terms containing r^2/ρ_C^2, and upon introducing

$$r\cos\phi = \frac{\boldsymbol{\rho}_C \cdot \mathbf{r}}{\rho_C} = x\cos\alpha + y\cos\beta + z\cos\gamma \tag{2.34}$$

where

$$\mathbf{r} = [\mathbf{e}_x \ \mathbf{e}_y \ \mathbf{e}_z] \begin{bmatrix} x \\ y \\ z \end{bmatrix} \tag{2.35}$$

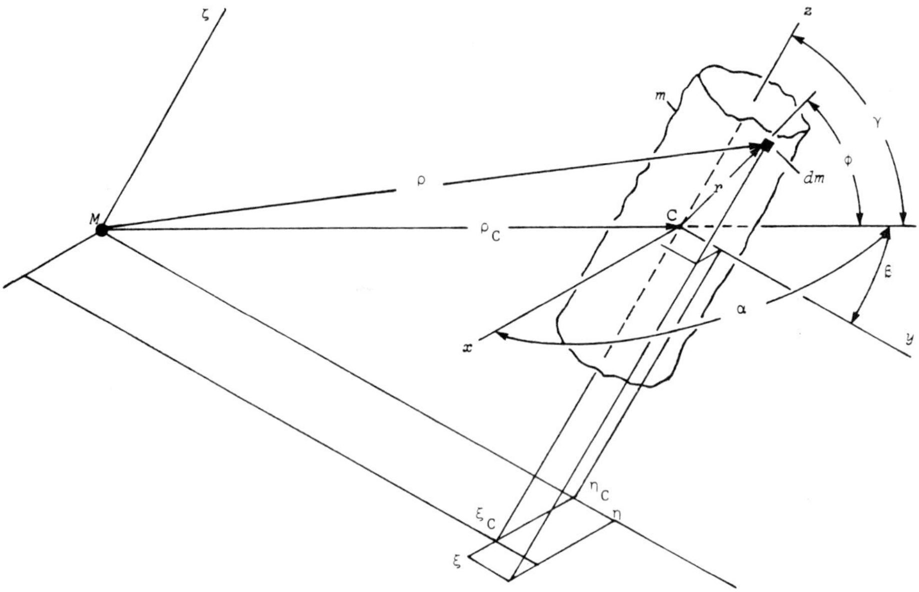

FIGURE 2.7. Location of the mass element dm of a satellite.

and

$$\boldsymbol{\rho}_{\mathrm{C}} = [\mathbf{e}_x \ \mathbf{e}_y \ \mathbf{e}_z] \begin{bmatrix} \rho_{\mathrm{C}} \cos\alpha \\ \rho_{\mathrm{C}} \cos\beta \\ \rho_{\mathrm{C}} \cos\gamma \end{bmatrix} \tag{2.36}$$

one obtains

$$\xi_{\mathrm{C}} + x_{\mathrm{G}} = \frac{\rho_{\mathrm{G}}^3}{\rho_{\mathrm{C}}^3} \frac{1}{m} \int_m (\xi_{\mathrm{C}} + x)[(1 - \frac{3}{\rho_{\mathrm{C}}}(x\cos\alpha + y\cos\beta + z\cos\gamma)$$

$$- \frac{3}{2\rho_{\mathrm{C}}^2}(x^2+y^2+z^2) + \frac{15}{2\rho_{\mathrm{C}}^2}(x\cos\alpha + y\cos\beta + z\cos\gamma)^2]dm. \tag{2.37}$$

Using the inertia tensor (2.29), equation (2.37) may be written

$$\xi_{\mathrm{C}} + x_{\mathrm{G}} = \frac{\rho_{\mathrm{G}}^3}{\rho_{\mathrm{C}}^3}\{\xi_{\mathrm{C}} - \frac{3\cos\alpha}{4m\rho_{\mathrm{C}}}[3(B+C) - A] + \frac{15}{4}\frac{\cos\alpha}{m\rho_{\mathrm{C}}}[(B+C-A)\cos^2\alpha$$

$$+ (C+A-B)\cos^2\beta + (A+B-C)\cos^2\gamma]\}. \tag{2.38}$$

The ratio $\rho_{\mathrm{G}}^3/\rho_{\mathrm{C}}^3$ can be treated by using equation (2.33) again. The law of cosines provides us with

$$\rho_{\mathrm{G}} = \rho_{\mathrm{C}}\sqrt{1 + 2\frac{r_{\mathrm{G}}}{\rho_{\mathrm{C}}}\cos\phi_{\mathrm{G}} + \frac{r_{\mathrm{G}}^2}{\rho_{\mathrm{C}}^2}} \tag{2.39}$$

which can be expanded into a series. We discontinue the series after terms containing r_G/ρ_C and form the third power of the ratio

$$\frac{\rho_G^3}{\rho_C^3} = 1 + \frac{3}{\rho_C}(x_G \cos\alpha + y_G \cos\beta + z_G \cos\gamma). \tag{2.40}$$

This result is entered into equation (2.38). The other two coordinates of the center G of gravity can be treated in a similar fashion. After some manipulation, we eventually obtain a system of three implicit equations with three unknowns.

$$[N]\begin{bmatrix} x_G \\ y_G \\ z_G \end{bmatrix} = -\frac{3}{4m\rho_C}\begin{bmatrix} \{3(B+C) - A - 5D\}\cos\alpha \\ \{3(C+A) - B - 5D\}\cos\beta \\ \{3(A+B) - C - 5D\}\cos\gamma \end{bmatrix} \tag{2.41}$$

where

$$D = (B+C-A)\cos^2\alpha + (C+A-B)\cos^2\beta + (A+B-C)\cos^2\gamma \tag{2.42}$$

and

$$[N] = \begin{bmatrix} 1-3\cos^2\alpha & -3\cos\alpha\cos\beta & -3\cos\alpha\cos\gamma \\ -3\cos\alpha\cos\beta & 1-3\cos^2\beta & -3\cos\beta\cos\gamma \\ -3\cos\alpha\cos\gamma & -3\cos\beta\cos\gamma & 1-3\cos^2\gamma \end{bmatrix}. \tag{2.43}$$

The inverse of the matrix $[N]$ is

$$[N]^{-1} = -\frac{1}{2}\begin{bmatrix} 3\cos^2\alpha - 2 & 3\cos\alpha\cos\beta & 3\cos\alpha\cos\gamma \\ 3\cos\alpha\cos\beta & 3\cos^2\beta - 2 & 3\cos\beta\cos\gamma \\ 3\cos\alpha\cos\gamma & 3\cos\beta\cos\gamma & 3\cos^2\gamma - 2 \end{bmatrix}. \tag{2.44}$$

Eventually, we obtain for the coordinates of the position vector $\mathbf{r}_G$ of the center G of gravity (Figure 2.8).

$$\begin{bmatrix} x_G \\ y_G \\ z_G \end{bmatrix} = \frac{3}{4m\rho_C}\begin{bmatrix} (3A - B - C - I_C) & \cos\alpha \\ (3B - C - A - I_C) & \cos\beta \\ (3C - A - B - I_C) & \cos\gamma \end{bmatrix} \tag{2.45}$$

or

$$\mathbf{r}_G = \frac{3}{4m\rho_C}[\mathbf{e}_x\ \mathbf{e}_y\ \mathbf{e}_z]\begin{bmatrix} (3A - B - C - I_C) & \cos\alpha \\ (3B - C - A - I_C) & \cos\beta \\ (3C - A - B - I_C) & \cos\gamma \end{bmatrix} \tag{2.46}$$

where

$$I_C = A\cos^2\alpha + B\cos^2\beta + C\cos^2\gamma \tag{2.47}$$

is the inertia moment of the satellite about an axis coinciding with the position vector ρ_C pointing to the mass center of the satellite.

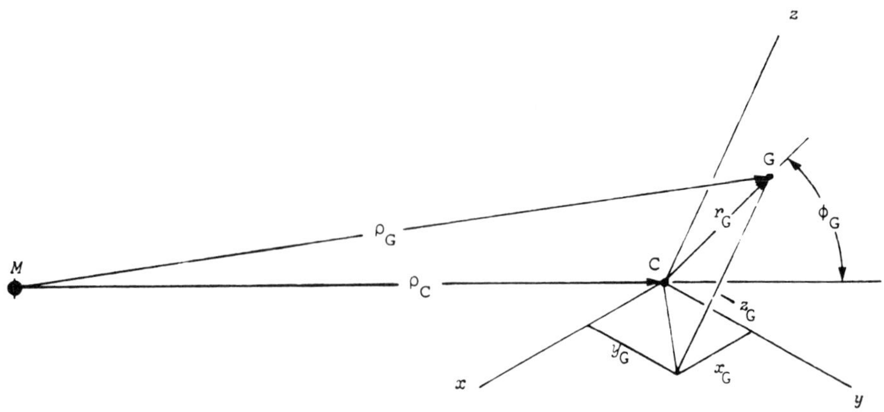

FIGURE 2.8. Location of the center G of gravity.

With the coordinates of the gravity center known, we can then combine equations (2.40) and (2.46) and obtain after a few substitutions

$$\frac{\rho_G^3}{\rho_C^3} = 1 + \frac{9}{4m\rho_C^2}(3I_C - A - B - C). \tag{2.48}$$

Recalling that the second term on the right side of equation (2.48) is very small, we can approximate the inverse by

$$\frac{\rho_C^3}{\rho_G^3} = 1 - \frac{9}{4m\rho_C^2}(3I_C - A - B - C). \tag{2.49}$$

2.4 Kepler Force for a Small Satellite

In general, the Kepler force of a satellite is given by (equation (2.22)):

$$\mathbf{K} = -\mu m \frac{\boldsymbol{\rho}_G}{\rho_G^3}.$$

From equations (2.46) and (2.49) we obtain for a *small* satellite

$$\frac{\boldsymbol{\rho}_G}{\rho_G^3} = \frac{\boldsymbol{\rho}_C + \mathbf{r}_G}{\rho_G^3} = \frac{1}{\rho_C^2}[\mathbf{e}_x\ \mathbf{e}_y\ \mathbf{e}_z]\begin{bmatrix} \left\{1 + \frac{3}{2}\frac{3A+B+C-5I_C}{m\rho_C^2}\right\}\cos\alpha \\ \left\{1 + \frac{3}{2}\frac{3B+C+A-5I_C}{m\rho_C^2}\right\}\cos\beta \\ \left\{1 + \frac{3}{2}\frac{3C+A+B-5I_C}{m\rho_C^2}\right\}\cos\gamma \end{bmatrix} \tag{2.50}$$

with

$$\frac{\boldsymbol{\rho}_C}{\rho_C} = [\mathbf{e}_x \ \mathbf{e}_y \ \mathbf{e}_z] \begin{bmatrix} \cos\alpha \\ \cos\beta \\ \cos\gamma \end{bmatrix}$$

one gets for the Kepler force either

$$\mathbf{K} = -\frac{\mu m}{\rho_C^2} \left\{ \left(1 + \frac{3(A+B+C-5I_C)}{2m\rho_C^2} \right) \frac{\boldsymbol{\rho}_C}{\rho_C} + \frac{3}{m\rho_C^2} (\mathbf{e}_x A \cos\alpha + \mathbf{e}_y B \cos\beta + \mathbf{e}_z C \cos\gamma \right\}. \tag{2.51}$$

or, using body-fixed unit vectors only,

$$\mathbf{K} = -\frac{\mu m}{\rho_C^2} [\mathbf{e}_x \ \mathbf{e}_y \ \mathbf{e}_z] \begin{bmatrix} \left\{ 1 + \frac{3}{2} \frac{3A+B+C-5I_C}{m\rho_C^2} \right\} \cos\alpha \\ \left\{ 1 + \frac{3}{2} \frac{3B+C+A-5I_C}{m\rho_C^2} \right\} \cos\beta \\ \left\{ 1 + \frac{3}{2} \frac{3C+A+B-5I_C}{m\rho_C^2} \right\} \cos\gamma \end{bmatrix}. \tag{2.52}$$

2.5 Torque about Mass Center, General Case

For the general case of a satellite in the potential field of a point master, the torque of the resultant Kepler force **K** about the mass center C of the satellite (Figure 2.9) is given by

$$\mathbf{M} = \mathbf{r}_G \times \mathbf{K} \tag{2.53}$$

Considering that $\boldsymbol{\rho}_G = \boldsymbol{\rho}_C + \mathbf{r}_G$ and considering equations (2.2) and (2.3), one obtains

$$\mathbf{M} = \mu \boldsymbol{\rho}_C \times \int_m \frac{\boldsymbol{\rho}}{\rho^3} dm. \tag{2.54}$$

Equation (2.54) cannot be solved in closed form for the general case. On the other hand, for the special case of a *small* satellite a closed-form solution can be obtained:

$$\mathbf{M} = 3\frac{\mu}{\rho_C^3} [\mathbf{e}_x \ \mathbf{e}_y \ \mathbf{e}_z] \begin{bmatrix} (C-B)\cos\gamma\cos\beta \\ (A-C)\cos\alpha\cos\gamma \\ (B-A)\cos\beta\cos\alpha \end{bmatrix} \tag{2.55}$$

where α, β, and γ are the direction angles between the position vector $\boldsymbol{\rho}_C$ and the principal axes of the satellite.

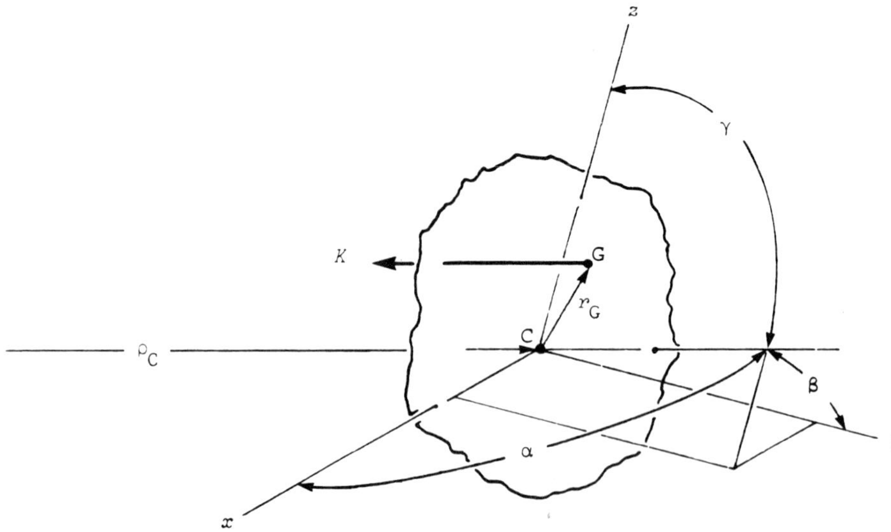

FIGURE 2.9. Torque of Kepler force about mass center.

2.6 Potential

One can establish the potential U of the Kepler force considering the finite size of a satellite. For a small satellite in the gravitational field of a point master, the potential is

$$U = -\frac{\mu m}{\rho_C}\left[1 + \frac{A+B+C-3I_C}{2m\rho_C^2}\right] \tag{2.56}$$

where an expansion into a series has been employed and smaller terms neglected.

The Kepler force is

$$\mathbf{K} = -\mathbf{grad}\ U \tag{2.57}$$

and the torque of the Kepler force about the mass center of the satellite is given by

$$\mathbf{M} = -\boldsymbol{\rho}_C \times \mathbf{K}. \tag{2.58}$$

Equation (2.56) indicates that the simplifying assumption of representing a satellite as a point mass, can, strictly speaking, only be made when the second term in the brackets vanishes, which is the case, e.g., when the satellite's massive shape is spherical, i.e., $A = B = C$, or when the direction cosines are equal.

For operational convenience, use

$$\mathbf{grad} = [\mathbf{e}_\xi \ \mathbf{e}_\eta \ \mathbf{e}_\zeta] \begin{bmatrix} \partial/\partial\xi \\ \partial/\partial\eta \\ \partial/\partial\zeta \end{bmatrix}$$

$$\rho_C = \sqrt{\xi^2 + \eta^2 + \zeta^2}$$

$$I_C = A\frac{\xi^2}{\rho_C^2} + B\frac{\eta^2}{\rho_C^2} + C\frac{\zeta^2}{\rho_C^2}$$

and

$$\cos\alpha = \frac{\xi}{\rho_C} \quad \cos\beta = \frac{\eta}{\rho_C} \quad \cos\gamma = \frac{\zeta}{\rho_C}$$

in equation (2.57).

2.7 Locked Rotation

If a satellite orbits its master on a circular path in such a manner that one and the same axis fixed in the satellite always points to the master, then one speaks of a *locked rotation* of the satellite about its master (Figure 2.10). For communications satellites, particularly those at synchronous altitude, locked rotation is an obvious design requirement. The Moon provides another well-known example of locked rotation: The same face of the Moon is always turned towards Earth. As will be shown later, there are, however, small deviations from a perfectly radial attitude, caused by the slight eccentricity of the Moon's orbit. The resulting small oscillatory motion that takes place about the radial vector is called *libration.*

2.8 Orbital Period for Locked Rotation

The circumstance that center of gravity and center of mass of a satellite of finite size do not coincide leads to orbital periods for these satellites that differ from that of a point satellite. Let us consider a rod satellite, orbiting on a circular path in locked rotation. We equate centrifugal force (applied at mass center) and Kepler force (applied at center of gravity)

$$m\rho_C\omega^2 = \frac{\mu m}{\rho_G^2}. \tag{2.59}$$

Considering that $\tau = 2\pi/\omega$, one obtains for the orbital period

$$\tau = 2\pi\sqrt{\frac{\rho_C^3}{\mu}\frac{\rho_G}{\rho_C}}. \tag{2.60}$$

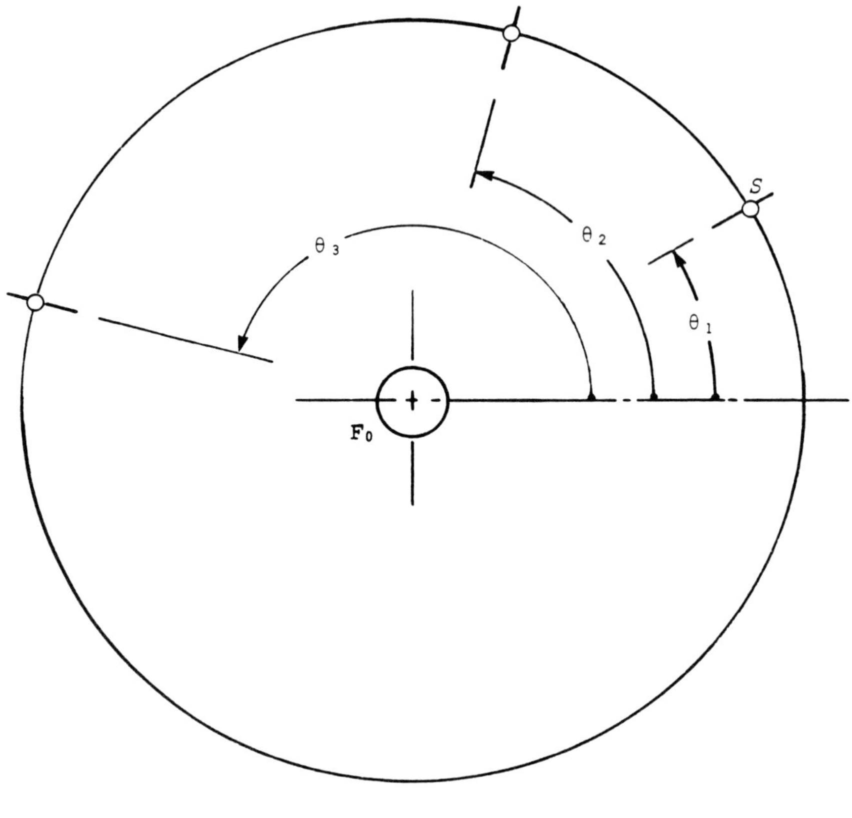

FIGURE 2.10. Locked rotation of a satellite.

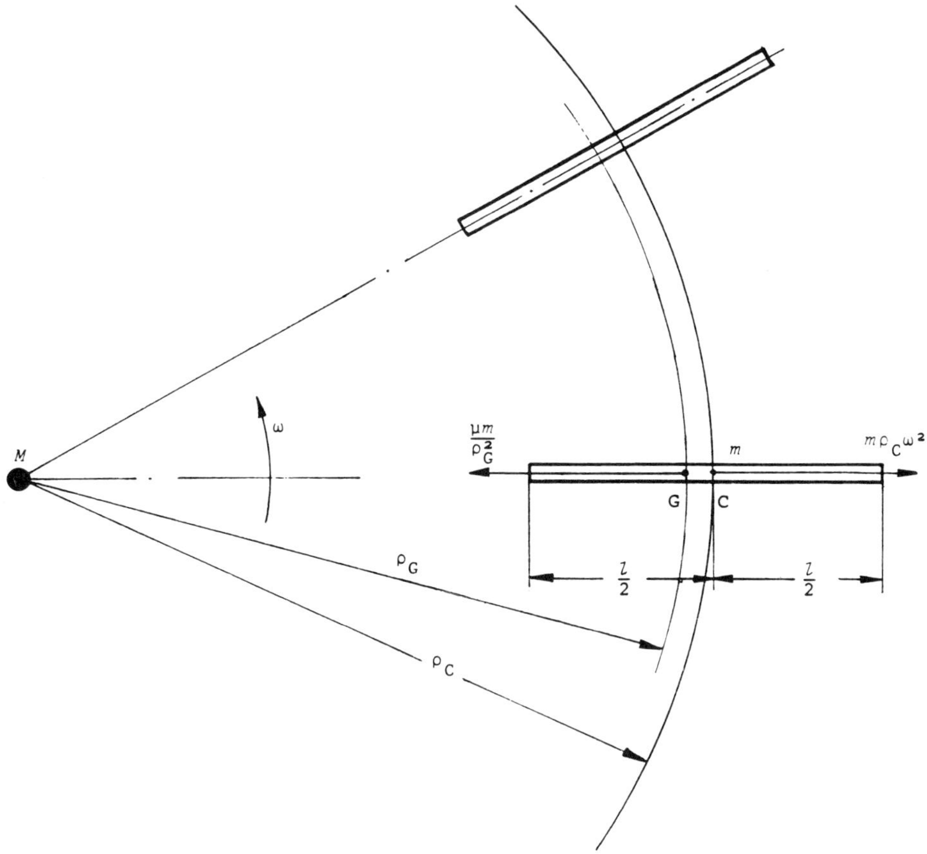

FIGURE 2.11. Rod satellite in radial attitude during locked rotation about master.

For the one-dimensional rod satellite with radial attitude (Figure 2.11) equation (2.6) is applicable and the orbital period (2.60) becomes

$$\tau = 2\pi\sqrt{\frac{\rho_C^3}{\mu}}\sqrt{1-\frac{l^2}{4\rho_C^2}} \tag{2.61}$$

which is smaller than the Kepler period (of a *point* satellite orbiting on the path of the mass center of the rod satellite). The radial attitude of the one-dimensional rod satellite happens to be stable. It will be shown later that the radial attitude is the one and *only* stable attitude of a rod satellite.

For a small satellite in locked rotation whose center of gravity is located at a position as given by equation (2.46), whose x-axis is pointing in radial direction, and whose z-axis is normal to the orbital plane, the orbital period

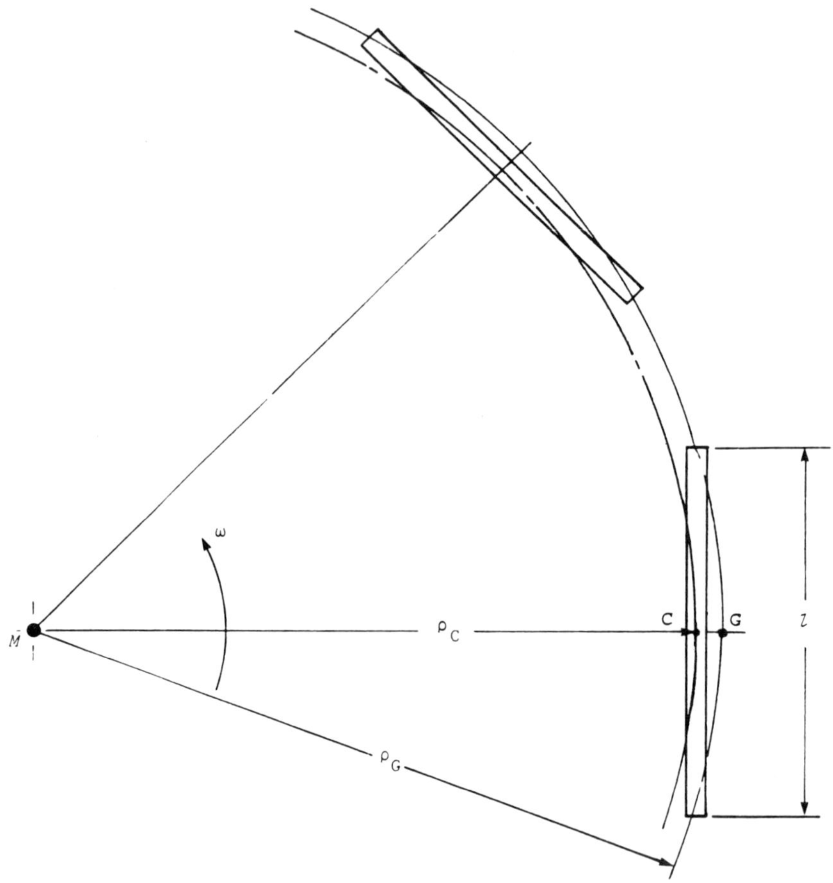

FIGURE 2.12. Orbital velocity in case of tangential attitude.

is

$$\tau = 2\pi\sqrt{\frac{\rho_C^3}{\mu}}\left\{1 + \frac{3}{4m\rho_C^2}(2A - B - C)\right\}. \tag{2.62}$$

This attitude of the satellite is also stable, provided that

$$A < B < C. \tag{2.63}$$

A more detailed discussion of the attitude stability of a satellite of arbitrary configuration is given later.

If a one-dimensional rod satellite would be orbiting in fixed rotation about the master at a (admittedly unstable) tangential attitude (Figure 2.12), then the period would be

$$\tau = 2\pi\sqrt{\frac{\rho_C^3}{\mu}}\sqrt[4]{1 + \frac{l^2}{4\rho_C^2}} \tag{2.64}$$

and would be larger than that of a point satellite orbiting the master on the orbit of the satellite's mass center. Equation (2.64) would also be obtained for a one-dimensional rod satellite with an attitude normal to the orbital plane.

The discussion of the orbital period of satellites in fixed rotation shows that the behavior of satellites of finite size can, strictly speaking, no longer be described using a point satellite. The reason for this is the fact that center of mass and center of gravity of a finite satellite no longer coincide. While the location of the mass center is merely a characteristic of the mass distribution of the satellite, we found that the location of the gravity center is, in addition, dependent upon altitude and attitude of the satellite.

An inspection of equation (2.62), and excluding the case $A = B = C$, shows that, for practical cases, a decoupling of a satellite's translatory (i.e., orbital) and rotary motion occurs only when $A <<< m\rho_C^2$, $B <<< m\rho_C^2$, and $C <<< m\rho_C^2$, i.e., when the satellite's principal inertia moments are negligible compared to the inertia moment $m\rho_C^2$ of the satellite (point) mass about the master mass center, or in other words, when the satellite can be considered a point mass.

2.9 Spacecraft with Booms

Long booms (e.g., antennas) on spacecraft are subject to gravity gradient loading that not only cause a torque but may be large enough to bring about a noticeable bending of the booms. The boom masses are typically small compared to the spacecraft central body mass, such that gravity center and mass center of the spacecraft can be assumed to coincide, $\rho_G = \rho_C$.

In the case of locked rotation, a boom mass element dm is subject to a centrifugal force

$$dm\,\rho\,\omega^2 \tag{2.65}$$

and to a gravitational attraction force

$$\frac{\mu dm}{\rho^2}. \tag{2.66}$$

The difference between the two is the gravity gradient force dF, which amounts to

$$dF = dm\left(\rho\omega^2 - \frac{\mu}{\rho^2}\right). \tag{2.67}$$

If spacecraft gravity center and spacecraft mass center coincide, then

$$\rho_C\omega^2 = \frac{\mu}{\rho_C^2} \tag{2.68}$$

leading to

$$dF = dm\,\rho_C\omega^2\left(\frac{\rho}{\rho_C} - \frac{\rho_C^2}{\rho^2}\right). \tag{2.69}$$

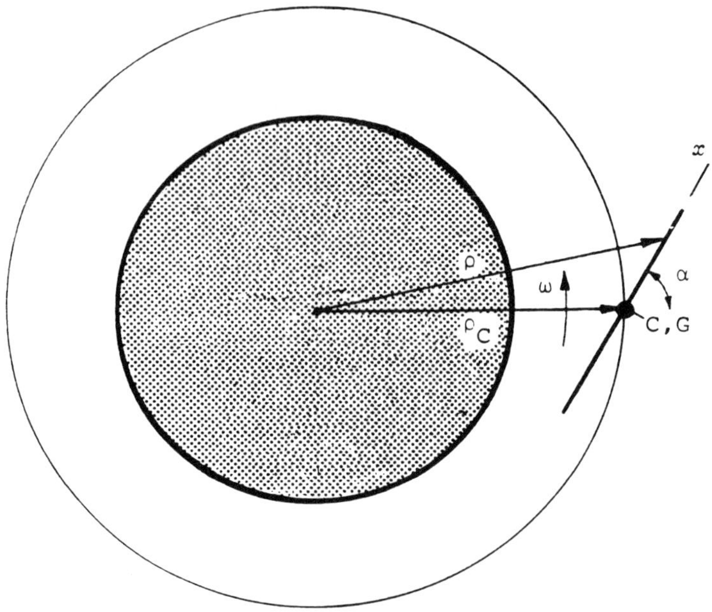

FIGURE 2.13. Spacecraft with symmetrical booms circling the earth.

An inspection of Figures 2.13 and 2.14 indicates that the radius ρ can be approximated by

$$\rho = \rho_C + x \cos\alpha \tag{2.70}$$

giving

$$\frac{\rho}{\rho_C} = 1 + \frac{x}{\rho_C} \cos\alpha \tag{2.71}$$

and consequently

$$\frac{\rho_C^2}{\rho^2} = 1 - 2\frac{x}{\rho_C} \cos\alpha + \dots . \tag{2.72}$$

Equations (2.71) and (2.72) can now be used to simplify equation (2.69), which becomes

$$dF = 3\, dm\, x\, \omega^2 \cos\alpha. \tag{2.73}$$

The mass element

$$dm = \frac{m}{l} dx \tag{2.74}$$

such that

$$\frac{dF}{dx} = 3\frac{m}{l} x\, \omega^2 \cos\alpha. \tag{2.75}$$

Upon integration between the limits $x = 0$ (in case of a point-size central body) and $x = l$, the total gravity gradient load is obtained:

$$F = \frac{3}{2} m\, l\, \omega^2 \cos\alpha. \tag{2.76}$$

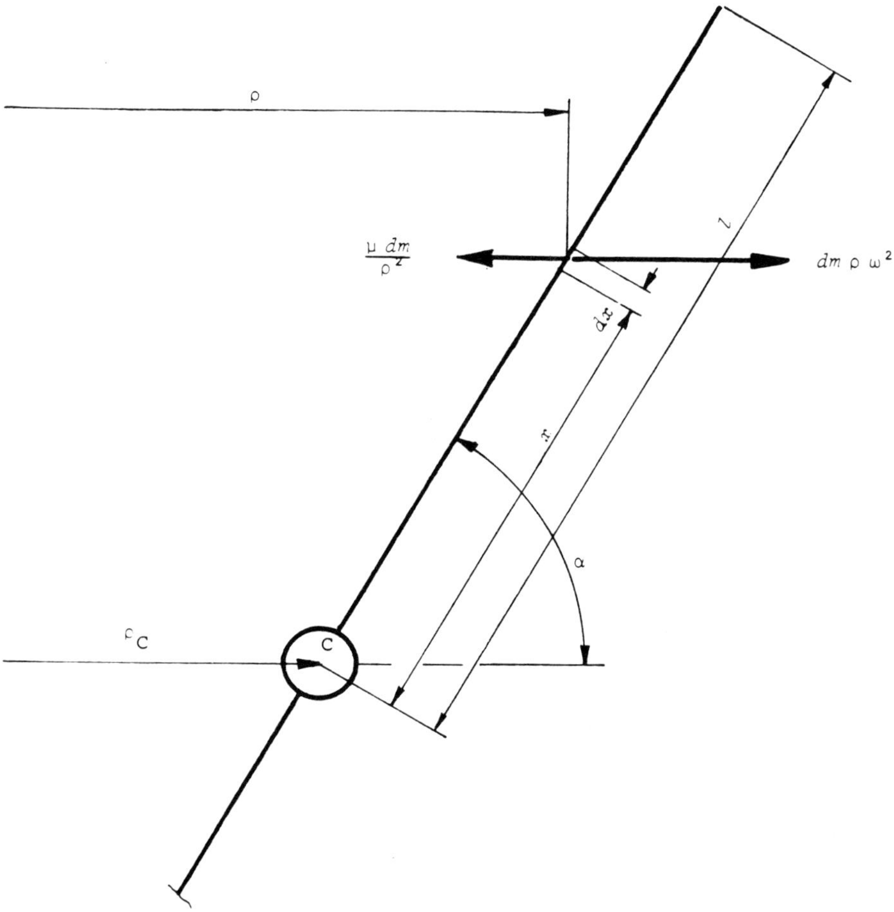

FIGURE 2.14. Centrifugal and gravitational force on boom element dm.

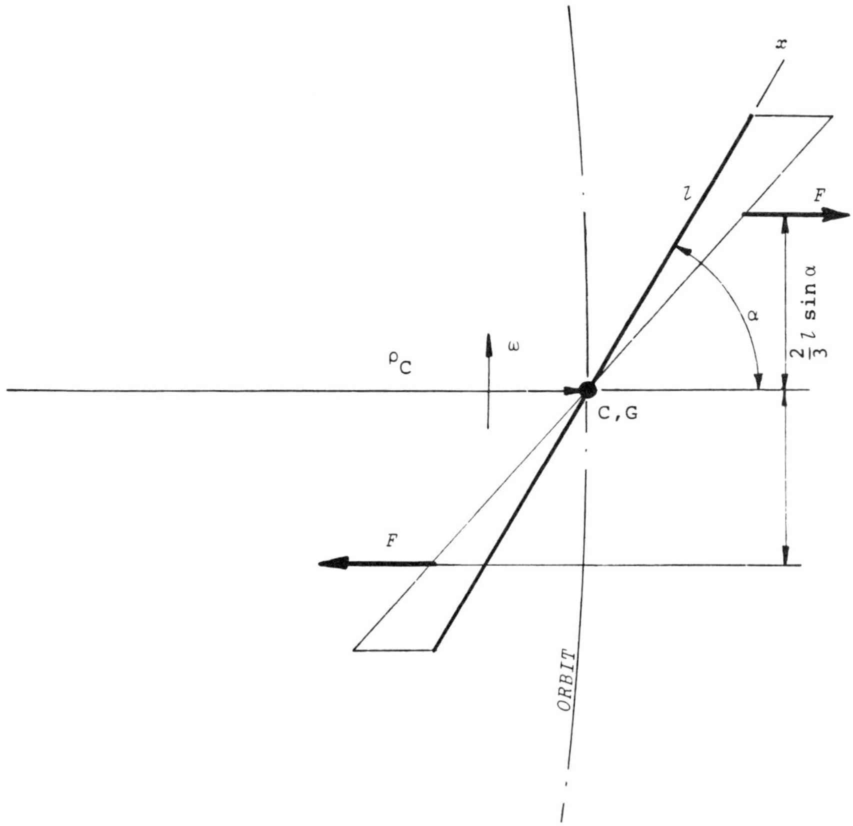

FIGURE 2.15. Torque about mass center.

The gravity gradient load is a distributed load, linearly rising from the built-in end (Figure 2.15).

The gravity gradient loads on the two booms also have a torque, directed such that it attempts to cause the spacecraft to assume a radial attitude (i.e., $\alpha = 0°$ or $180°$).

The torque is equal to

$$M = F\left(2\frac{2}{3}l \sin\alpha\right) = m\,l^2\,\omega^2 \sin 2\alpha. \tag{2.77}$$

Note that the torque has a maximum for $\alpha = 45°$, $135°$, $225°$, and $315°$, and vanishes for $\alpha = 0°$, $90°$, $180°$, and $270°$.

Take as an example an orbiter (m = 100 000 kg) on a circular orbit of 200-km height, with two thin-walled tubular booms of 50-kg mass and 150-m length each, symmetrically installed at a 45° attitude from radial.

The gravity gradient torque, then, is

$$M = (50)\,150^2 \frac{398\,601}{(6371+200)^3} 1 = 1.58 \text{ N m}. \tag{2.78}$$

2.10 Microgravity

In Figure 2.16 a hollow spacecraft of mass m_s is shown with a very small point mass m inside, located at

$$\mathbf{r} = [\mathbf{e}_x\ \mathbf{e}_y\ \mathbf{e}_z] \begin{bmatrix} \rho_C + x \\ y \\ z \end{bmatrix} \tag{2.79}$$

from the center of the master mass M.

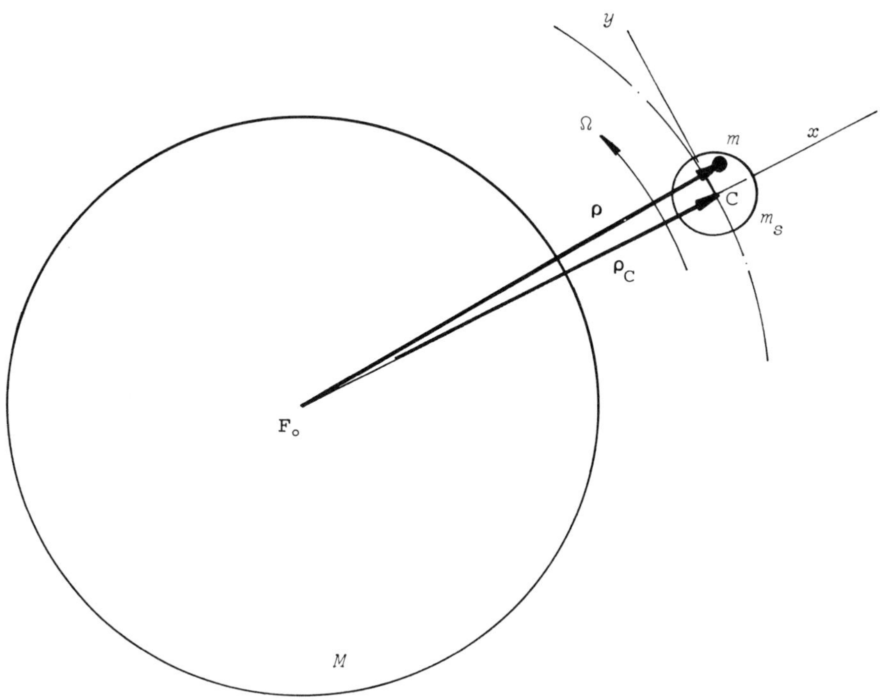

FIGURE 2.16. Point mass m inside spacecraft in locked rotation.

The spacecraft orbits the master in locked rotation. The $Cxyz$ coordinate system is spacecraft-fixed. The attraction force exerted upon the point mass m from the spacecraft is considered negligible. The centrifugal force acting on the point mass m is in the xy-plane. We make an ansatz

$$\mathbf{W}_\mu = \mathbf{C} + \mathbf{K} \tag{2.80}$$

where $\mathbf{W}_\mu$ = microweight force; $\mathbf{K}$ = Kepler force; $\mathbf{C}$ = centrifugal force; and with

$$\frac{\mu m}{(\rho_C + x)^2} = \frac{\mu m}{\rho_C^2}\left(1 - 2\frac{x}{\rho_C} + \ldots\right) \tag{2.81}$$

and consequently

$$\mathbf{K} = [\mathbf{e}_x \ \mathbf{e}_y \ \mathbf{e}_z]\begin{bmatrix} -\frac{\mu m}{\rho_C^2}(1 - 2\frac{x}{\rho_C}) \\ -\frac{\mu m}{\rho_C^2}\frac{y}{\rho_C} \\ -\frac{\mu m}{\rho_C^2}\frac{z}{\rho_C} \end{bmatrix} \tag{2.82}$$

and

$$\mathbf{C} = [\mathbf{e}_x \ \mathbf{e}_y \ \mathbf{e}_z]\begin{bmatrix} m(\rho_C + x)\Omega^2 \\ m(\rho_C + x)\Omega^2\frac{y}{\rho_C + x} \\ 0 \end{bmatrix} \tag{2.83}$$

and considering that

$$\Omega^2 = \frac{\mu}{\rho_C^3}\frac{\rho_C^2}{\rho_G^2} \simeq \frac{\mu}{\rho_C^3} \tag{2.84}$$

the linearized microweight force becomes

$$\mathbf{W}_\mu = [\mathbf{e}_x \ \mathbf{e}_y \ \mathbf{e}_z]\begin{bmatrix} \frac{3\mu m}{\rho_C^3}x \\ 0 \\ -\frac{\mu m}{\rho_C^3}z \end{bmatrix}. \tag{2.85}$$

From equation (2.85) it is seen that the microweight force's z-component is directed *toward* the spacecraft's mass center, while the three-times-larger x-component is directed *away*. Expressed in other words, the linearized microgravity field is *repelling* in radial direction, *indifferent* in tangential direction, and *attracting* in normal direction. Division by m gives a *microgravitational attraction coefficient:*

$$\mathbf{g}_\mu = [\mathbf{e}_x \ \mathbf{e}_y \ \mathbf{e}_z]\begin{bmatrix} \frac{3\mu}{\rho_C^3}x \\ 0 \\ -\frac{\mu}{\rho_C^3}z \end{bmatrix}. \tag{2.86}$$

True zero gravity is only achieved on the y-axis, as an inspection of equation (2.85) or (2.86) shows.

As an example, consider a spacecraft in locked rotation on an orbit of 200-km altitude above the Earth sphere. The ceiling of the spacecraft is 5 m from the spacecraft's mass center. A small experiment is installed at the ceiling, i.e., at $x = 5$ m, $y = z = 0$. What microgravity is the experiment subjected to?

The solution is

$$g_\mu = \frac{3(398\,601)5}{(6371+200)^3} = 0.000\,021\,\mathrm{m/s}^2$$

in x-direction, i.e., *away* from the spacecraft mass center.

If the spacecraft is not rotating, then the centrifugal force throughout the spacecraft is merely

$$\mathbf{C} = [\mathbf{e}_x \; \mathbf{e}_y \; \mathbf{e}_z] \begin{bmatrix} m\Omega^2\rho_\mathrm{C} \\ 0 \\ 0 \end{bmatrix} \tag{2.87}$$

and the microweight force consequently

$$\mathbf{W}_\mu = [\mathbf{e}_x \; \mathbf{e}_y \; \mathbf{e}_z] \begin{bmatrix} \frac{2\mu m}{\rho_\mathrm{C}^3} x \\ -\frac{\mu m}{\rho_\mathrm{C}^3} y \\ -\frac{\mu m}{\rho_\mathrm{C}^3} z \end{bmatrix} . \tag{2.88}$$

In addition to the microgravity field discussed here, there are numerous other effects that exert small forces on a mass located inside a spacecraft, such as spacecraft maneuvers, crew motion within the spacecraft, the oblateness of the master mass, a noncircular orbit, Coriolis and other forces due to motion of the mass within the spacecraft, air currents in the spacecraft, etc. For a more complete description of microgravity effects on experiments inside a spacecraft see, e.g., Bauer (1982).

Suggested Reading

1. Bauer, H.F. "Environmental Effects on Micro-Gravity Experiments," *Zeitschrift für Flugwissenschaft und Weltraumforschung,* 6, 3, 1982, 184–194.

2. Halfman, L. *Dynamics* Volume I, Addison-Wesley, 1962, 379 pp.

3. Schiehlen, W.; O. Kolbe. "A Generalization of Kepler Orbits for Large Satellites," *Applied Mechanics* (Ed. M. Hetényi and W.G. Vincenti) Springer-Verlag, 1969, 364–381.

Problems

2.1. Given is a rod satellite of uniform mass distribution and length l whose attitude is perpendicular to the line connecting satellite mass center C and point master. Determine the distance ρ_G between satellite gravity center and point master.

2.2. Invert the matrix

$$[N] = \begin{bmatrix} 1 - 3\cos^2\alpha & -3\cos\alpha\cos\beta & -3\cos\alpha\cos\gamma \\ -3\cos\alpha\cos\beta & 1 - 3\cos^2\beta & -3\cos\beta\cos\gamma \\ -3\cos\alpha\cos\gamma & -3\cos\beta\cos\gamma & 1 - 3\cos^2\gamma \end{bmatrix}.$$

2.3. A satellite ($m = 1200$ kg) is located 7000 km from Earth center. It was the shape of a rectangular parallelepiped and its mass is uniformly distributed. Determine (a) its moments of inertia; (b) the location $\mathbf{r}_G$ of its gravity center G if $\alpha = \beta = \gamma$ and $a = 3$ m, $b = 4$ m, and $c = 12$ m.

2.4. Show how

$$\xi_C + x_G = \frac{\rho_G^3}{\rho_C^3}\frac{1}{m}\int_m (\xi_C + x)\left(1 - \frac{3}{\rho_C}(x\cos\alpha + y\cos\beta + z\cos\gamma)\right.$$

$$\left. - \frac{3}{2\rho_C^2}(x^2 + y^2 + z^2) + \frac{15}{2\rho_C^2}(x\cos\alpha + y\cos\beta + z\cos\gamma)^2\right) dm$$

becomes

$$\xi_C + x_G = \frac{\rho_G^3}{\rho_C^3}\left\{\xi_C - \frac{3\cos\alpha}{4m\rho_C}[3(B + C) - A + \frac{15\cos\alpha}{4\ m\rho_C}[(B + C - A)\cos^2\alpha\right.$$

$$+ \ (C + B - A)\cos^2\beta + (A + B - C)\cos^2\gamma]\}\,.$$

2.5. Express the Kepler force

$$\mathbf{K} = -\frac{\mu m}{\rho_C^2}\left\{\left(1 + \frac{3(A + B + C - 5I_C)}{2m\rho_C^2}\right)\frac{\rho_C}{\rho_C} + \frac{3}{m\rho_C^2}(\mathbf{e}_x A\cos\alpha\right.$$

$$+ \ \mathbf{e}_y B\cos\beta + \mathbf{e}_z C\cos\gamma)\}\,.$$

in terms of the body-fixed unit vectors $\mathbf{e}_x$, $\mathbf{e}_y$, and $\mathbf{e}_z$ only.

2.6. A brick-shaped satellite body of uniform mass distribution and total mass of 500 kg is located 2000 km from the mass center of its point master, and the body's principal axes make angles of $\alpha = 45°$, $\beta = 60°$, and $\gamma = 60°$, with the (extended) radius vector linking the master mass center and satellite mass center C. (a) Determine the principal inertia tensor of the satellite body; (b) determine the location of the gravity center G of the satellite body; and (c) make a sketch. The dimensions of the satellite body are $a = 12$ m, $b = 3$ m, and $c = 4$ m.

2.7. The mass center of a parallelepiped-shaped satellite body ($a = 12$ m, $b = 3$ m, and $c = 4$ m) of the uniform mass distribution and total mass of 1200 kg is located 3000 km from the mass center of its point master. The satellite body's principal axes make angles $\alpha = 80°$, $\beta = 90°$, and $\gamma = 10°$ with the (extended) radius vector linking the master mass center C. (a) Determine the principal inertia tensor of the satellite body; (b) determine the location of the gravity center G of the satellite body; and (c) make a sketch.

2.8. A thin satellite of large size is $5a$ long and has a shape as shown. Determine (a) the location of the mass center C; (b) the location of the gravity center G; and (c) the resultant Kepler force.

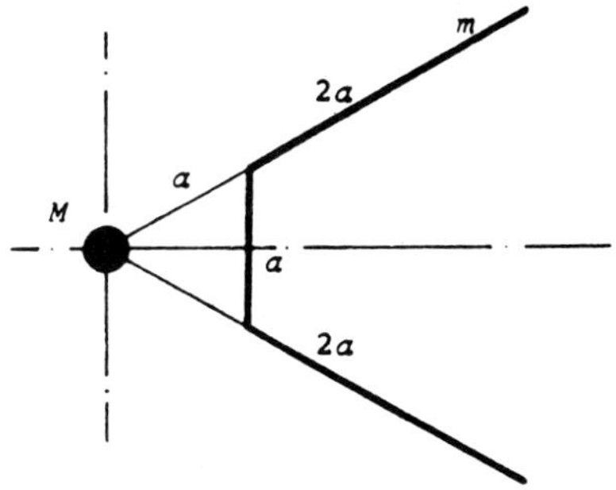

2.9. A one-dimensional satellite has a toroidal shape, whose radius of curvature is equal to the distance from the master as shown. (a) Determine the location of the mass center; (b) determine the location of the gravity center; (c) determine the resultant Kepler force; and (d) complete Table 2.1.

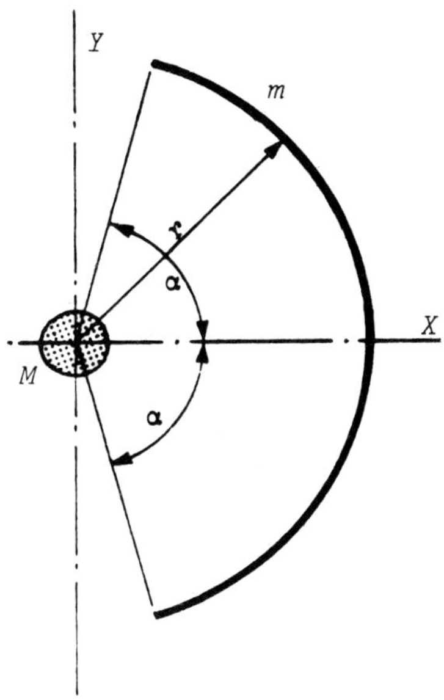

Table 2.1

α	r_C	r_G		K
0°				$\mu m/r^2$
90°	$0.637r$	$1.253r$	0.637	$\mu m/r^2$
150°				
180°				
210°				

2.10. A satellite consisting of one-dimensional rods has a shape and position as shown. Determine (a) the location of the mass center; (b) the location of the gravity center; and (c) the resultant Kepler force.

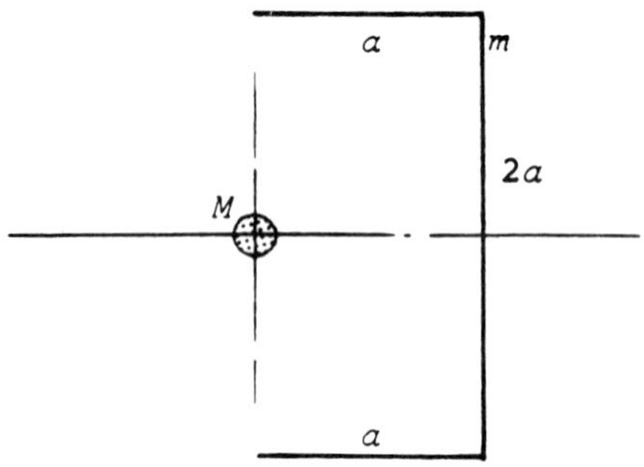

2.11. A huge satellite of mass $m << M$ and consisting of rigid one-dimensional rods has a shape and position as shown. Determine (a) the location of the mass center C; (b) the location of the gravity center G, and (c) the resultant concentrated Kepler force. (d) Is the satellite attitude shown stable?

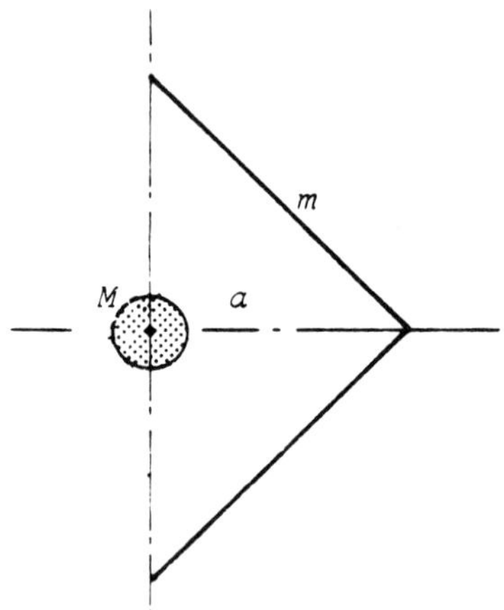

2.12. Derive the following equation:

$$\tau = 2\pi\sqrt{\frac{\rho_C^3}{\mu}}\left(1 + \frac{3}{4m\rho_C^2}(2A - B - C)\right)$$

and discuss it.

2.13. Given is a very long dumbbell satellite of total mass m and total length l, which moves in locked rotation about its master. (a) Calculate the magnitude, determine the direction, and locate the line of action of the resultant centrifugal force. (b) Repeat the calculation for the same satellite on a circular orbit, but moving in translatory fashion. (c) Locate the center of gravity for the attitude shown and $l = 2r$.

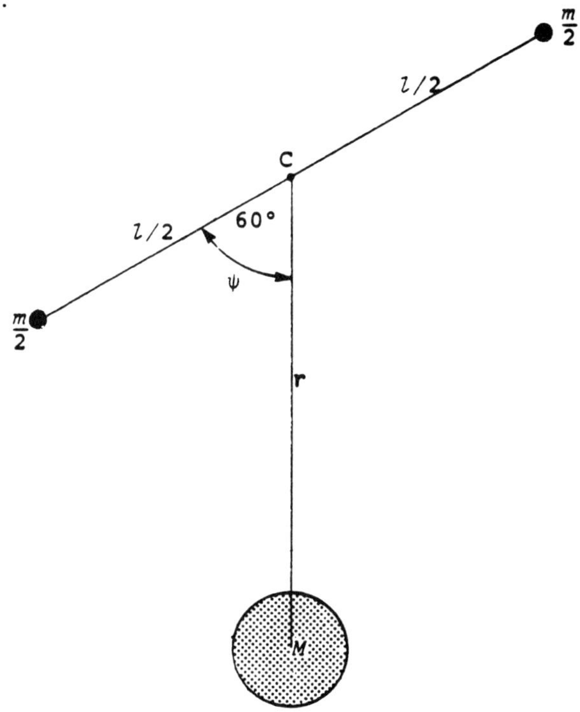

2.14. Expand the expression

$$y = \sqrt{a^2 + \left(r + \frac{l}{2}\right)^2} - \sqrt{a^2 + \left(r - \frac{l}{2}\right)^2}$$

into a series, for $r >> a$ and $r >> l$. The series is to be broken off after terms with l^2/r^2 and a^2/r^2.

2.15. The gravitational torque about the mass center of a finite-size satellite is

$$\mathbf{M} = \mu \boldsymbol{\rho}_C \times \int_m \frac{\boldsymbol{\rho}}{\rho^3} dm.$$

Show, by using

$$\boldsymbol{\rho} = \boldsymbol{\rho}_C + \mathbf{r}$$

and by a series expansion, that for a small satellite the torque becomes

$$\mathbf{M} = 3\frac{\mu}{\rho_C^3}[\mathbf{e}_x \; \mathbf{e}_y \; \mathbf{e}_z] \begin{bmatrix} (C-B)\cos\gamma\cos\beta \\ (A-C)\cos\alpha\cos\gamma \\ (B-A)\cos\beta\cos\alpha \end{bmatrix}$$

where the $Cxyz$ coordinate system is body-fixed, originating at the mass center C, and coinciding with the principal axes of the satellite.

2.16. Given is a very large, thin, rigid ring of radius x, with uniform mass distribution, and total mass $m << M$. The axis of the ring points towards the point master mass M. Determine (a) the attraction force on the ring; and (b) the location of the gravity center G.

2.17. Given is a very large, thin, rigid disk of radius a, with uniform mass distribution, and total mass $m << M$. The axis of the disk points towards the point master mass M. Determine (a) the attraction force acting on the disk; and (b) the location of the disk's gravity center G.

2.18. Given is a very large homogeneous drum of radius a and height l. The mass of the drum is, however, small such that $m << M$. The longitudinal axis of the drum points towards the center of the master. Determine the total attraction force acting on the drum.

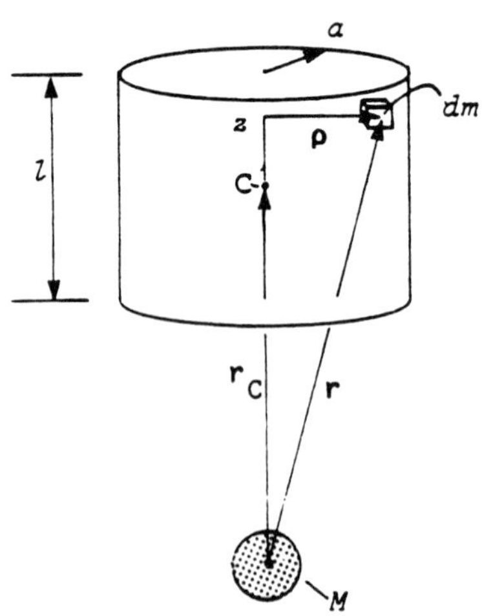

2.19. A rod satellite of length l, inertia tensor

$$[I] = \begin{bmatrix} A & 0 & 0 \\ 0 & A & 0 \\ 0 & 0 & 0 \end{bmatrix}$$

and mass m is moving on a circular orbit about its point master in locked rotation, with a tangential attitude. The distance between center of master and mass center of satellite is r. No force, other than the master's gravitational attraction, is acting on the satellite. Calculate (a) the orbital period; (b) the total angular momentum about the master; (c) the potential energy; (d) the kinetic energy; and (e) the total energy of the satellite.

After a long time interval, say one year, the attidude of the satellite has become radial. Libration has ceased. Calculate (f) the new orbital period (using an (unknown) orbital radius r_2); (g) the total angular momentum, which equals the original total angular momentum and thus provides an equation for the determination of r_2 in terms of r; (h) the potential energy; (i) the kinetic energy; and (j) the total energy of the satellite. (k) How much energy has been lost? (l) If all the energy dissipated is absorbed by the satellite's hysteresis, what is the satellite's absorption power (in μW) if it is assumed to have been constant over the year and if $m = 500$ kg, $l = 1$ km, $r = 7000$ km, and $\mu = 398601\ \text{km}^3/\text{s}^2$. (m) By how much has the orbital radius been reduced? [Hint: Approximate throughout by using the first two terms of the binomial expansion, with $A = ml^2/12$.]

2.20. A small satellite of spatial cross-boom configuration with uniform mass distribution ($a = 100$ m, $b = 50$ m, $c = 10$ m) and total mass $m = 1200$ kg is located 229 km above the Earth sphere's surface. The satellite's x-axis makes an angle of 60° with the vector connecting Earth mass center and satellite mass center, while the satellite's y-axis is at 45°. Determine (a) the potential energy of the satellite; and (b) the torque of the Kepler force about the satellite's mass center.

2.21. During assembly of a space station, the partially finished structure ($m << M$) has a shape that resembles a U consisting of thin rigid rods ($a << \rho_C$) with the open side facing the Earth, as shown. Determine (a) the principal inertia moments of the structure; and (b) the distance between gravity center and mass center. (c) Use $\rho_C = 7000$ km and $a = 1$ km to get a numerical value. (d) State whether the attitude shown is stable or not.

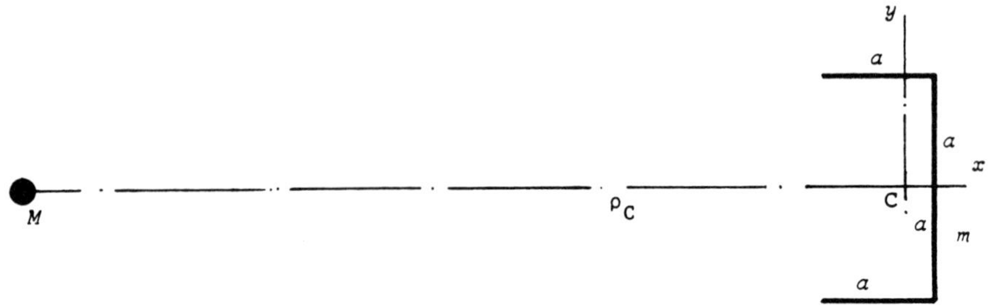

2.22. A very heavy point mass M moves on a circular orbit of radius r about its master. Two massless, perfectly straight booms extend a distance l from the point mass M, each carrying a small point mass m at the tip. (The total mass of the satellite is $M + 2m$.) The boom axes are normal to the orbital plane. Determine the bending moment transmitted through the booms.

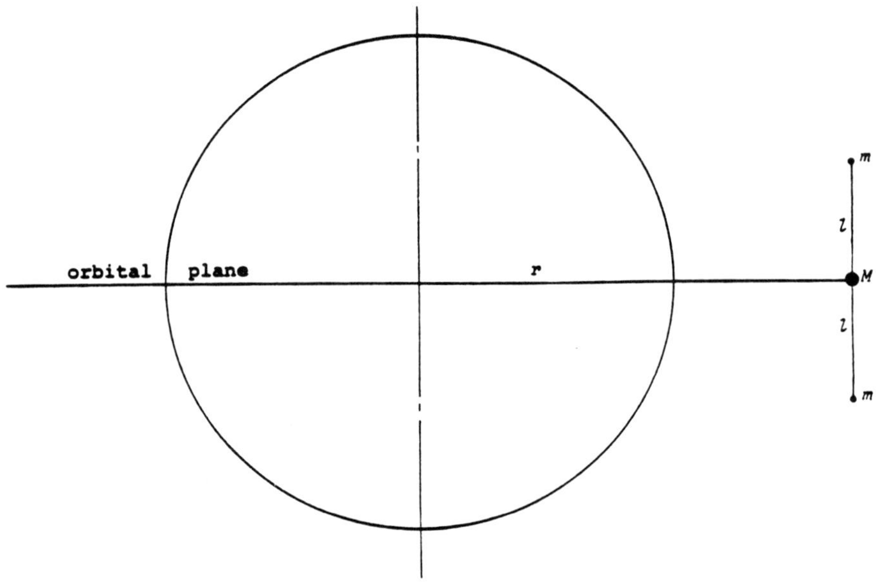

2.23. A large dumbbell satellite orbits its master in locked rotation and radial attitude. The two point masses m are separated by a massless rod of length l. Determine (a) the common Kepler force; and (b) the location of the gravity center G. (c) What is the orbital period τ? (d) Which force is transmitted through the rod? After the derivations of general expressions, calculate the numerical values for $m = 200$ kg, $l = 1$ km, $r = 7000$ km, and $\mu = 398\,601 \mathrm{km}^3/\mathrm{s}^2$.

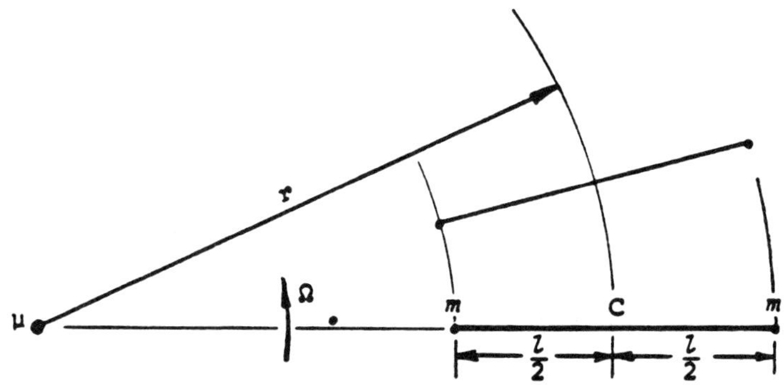

2.24. Sometimes a cable is proposed that would stretch from a point on the (spherical) Earth's equator to a geostationary satellite (and beyond if necessary), along which a cablecar would travel. To analyze the proposition, begin by assuming that a straight mast (no bending, no buckling) is extended vertically from a suitable base at the equator to a height l. Locate (a) the mass center C; and (b) gravity center G, in terms of l. (c) Calculate the reaction force applied to the mast at the base. In order to be replaceable by a cable, the mast must be free of compression. (d) How long would the mast have to be for the reaction force to vanish? (e) Make a sketch to scale and show G and C. (f) For the case of zero reaction force at the base, what is the stress transmitted at a cross section located at a distance x from the base? (g) Plot the stress distribution along the cable. (h) At which cross section is the stress transmitted a maximum; and (i) what is the maximum stress transmitted? If Kevlar ($\rho = 1440\,\mathrm{kg/m^3}$ and $\sigma_{ult} = 2760$ MPa) is chosen, (j) what is the maximum stress (in MPa); and (k) what is your assessment of the proposition?

2.25. Form the negative gradient of

$$U = -\frac{\mu m}{\rho_C}\left(1 + \frac{A + B + C - 3I_C}{2m\rho_C^2}\right)$$

to find the Kepler force **K**. [Hint: Write

$$\mathbf{grad} = [\mathbf{e}_\xi\ \mathbf{e}_\eta\ \mathbf{e}_\zeta]\begin{bmatrix} \partial/\partial\xi \\ \partial/\partial\eta \\ \partial/\partial\zeta \end{bmatrix}$$

$$\cos\alpha = \frac{\xi}{\rho_C}, \quad \cos\beta = \frac{\eta}{\rho_C}, \quad \cos\gamma = \frac{\zeta}{\rho_C}, \quad \text{and} \quad \rho_C = \sqrt{\xi^2 + \eta^2 + \zeta^2}].$$

2.26. For purposes of this problem, let lower case letters refer to satellite-related quantities, and capital letters to master-related ones. The potential field can then be expressed by

$$U = \frac{GMm}{\rho}\left(1 + \frac{A+B+C-3I_C}{2M\rho^2} + \frac{a+b+c-3i_c}{2m\rho^2}\right)$$

where ρ is the distance between master mass center and satellite mass center, and

$$I_C = A\cos^2 \mathrm{A} + B\cos^2 \mathrm{B} + C\cos^2 \Gamma$$
$$i_c = a\cos^2\alpha + b\cos^2\beta + c\cos^2\gamma.$$

Let $B = A$ and $b = a$. (a) Express the potential U in terms of the declination angles Δ and δ, measured from the respective equatorial planes. (b) Sketch the case $\Delta = 0$ and $\delta = 0$ and calculate the attraction force. (c) Sketch the case $\Delta = 0$ and $\delta = 90°$ and calculate the attraction force. (d) Sketch the case $\Delta = 90°$ and $\delta = 90°$ and calculate the attraction force. (e) Sketch the case $\Delta = 45°$ and $\delta = 0$ and calculate the attraction force and moment of the attraction force about the master's mass center. (f) Sketch the case $\Delta = 0$ and $\delta = 45°$ and calculate the attraction force and the moment of the attraction force about the satellite's mass center.

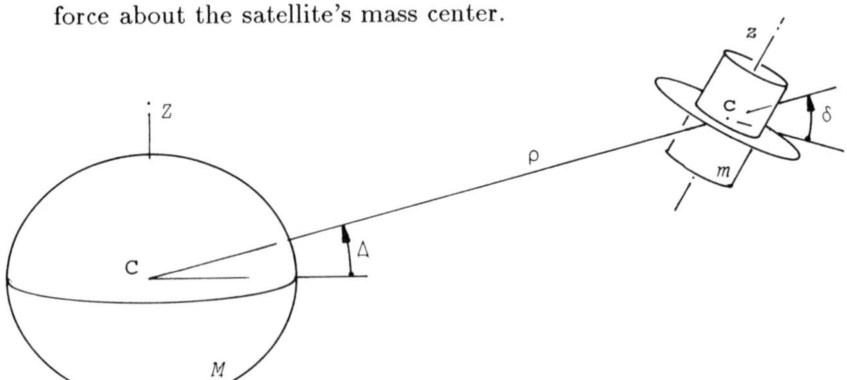

2.27. Locate (a) the gravity center G of a small flat disk of mass m and radius a, whose symmetry axis (= z-axis) and whose y-axis are normal to the line connecting the mass center of the disk with the point master; and (b) when the symmetry axis points in the direction of the master.

2.28. A thin rod satellite of length l and mass m is moving on a circular orbit about its point master in locked rotation, with an attitude tangential to its orbit. The distance between center of master and mass center of satellite is r. No force, other than the master's gravitational attraction, is acting on the satellite. Calculate (a) the orbital angular velocity Ω; (b) the potential energy; and (c) the kinetic energy.

As the tangential attitude is unstable, libration (of 90° amplitude) begins very soon. Assume that the satellite's mass center remains on the same circular orbit, with r constant, throughout. (d) Determine the satellite's potential energy when its attitude is radial, and (e) use energy conservation to obtain an approximate value for the absolute angular velocity of the librating satellite as it passes through the radial attitude. [Hint: Write $T_1 = \frac{1}{2}mv^2 + \frac{1}{2}A\Omega^2$ and $T_2 = \frac{1}{2}mv^2 + \frac{1}{2}A(\Omega + \omega)^2$ and find ω in terms of Ω.]

2.29. The mass center of a rod satellite of 1-km length and 1000-kg mass is located at a distance of 7000 km from Earth center. What is the compression force inside the rod at the mass center, if the rod's longitudinal axis is perpendicular to the radius vector connecting master mass center and satellite mass center, and the rod satellite is at rest, i.e., just commencing a free fall? Make a sketch.

2.30. A spacecraft has the shape of a hollow box ($a = 12$ m, $b = 4$ m, and $c = 3$ m). It moves in locked rotation around the Earth at 200-km altitude. In one corner (at $x = 6$ m, $y = 2$ m, and $z = 1.5$ m) an experiment is carried out, involving a mass of 2 kg. What is the total microweight force (in newton) acting on this mass?

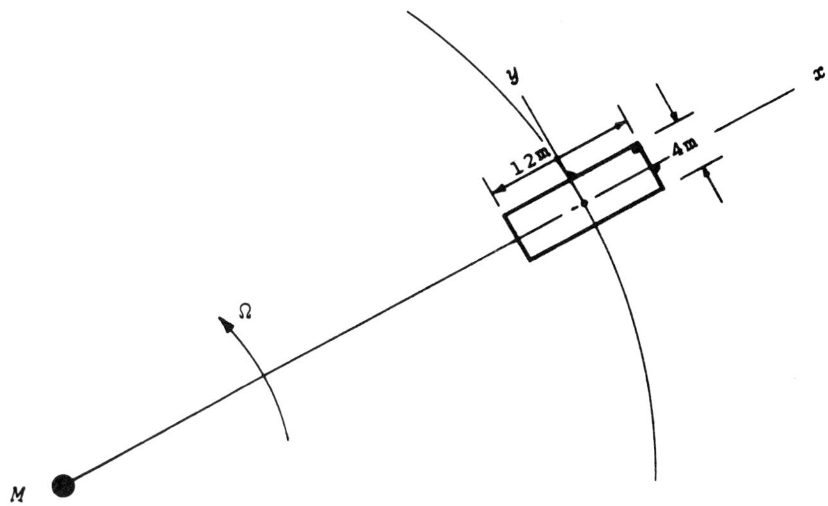

3

Libration

Vibratory motion superimposed upon locked rotatory motion of a satellite is called *libration*. In order to obtain a basic understanding of the libration behavior of satellites, we shall investigate a few simple special cases. Because it exhibits strong sensitivity to the gravity gradient, among all possible satellite configurations, we select a one-dimensional rod satellite, with an inertia tensor of, say,

$$[I] = \begin{bmatrix} 0 & 0 & 0 \\ 0 & B & 0 \\ 0 & 0 & B \end{bmatrix}. \tag{3.1}$$

We assume that the satellite is very small, i.e., $B << m\rho_C^2$.

Approximating $\mu/\rho_C\rho_G^2 = \Omega^2$ by $\mu/\rho_C^3 = \Omega^2$, where Ω is the angular velocity of the satellite's position vector, we can rewrite equation (2.55) for the torque of the Kepler force about the satellite's mass center C and obtain

$$\mathbf{M} = 3\Omega^2[\mathbf{e}_x\ \mathbf{e}_y\ \mathbf{e}_z] \begin{bmatrix} (C-B)\cos\gamma\cos\beta \\ (A-C)\cos\alpha\cos\gamma \\ (B-A)\cos\beta\cos\alpha \end{bmatrix}. \tag{3.2}$$

3.1 Libration about Position Vector and within Orbital Plane

Figure 3.1 shows a rod satellite of inertia tensor (3.1), deviating by a small pitch angle ζ from a perfectly radial attitude. The satellite's angular velocity then contains the amount Ω for locked rotation and a small pitch velocity $\dot{\zeta}$ superimposed:

$$\boldsymbol{\omega} = [\mathbf{e}_x\ \mathbf{e}_y\ \mathbf{e}_x] \begin{bmatrix} 0 \\ 0 \\ \Omega + \dot{\zeta} \end{bmatrix}. \tag{3.3}$$

The satellite's angular momentum is

$$\mathbf{H} = [\mathbf{e}_x\ \mathbf{e}_y\ \mathbf{e}_z] \begin{bmatrix} 0 \\ 0 \\ B(\Omega + \dot{\zeta}) \end{bmatrix} \tag{3.4}$$

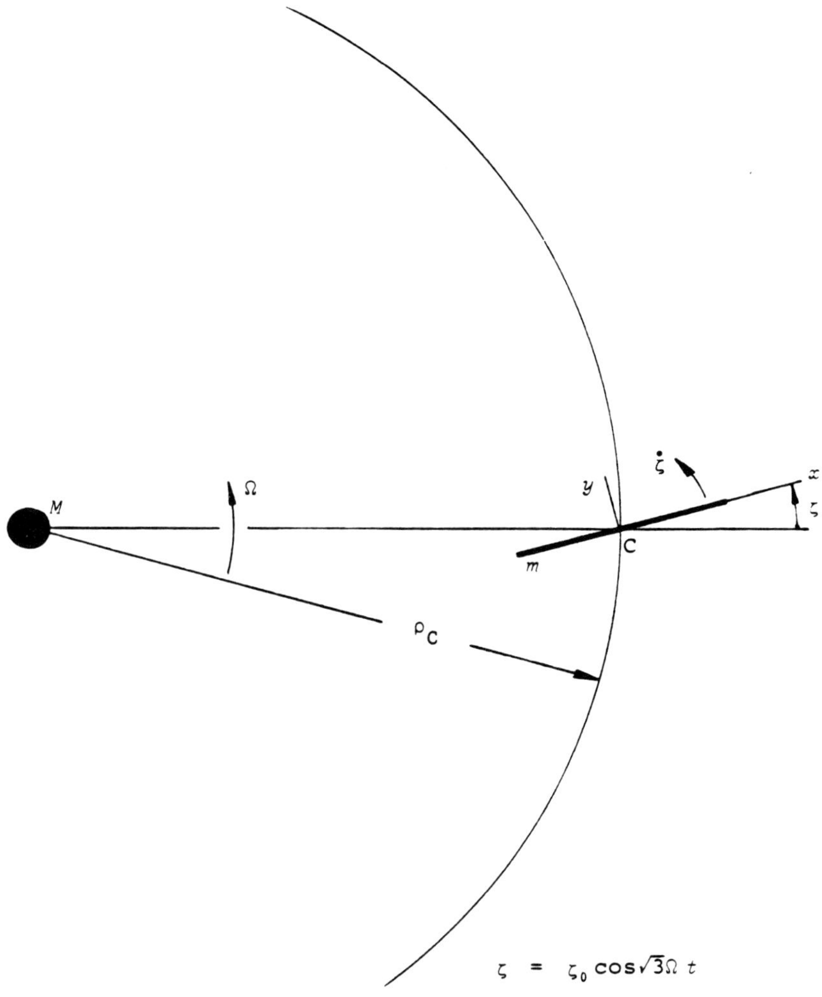

FIGURE 3.1. Point master with rod satellite librating about radial attitude within orbital plane.

and its time derivative

$$\dot{\mathbf{H}} = [\mathbf{e}_x\ \mathbf{e}_y\ \mathbf{e}_z] \begin{bmatrix} 0 \\ 0 \\ B\ddot{\zeta} \end{bmatrix}. \tag{3.5}$$

An inspection of Figure 3.1 indicates that the direction angles are $\alpha = \zeta$, $\beta = 90^\circ + \zeta$, and $\gamma = 90^\circ$. For small values of ζ, equation (3.2) then yields

$$\mathbf{M} = 3\Omega^2 [\mathbf{e}_x\ \mathbf{e}_y\ \mathbf{e}_z] \begin{bmatrix} 0 \\ 0 \\ -B\zeta \end{bmatrix}. \tag{3.6}$$

Forming $\mathbf{M} = \dot{\mathbf{H}}$, and using equations (3.5) and (3.6), leads to

$$\ddot{\zeta} + 3\Omega^2\zeta = 0 \tag{3.7}$$

which is quickly solved and gives

$$\zeta = \zeta_0 \cos\sqrt{3}\Omega t \tag{3.8}$$

indicating that the libration frequency of the satellite is equal to the satellite's Kepler frequency multiplied by $\sqrt{3}$.

3.2 Libration about Position Vector and Normal to Orbital Plane

Figure 3.2 shows the rod satellite, with inertia tensor (3.1), oriented along the position vector, this time with a slight deviation normal to the orbital plane, as expressed by roll angle η. Its angular velocity consists of the amount Ω for locked rotation, with a roll frequency $\dot{\eta}$ superimposed, resulting in

$$\boldsymbol{\omega} = [\mathbf{e}_x\ \mathbf{e}_y\ \mathbf{e}_z] \begin{bmatrix} -\Omega\cos\eta \\ \dot{\eta} \\ \Omega\cos\eta \end{bmatrix}.$$

The angular momentum is

$$\mathbf{H} = [\mathbf{e}_x\ \mathbf{e}_y\ \mathbf{e}_z] \begin{bmatrix} 0 \\ B\dot{\eta} \\ B\Omega\cos\eta \end{bmatrix}.$$

The angular momentum time derivative is obtained by invoking $\dot{\mathbf{H}} = \overset{\circ}{\mathbf{H}} + \boldsymbol{\omega} \times \mathbf{H}$, resulting in

$$\dot{\mathbf{H}} = [\mathbf{e}_x\ \mathbf{e}_y\ \mathbf{e}_z] \begin{bmatrix} 0 \\ B(\ddot{\eta} + \Omega^2\eta) \\ 0 \end{bmatrix} \tag{3.9}$$

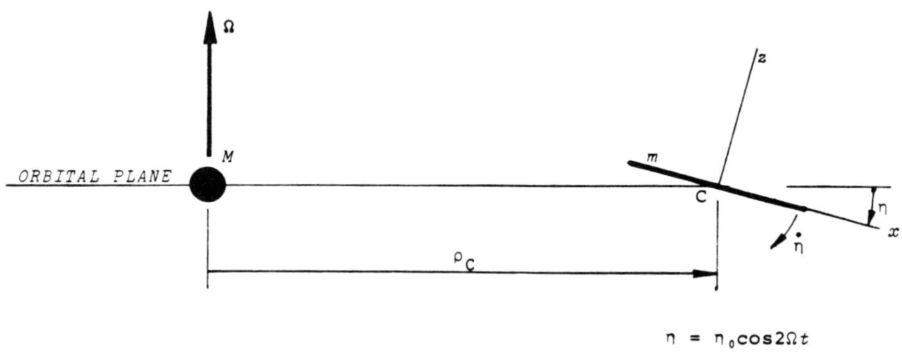

FIGURE 3.2. Libration about radial attitude normal to orbital plane.

for small roll angles η and small roll frequencies $\dot{\eta}$. The externally applied torque (equation (3.2)) with $\alpha = \eta$, $\beta = 90°$, and $\gamma = 90° - \eta$, becomes, for small roll angles η,

$$\mathbf{M} = [\mathbf{e}_x \ \mathbf{e}_y \ \mathbf{e}_z] \begin{bmatrix} 0 \\ -3\Omega^2 B\eta \\ 0 \end{bmatrix}. \tag{3.10}$$

Equating equations (3.9) and (3.10), we obtain

$$\ddot{\eta} + 4\Omega^2\eta = 0 \tag{3.11}$$

leading to

$$\eta = \eta_0 \cos 2\Omega t \tag{3.12}$$

and the conclusion that libration about the radial vector normal to the orbital plane takes place at twice the Kepler frequency of the satellite's orbital motion.

3.3 Libration about Tangent to Orbit and Normal to the Orbital Plane

Figure 3.3 shows the rod satellite oriented along the tangent of the orbit, but deviating by a small angle ξ normal to the orbital plane. What will the librational behavior be this time?

The satellite's inertia tensor is

$$[I] = \begin{bmatrix} A & 0 & 0 \\ 0 & 0 & 0 \\ 0 & 0 & A \end{bmatrix}. \tag{3.13}$$

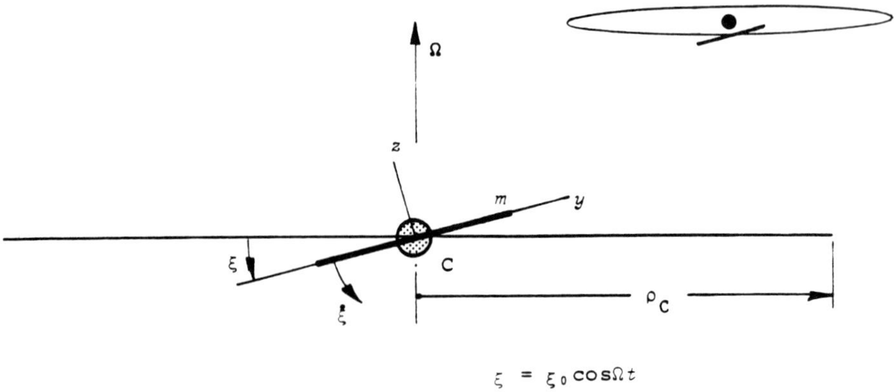

FIGURE 3.3. Theoretical case of libration about (unstable) tangential attitude normal to orbital plane.

The angular velocity is then

$$\boldsymbol{\omega} = [\mathbf{e}_x \ \mathbf{e}_y \ \mathbf{e}_z] \begin{bmatrix} \dot{\xi} \\ \Omega \sin \xi \\ \Omega \cos \xi \end{bmatrix} \tag{3.14}$$

and the angular momentum of the rod satellite with respect to its own mass center is

$$\mathbf{H} = [\mathbf{e}_x \ \mathbf{e}_y \ \mathbf{e}_z] \begin{bmatrix} A\dot{\xi} \\ 0 \\ A\Omega \cos \xi \end{bmatrix} \tag{3.15}$$

and (with $\sin \xi \simeq \xi$, and $\xi\dot{\xi} \simeq 0$)

$$\dot{\mathbf{H}} = [\mathbf{e}_x \ \mathbf{e}_y \ \mathbf{e}_z] \begin{bmatrix} A\ddot{\xi} + A\Omega^2 \xi \\ 0 \\ 0 \end{bmatrix}. \tag{3.16}$$

Because of $\gamma = 90°$ we have $\mathbf{M} = 0$. Consequently $\dot{\mathbf{H}} = 0$ and

$$\ddot{\xi} + \Omega^2 \xi = 0. \tag{3.17}$$

We might thus expect a rod satellite to librate about the tangent with exactly the Kepler frequency. However, before we accept the tangential attitude of a rod satellite as inherently stable, we want to look at the possibility of libration *within* the orbital plane.

3.4 Attitude Change about Tangent and within Orbital Plane

The situation to be investigated is depicted in Figure 3.4. The satellite is a rod with a body-fixed $Cxyz$ coordinate system and an inertia tensor (equation (2.13)):

$$[I] = \begin{bmatrix} A & 0 & 0 \\ 0 & 0 & 0 \\ 0 & 0 & A \end{bmatrix}.$$

The small deviation from a perfectly tangential attitude is expressed by the pitch angle ζ. The satellite's angular velocity is

$$\boldsymbol{\omega} = [\mathbf{e}_x \ \mathbf{e}_y \ \mathbf{e}_z] \begin{bmatrix} 0 \\ 0 \\ \Omega + \dot{\zeta} \end{bmatrix} \tag{3.18}$$

and its angular momentum about its mass center C is

$$\mathbf{H} = [\mathbf{e}_x \ \mathbf{e}_y \ \mathbf{e}_z] \begin{bmatrix} 0 \\ 0 \\ A(\Omega + \dot{\zeta}) \end{bmatrix}. \tag{3.19}$$

The time derivative is

$$\dot{\mathbf{H}} = [\mathbf{e}_x \ \mathbf{e}_y \ \mathbf{e}_z] \begin{bmatrix} 0 \\ 0 \\ A\ddot{\zeta} \end{bmatrix}. \tag{3.20}$$

The external torque acting on the satellite is given by equation (2.55). With direction angles $\alpha = \zeta$, $\beta = 90° + \zeta$, and $\gamma = 90°$, with $\mu/\rho_C^3 \simeq \Omega^2$, and for small values of ζ, it yields

$$\mathbf{M} = 3\Omega^2[\mathbf{e}_x \ \mathbf{e}_y \ \mathbf{e}_z] \begin{bmatrix} 0 \\ 0 \\ A\zeta \end{bmatrix}. \tag{3.21}$$

Euler's law $\mathbf{M} = \dot{\mathbf{H}}$ gives

$$\ddot{\zeta} - 3\Omega^2\zeta = 0$$

a differential equation whose solution is of the type

$$\zeta = \zeta_0 \operatorname{Cosh} \sqrt{3}\Omega t \tag{3.22}$$

that indicates that no libration about the orbit tangent will take place, but that ζ will become greater and greater. We conclude that the tangential attitude is *unstable,* and that the satellite tends to move towards the (stable) radial attitude. Equation (3.22) cannot, of course, be used to study the satellite's motion into the new attitude, because it was obtained making the assumption that the pitch angle ζ remains small.

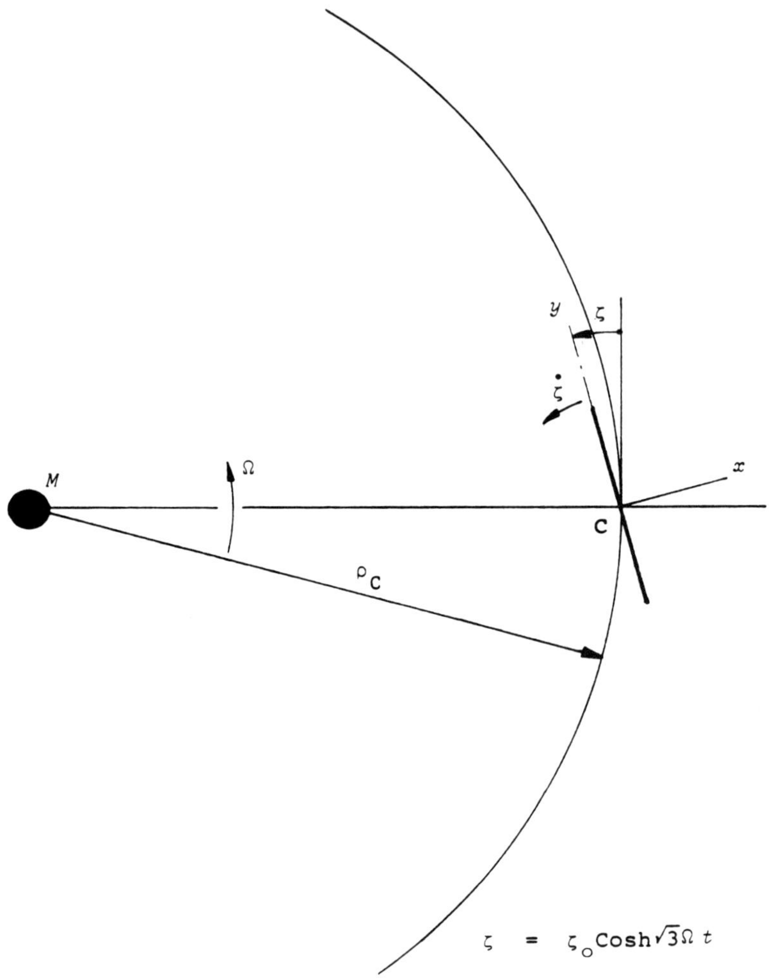

FIGURE 3.4. Rod satellite moving away from the (unstable) tangential attitude into the (stable) radial attitude.

3.5 Attitude Stability

In the preceding it has been shown that a one-dimensional rod satellite orbiting its master in locked rotation is stable only when it has a radial attitude. If the satellite experiences a small deviation from the radial attitude, it will subsequently carry out a small vibratory motion, called *libration* (from Latin libra = balance), about the radial orientation. The libration can be brought to die away by suitable means of damping.

Effects on the satellite that cause libration include the torque produced by the gravitational force in the case of elliptical orbits and/or a not perfectly spherical equipotential field surrounding the master, the magnetic torque acting on a not completely nonmagnetic satellite, the solar wind, the atmospheric resistance, and short impulses such as could be caused by meteorite strikes. Effects that contribute to a damping of any libration motion are the very tenuous atmosphere surrounding a satellite and the internal hysteresis of satellite structures, both, however, only to a very limited extent. More effective are specially designed libration dampers that can be installed in satellites. The stability of the radial attitude of a rod satellite (or more generally of a satellite whose minimum principal inertia moment is about an axis that is aligned with the radial vector) is due to the fact that the gradient of the gravitational attraction force acting on the satellite is not constant.

A stabilization making use of this effect is called *gravity gradient stabilization* (*GG stabilization* for short). GG stabilization is passive, i.e., no power source on board the satellite is required.

For practical purposes, a rod configuration for a satellite is not too feasible. GG stabilization can also be achieved for satellites with a dumbbell configuration, or with very long booms (Figure 3.5). Since satellites with long booms cannot be stowed within the fairing of a rocket nose cone, such booms must be retractable and extendable. Furlable overlapped tubing (Rimrott, 1965) has found wide application for extendable booms. Engineers refer to such tubing as STEM or TEE tubing.

3.6 Libration of the Moon

Astronomers observed long ago that the Moon carries out a libratory motion superimposed on the locked rotation about the Earth. The Moon's libration consists of two portions, a very small *true* libration and a much larger *apparent* libration, which in turn combines several effects. Because of the libration, one can see more than half of the Moon's surface. In the course of time this amounts to some 59% of the total Moon surface.

The apparent libration consists of a longitudinal, a latitudinal, and a diurnal component. The *longitudinal* libration is brought about by the fact that the Moon orbits the Earth on a (slightly) elliptical orbit. If we

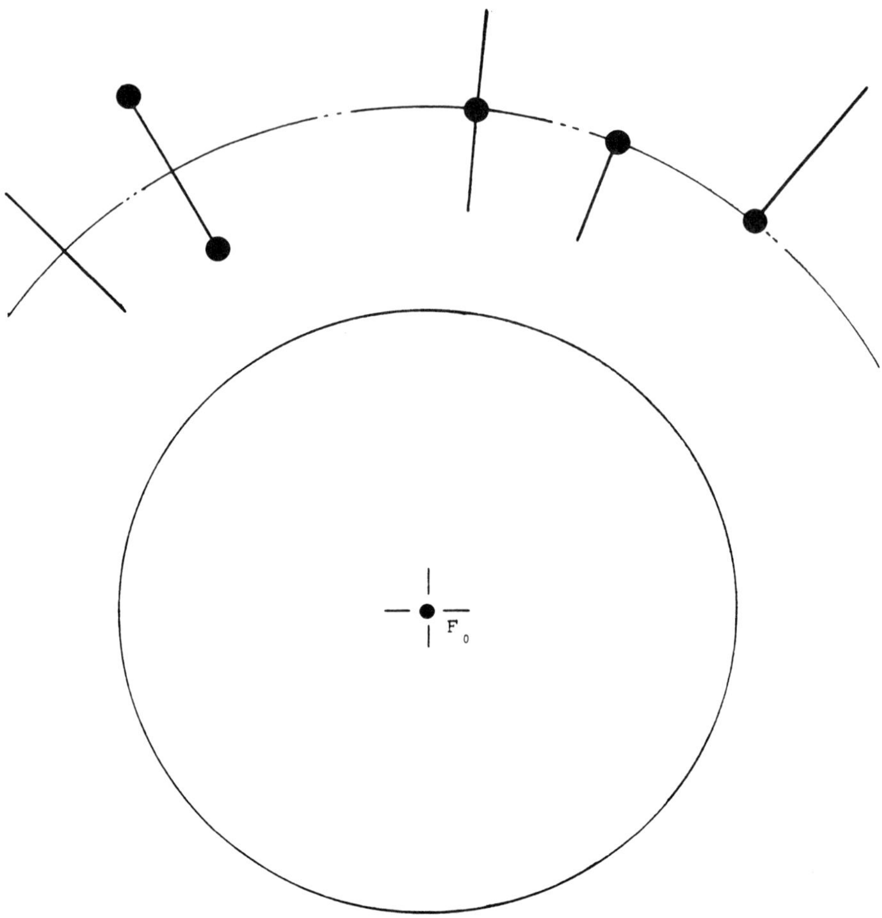

FIGURE 3.5. GG stabilization: rod satellite, dumbbell satellite, satellite with extended rods.

disregard the (very small) true libration for the subsequent argument, then we can assume that the Moon spins about its own axis at a constant angular velocity. Its translation speed along the orbit, on the other hand, depends on its distance from the Earth, and thus varies. As the Moon moves through the perigee, the spin is a little too slow for an exact locking. At the apogee the spin is a little too fast. The consequence of these differences is that an observer on the Earth gains the impression that the Moon oscillates.

There are some additional oscillations. The spin axis of the Moon is not exactly normal to the orbital plane of its motion about the Earth, it deviates some 6.68° from the normal. Therefore, one can see, in the course of one month, first the region around one of the Moon's poles and later the region around the other pole. This effect is the *latitudinal* libration.

The Moon is also close enough to the Earth to display a somewhat different portion of its surface to an observer located on the Earth such that the Moon appears in the zenith, than to an observer located such that the Moon appears near the horizon. This effect is called the *diurnal* libration.

During the long, long history of the study of the Moon no human being had ever seen the remaining 41% of the Moon's surface on the far side. This changed abruptly when the unmanned spacecraft *Luna 3* (1959-October-04) took photographs of the Moon's far side and subsequently returned with them to Earth.

Locked rotation, with libration superimposed, about one's master, turns out to be a fairly common occurrence among natural satellites of planets. There is evidence (Beletskii, 1975; Goldeich and Peale, 1968) that Jupiter's trabants Io, Europa, and Callisto, that Saturn's trabants Japetus, Rhea, Tethys, and Dione, and that Mars' minute trabants Phobos and Deimos orbit their master in locked rotation.

Large satellites are subject to libration, and so are tethered bodies (such as space shuttle and tethered subsatellite). The dynamics of tethered satellite systems is very complex. However, reasonably good models have been developed by Kohler (1978) and Xu et al. (1981).

Suggested Reading

1. Beletskii, V.V. "Resonance Phenomena at Rotations of Artificial and Natural Celestial Bodies," *Satellite Dynamics* (Ed. G.E.O. Giacaglia) Springer-Verlag, 1975, 191–232.

2. Goldeich, P.; S.J. Peale. "The Dynamics of Planetary Rotations," *Annual Review of Astronomy and Astrophysics,* 6, Palo Alto, 1968, 287–320.

3. Kohler, P. "Dynamics of a System of Two Satellites Connected by a Deployable and Extensible Tether of Finite Mass," *ESTEC,* 2992/76/NL/AK (SC), 1 and 2, 1978.

4. McCord, T.B.; T.V. Johnson; J.H. Elias. "Saturn and Its Satellites," *Astrophysics Journal,* 165, 1971, 413–424.

5. Rimrott, F.P.J. "Storable Tubular Extendible Member—A Unique Machine Element," *Machine Design,* 37, 28, 1965, 156–165.

6. Xu, D.M.; V.J. Modi; A.K. Misra. "On Three-Dimensional Librational Control of the Shuttle Tethered Subsatellite System during Deployment and Retrieval," *Proceedings,* CANCAM 81, 1981, 389–391.

Problems

3.1. A rod satellite circles the Earth in locked rotation, carrying out a small libration in the orbital plane about the position vector. The satellite is 629 km above the surface of the Earth sphere. Determine the libration period.

3.2. A librating rod satellite, remaining in its orbital plane during libration, is at its maximum amplitude of $\zeta_0 = 12°$ at $t = 0$. After three orbits, at what libration amplitude will the satellite be?

3.3. A rod satellite librates normal to the orbital plane about its radial attitude. Determine the libration frequency.

4

Stability of Satellite Attitude in a Central Force Field

The subsequent stability analyses are for absolutely rigid satellites of finite size in the gravitational field of a point master. The term *gravity gradient stabilization* (or *GG stabilization*) is used frequently by space engineers for satellites stabilized by taking advantage of the gradient of the gravitational field surrounding the master, the gradient of the field being none other than the Kepler force **K**, which, because of the finite size of the satellite, also exerts a torque on the satellite.

4.1 Cardan Angles

The system of Euler angles introduced in Chapter 1 is sometimes very awkward, for instance because of the singularity at $\nu = 0$, as a consequence of which ψ and σ become coplanar and hence indistinguishable, or in situations where small deviations from a reference value of 90° of a Euler angle are to be investigated. From the many angle systems one can devise, each with certain advantages and disadvantages, we shall select the Cardan[1] angles ξ, η, and ζ, as shown in Figure 4.1, for an analysis of the attitude stability of satellites in a central force field. (See also Appendix C.)

The Cardan components of the angular velocity are integrable, i.e., Cardan angles can be obtained as function time:

$$\xi = \xi_0 + \int \dot{\xi}\, dt$$

$$\eta = \eta_0 + \int \dot{\eta}\, dt$$

$$\zeta = \zeta_0 + \int \dot{\zeta}\, dt.$$

In Figure 4.1 a disk gyro is shown, with a body-fixed axis system $Cxyz$. The attitude of the disk gyro with respect to the reference coordinate system $Cx_0y_0z_0$ is described by first rotating the gyro about the (original)

---

[1] After Geronimo Cardano (1501–1576), who described the suspension of gyros in gimbals (*cardanic suspension*). Cardan angles are also referred to as *Bryant* angles by some authors, e.g., Wittenburg (1977).

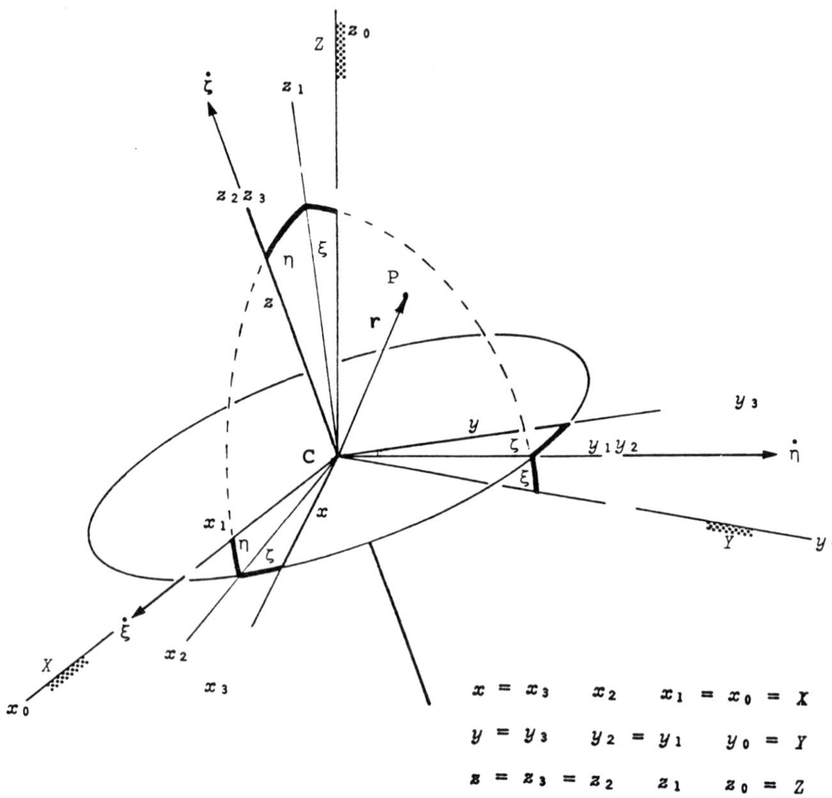

FIGURE 4.1. Gyro axes $Cxyz$ rotated through Cardan angles of the first kind.

x_0-axis through an angle ξ, resulting in a first *carried* coordinate system $Cx_1y_1z_1$, whose x_1-axis coincides with the x_0-axis. Subsequently, the gyro is rotated about the carried y_1-axis through an angle η, resulting in a second carried coordinate system $Cx_2y_2z_2$, whose y_2-axis coincides with the y_1-axis. Lastly, the gyro is rotated about the z_2-axis through an angle ζ, resulting in a third carried coordinate system $Cx_3y_3z_3$, whose z_3-axis coincides with the z_2-axis. The $Cx_3y_3z_3$ coordinate system is identical with the body-fixed $Cxyz$ system of the gyro body whose attitude is given by the Cardan angles ξ, η, and ζ. The sequence of the rotations is significant, and as a result six kinds of Cardan angle systems can be defined. The system used herein is a Cardan angle system of the first kind.

An arbitrary vector $\mathbf{r}$, connecting two points C and P, may be expressed in different coordinate systems:

$$\mathbf{r} = [\mathbf{e}_{x_0}\ \mathbf{e}_{y_0}\ \mathbf{e}_{z_0}] \begin{bmatrix} x_0 \\ y_0 \\ z_0 \end{bmatrix} = [\mathbf{e}_{x_1}\ \mathbf{e}_{y_0}\ \mathbf{e}_{z_0}] \begin{bmatrix} x_1 \\ y_1 \\ z_1 \end{bmatrix}.$$

The vector's components in the 1-system are obtained from its components in the 0-system by

$$\begin{bmatrix} x_1 \\ y_1 \\ z_1 \end{bmatrix} = [T]_{1\leftarrow 0} \begin{bmatrix} x_0 \\ y_0 \\ z_0 \end{bmatrix}$$

where $[T]_{1\leftarrow 0}$ is the *transition matrix.*

The transition matrix for the first coordinate rotation, which takes place about the x_0-axis and sweeps out the first Cardan angle ξ, is

$$[T]_{1\leftarrow 0} = \begin{bmatrix} 1 & 0 & 0 \\ 0 & \cos\xi & \sin\xi \\ 0 & -\sin\xi & \cos\xi \end{bmatrix}. \tag{4.1}$$

The transition matrix for the subsequent coordinate rotation, which takes place about the y_1-axis and sweeps out the second Cardan angle η, is

$$[T]_{2\leftarrow 1} = \begin{bmatrix} \cos\eta & 0 & -\sin\eta \\ 0 & 1 & 0 \\ \sin\eta & 0 & \cos\eta \end{bmatrix}. \tag{4.2}$$

The transition matrix for the final coordinate rotation, which take place about the z_2-axis and sweeps out the third Cardan angle ζ, is

$$[T]_{3\leftarrow 2} = \begin{bmatrix} \cos\zeta & \sin\zeta & 0 \\ -\sin\zeta & \cos\zeta & 0 \\ 0 & 0 & 1 \end{bmatrix}. \tag{4.3}$$

The three transition matrices may be combined into one, i.e.,

$$[T]_{3\leftarrow 0} = [T]_{3\leftarrow 2}[T]_{2\leftarrow 1}[T]_{1\leftarrow 0} \tag{4.4}$$

which is

$$[T]_{3\leftarrow 0} = \begin{bmatrix} \cos\eta\cos\zeta & \cos\xi\sin\zeta+\sin\xi\sin\eta\cos\zeta & \sin\xi\sin\zeta-\cos\xi\sin\eta\cos\zeta \\ -\cos\eta\sin\zeta & \cos\xi\cos\zeta-\sin\xi\sin\eta\sin\zeta & \sin\xi\cos\zeta+\cos\xi\sin\eta\sin\zeta \\ \sin\eta & -\sin\xi\cos\eta & \cos\xi\cos\eta \end{bmatrix}. \tag{4.5}$$

The angular velocity $\boldsymbol{\omega}$ of the satellite can either be expressed in terms of components along the $Cxyz$ coordinate system being used in the problem solution, or in terms of the *Cardan frequencies* along the pertinent (nonorthogonal) carried coordinates:

$$\boldsymbol{\omega} = [\mathbf{e}_x\ \mathbf{e}_y\ \mathbf{e}_z] \begin{bmatrix} \omega_x \\ \omega_y \\ \omega_z \end{bmatrix} = [\mathbf{e}_{x_0}\ \mathbf{e}_{y_1}\ \mathbf{e}_{z_2}] \begin{bmatrix} \dot{\xi} \\ \dot{\eta} \\ \dot{\zeta} \end{bmatrix}.$$

The carried coordinates used here are for Cardan angles of the first kind.

An inspection of Figure 4.1 shows that the transition between the two coordinate systems can be effected by

$$\begin{bmatrix} \omega_x \\ \omega_y \\ \omega_z \end{bmatrix} = \begin{bmatrix} \cos\eta\cos\zeta & \sin\zeta & 0 \\ -\cos\eta\sin\zeta & \cos\zeta & 0 \\ \sin\eta & 0 & 1 \end{bmatrix} \begin{bmatrix} \dot{\xi} \\ \dot{\eta} \\ \dot{\zeta} \end{bmatrix}. \tag{4.6}$$

In the transition matrix of equation (4.6) the first column is seen to be taken from $[T]_{3\leftarrow 0}$, the second from $[T]_{3\leftarrow 1}$ (or $[T]_{3\leftarrow 2}$), and the third form $[T]_{3\leftarrow 3}$ (i.e., the identity matrix).

For sufficiently small Cardan angles, i.e., when the cosine can be approximated by unity, and the sine by the angle itself, and products of angles are negligibly small, the transition matrix (4.5) yields

$$[T]_{3\leftarrow 0} = \begin{bmatrix} 1 & \zeta & -\eta \\ -\zeta & 1 & \xi \\ \eta & -\xi & 1 \end{bmatrix} \tag{4.7}$$

and the angular velocity components in terms of Cardan frequencies become

$$\begin{bmatrix} \omega_x \\ \omega_y \\ \omega_z \end{bmatrix} = \begin{bmatrix} 1 & \zeta & 0 \\ -\zeta & 1 & 0 \\ \eta & 0 & 1 \end{bmatrix} \begin{bmatrix} \dot{\xi} \\ \dot{\eta} \\ \dot{\zeta} \end{bmatrix}. \tag{4.8}$$

For the case of a gyro spinning about its z-axis, with sufficiently small spin axis attitude deviations ξ and η, the transition matrix (4.5) yields

$$[T]_{3\leftarrow 0} = \begin{bmatrix} \cos\zeta & \sin\zeta & \xi\sin\zeta - \eta\cos\zeta \\ -\sin\zeta & \cos\zeta & \xi\cos\zeta + \eta\sin\zeta \\ \eta & -\xi & 1 \end{bmatrix} \tag{4.9}$$

and the transition of angular velocity components assumes the form

$$\begin{bmatrix} \omega_x \\ \omega_y \\ \omega_z \end{bmatrix} = \begin{bmatrix} \cos\zeta & \sin\zeta & 0 \\ -\sin\zeta & \cos\zeta & 0 \\ \eta & 0 & 1 \end{bmatrix} \begin{bmatrix} \dot{\xi} \\ \dot{\eta} \\ \dot{\zeta} \end{bmatrix}. \tag{4.10}$$

4.2 Spinning Axisymmetric Satellite

For very small satellites, the rotational and orbital motions are decoupled. We use the Euler equations (1.68–1.70) and the torque of the gravitational field as given by equation (2.55), and thus obtain

$$A\dot{\omega}_x - (B - C)\omega_y\omega_z = \frac{3\mu}{\rho_C^3}(C - B)\cos\gamma\cos\beta \tag{4.11a}$$

$$B\dot{\omega}_y - (C - A)\omega_z\omega_x = \frac{3\mu}{\rho_C^3}(A - C)\cos\alpha\cos\gamma \tag{4.11b}$$

$$C\dot{\omega}_z - (A - B)\omega_x\omega_y = \frac{3\mu}{\rho_C^3}(B - A)\cos\beta\cos\alpha. \tag{4.11c}$$

If we restrict further work to a circular orbit, then we can relate the centrifugal force to the Kepler force by

$$m\rho_C\Omega^2 = \frac{\mu m}{\rho_G^2}$$

and use $\rho_G = \rho_C$ as an approximation, to get

$$\frac{\mu}{\rho_C^3} = \Omega^2. \tag{4.12}$$

We also will restrict ourselves to satellites of rotational symmetry, i.e., those for which $B = A$, a condition that immediately leads to a solution of equation (4.11c), i.e.,

$$\omega_z = \omega_0 = \text{constant}. \tag{4.13}$$

With the help of the constraints just introduced, the differential equations (4.11a) and (4.11b) of motion become

$$A\dot{\omega}_x - (A - C)\omega_0\omega_y - 3\Omega^2(C - A)\cos\gamma\cos\beta = 0 \tag{4.14a}$$

$$A\dot{\omega}_y - (C - A)\omega_0\omega_x - 3\Omega^2(A - C)\cos\alpha\cos\gamma = 0. \tag{4.14b}$$

For a spinning satellite, the axis $\mathbf{e}_z$ of symmetry of a stable satellite will be at or near the axis $\mathbf{e}_Z$ of the orbit-oriented coordinate system with the unit vectors $\mathbf{e}_r$, $\mathbf{e}_\theta$, and $\mathbf{e}_Z$ (Figure 4.2).

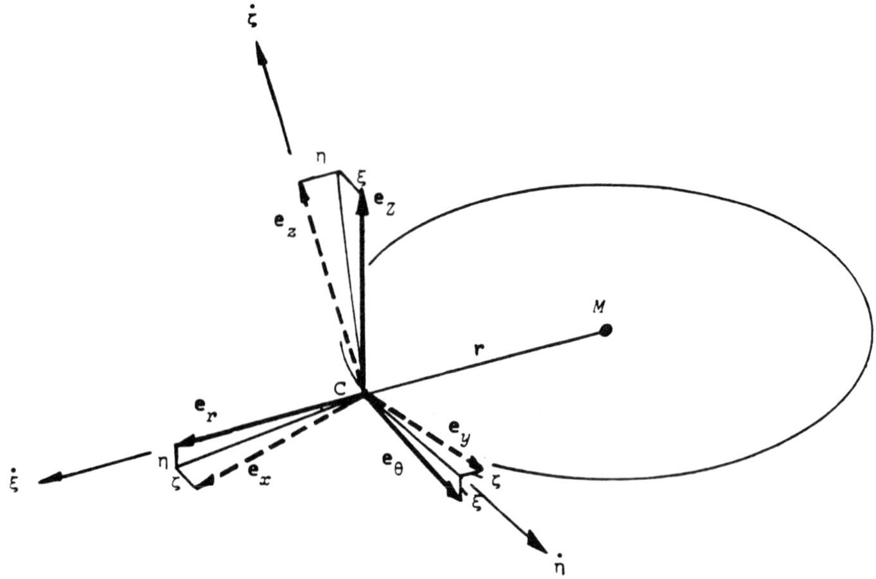

FIGURE 4.2. Orbital cylindrical coordinates r, θ, and Z; Cardan angles ξ, η, and ζ; and body-fixed $Cxyz$ coordinates.

Using the Cardan angles of Figure 4.2, and equation (4.6) for the angular velocity components, one obtains

$$\begin{bmatrix} \omega_x \\ \omega_y \\ \omega_z \end{bmatrix} = \begin{bmatrix} \cos\eta\cos\zeta & \sin\zeta & 0 \\ -\cos\eta\sin\zeta & \cos\zeta & 0 \\ \sin\eta & 0 & 1 \end{bmatrix} \begin{bmatrix} \dot\xi \\ \dot\eta \\ \dot\zeta \end{bmatrix} + \Omega \begin{bmatrix} \sin\xi\sin\zeta - \cos\xi\sin\eta\cos\zeta \\ \sin\xi\cos\zeta + \cos\xi\sin\eta\sin\zeta \\ \cos\xi\cos\eta \end{bmatrix}. \tag{4.15}$$

We allow very small angular deviations ξ and η of the satellite's axis of symmetry from the space-fixed Z-direction. This entitles us to assume that

$$\sin\xi = \xi \tag{4.16a}$$

$$\cos\xi = 1 \tag{4.16b}$$

$$\sin\eta = \eta \tag{4.16c}$$

$$\cos\eta = 1. \tag{4.16d}$$

If we furthermore neglect terms containing the product $\xi\eta$, the angular velocity components can be approximated by

$$\omega_x = \dot\xi\cos\eta + \dot\eta\sin\zeta + \Omega(\xi\sin\zeta - \eta\cos\zeta) \tag{4.17a}$$

$$\omega_y = -\dot{\xi}\sin\zeta + \dot{\eta}\cos\zeta + \Omega(\xi\cos\zeta + \eta\sin\zeta) \tag{4.17b}$$

$$\omega_z = \dot{\zeta} + \Omega = \omega_0. \tag{4.17c}$$

For the direction cosines we have, from Appendix C,

$$\begin{bmatrix} \cos\alpha \\ \cos\beta \\ \cos\gamma \end{bmatrix} = \begin{bmatrix} \cos\eta\cos\zeta \\ -\cos\eta\sin\zeta \\ \sin\eta \end{bmatrix} \tag{4.18}$$

or for small angles η

$$\begin{bmatrix} \cos\alpha \\ \cos\beta \\ \cos\gamma \end{bmatrix} = \begin{bmatrix} \cos\zeta \\ -\sin\zeta \\ \eta \end{bmatrix}. \tag{4.19}$$

With the help of equations (4.11) and (4.19) the equations of motion become very lengthy expressions in which each term contains a $\sin\zeta$ or $\cos\zeta$. We can multiply through, one equation with $\sin\zeta$ and the other with $\cos\zeta$, and add, then multiply by $\cos\zeta$ and $\sin\zeta$ and subtract, to arrive at a system of two simultaneous linear differential equations with constant coefficients having the comparatively simple form

$$A\ddot{\xi} + (C\omega_0 - 2A\Omega)\dot{\eta} + (C\omega_0\Omega - A\Omega^2)\xi = 0 \tag{4.20a}$$

$$A\ddot{\eta} - (C\omega_0 - 2A\Omega)\dot{\xi} + (C\omega_0\Omega + 3C\Omega^2 - 4A\Omega^2)\eta = 0. \tag{4.20b}$$

With a solution ansatz $\xi = \xi_0 e^{\lambda t}$ and $\eta = \eta_0 e^{\lambda t}$, the characteristic equation of this system is

$$\begin{vmatrix} A\lambda^2 + (C\omega_0\Omega - A\Omega^2) & (C\omega_0 - 2A\Omega)\lambda \\ -(C\omega_0 - 2A\Omega)\lambda & A\lambda^2 + (C\omega_0\Omega + 3C\Omega^2 - 4A\Omega^2) \end{vmatrix} = 0 \tag{4.21}$$

giving

$$\lambda^4 + p\lambda^2 + q = 0 \tag{4.22}$$

with

$$p = \left(3\frac{C}{A} - 1\right)\Omega^2 - 2\frac{C}{A}\omega_0\Omega + \left(\frac{C}{A}\omega_0\right)^2 \tag{4.23}$$

$$q = \left(4 - 3\frac{C}{A}\right)\Omega^4 + \frac{C}{A}\left(3\frac{C}{A} - 5\right)\omega_0\Omega^3 + \left(\frac{C}{A}\omega_0\right)^2\Omega^2. \tag{4.24}$$

The satellite's attitude will be stable, i.e, the amplitudes of the angles ξ and η will remain bounded, if

$$p > 0 \tag{4.25a}$$

$$q > 0 \tag{4.25b}$$

$$p^2 - 4q > 0. \tag{4.25c}$$

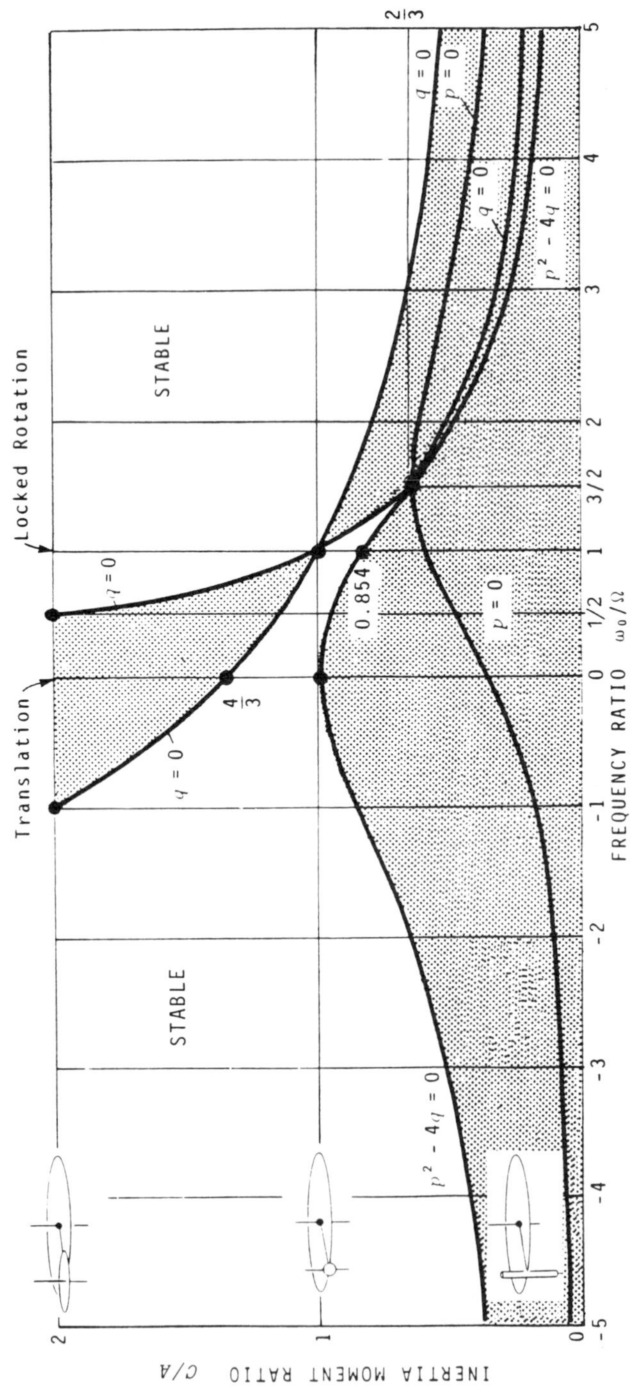

FIGURE 4.3. Stability regions for rigid axisymmetric satellites spinning around the z-axis on a circular orbit (according to Thomson (1962) and Kane (1962)).

The three curves delimiting the areas of stability resulting from inequalities (4.25a–c) are plotted in Figure 4.3 (Thomson, 1962).

It is interesting to note that, e.g., a disk satellite in pure translation is unstable, in spite of the fact that its gravity center is closer to the master than its mass center. And indeed, there is a restoring torque when the satellite assumes an angle η. There is, however, no restoring torque against any drift ξ, and consequently there is attitude instability. The area in the stability diagram (Figure 4.3) under the $C/A = 1$ line is of interest only for short-lived satellites if they can be considered rigid. Long-lived satellites with $C/A < 1$ will eventually succumb to internal energy dissipation, i.e., they are unstable in the long run, as will be shown later.

The effect of the presence of a nutation damper with a small mass of finite size has been studied in detail by Tyc (1987), who shows that the stability region begins at the $C/A = 1$ line for zero damper mass and that it moves further above the $C/A = 1$ line, the larger the damper mass.

If equations (4.20) are written in the form

$$\begin{bmatrix} A & 0 \\ 0 & A \end{bmatrix} \begin{bmatrix} \ddot{\xi} \\ \ddot{\eta} \end{bmatrix} + \begin{bmatrix} 0 & (C\omega_0 - 2A\Omega) \\ -(C\omega_0 - 2A\Omega) & 0 \end{bmatrix} \begin{bmatrix} \dot{\xi} \\ \dot{\eta} \end{bmatrix}$$

$$+ \begin{bmatrix} (C\omega_0\Omega - A\Omega^2) & 0 \\ 0 & (C\omega_0\Omega + 3C\Omega^2 - 4A\Omega^2) \end{bmatrix} \begin{bmatrix} \xi \\ \eta \end{bmatrix} = \begin{bmatrix} 0 \\ 0 \end{bmatrix} \quad (4.26)$$

then the matrix

$$[G] = \begin{bmatrix} 0 & (C\omega_0 - 2A\Omega) \\ -(C\omega_0 - 2A\Omega) & 0 \end{bmatrix} \quad (4.27a)$$

is referred to as the *gyric* (or *gyroscopic*) *matrix*. In the present case it provides the coupling between the two coordinates.

The matrix

$$[K] = \begin{bmatrix} C\omega_0\Omega - A\Omega^2 & 0 \\ 0 & C\omega_0\Omega + 3C\Omega^2 - 4A\Omega^2 \end{bmatrix} \quad (4.27b)$$

is, in analogy to elastodynamic systems, referred to as the *stiffness* matrix. It is positive definite when $(C/A)(\omega_0/\Omega) > 1$ and negative definite when $(C/A)/(\omega_0/\Omega) < 1$. This leads engineers to refer to the right stable region in Figure 4.3 as *statically* stable, and the left stable region as only *gyroscopically* stable.

4.3 Arbitrary Satellite in Locked Rotation

Let us now consider a satellite with an inertia tensor

$$\begin{bmatrix} A & 0 & 0 \\ 0 & B & 0 \\ 0 & 0 & C \end{bmatrix} \quad (4.28)$$

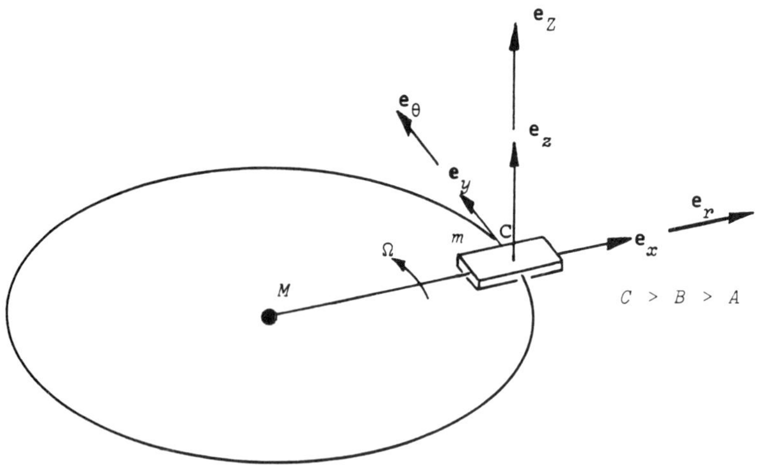

FIGURE 4.4. Satellite on circular orbit in locked rotation.

where $A \neq B \neq C$, on a circular orbit (Figure 4.4) in locked rotation. The satellite's z-axis is deviating only slightly from the Z-axis of the orbit coordinate system. The same is true for the other two axes, the x-axis remains close to the radial direction, and the y-axis remains close to the orbit tangent. As long as the deviations remain small we can write for the velocity vector

$$\boldsymbol{\omega} = [\mathbf{e}_x \ \mathbf{e}_y \ \mathbf{e}_z] \begin{bmatrix} \dot{\xi} - \Omega\eta \\ \dot{\eta} + \Omega\xi \\ \dot{\zeta} + \Omega \end{bmatrix} \tag{4.29}$$

and for the direction cosines

$$\begin{bmatrix} \cos\alpha \\ \cos\beta \\ \cos\gamma \end{bmatrix} = \begin{bmatrix} 1 \\ -\zeta \\ \eta \end{bmatrix}. \tag{4.30}$$

Considering that ω_x and ω_y are each so small that the product $\omega_x\omega_y$ can be neglected, equation (4.9) assumes the form

$$C\ddot{\zeta} + 3(B - A)\Omega^2\zeta = 0 \tag{4.31}$$

allowing a solution for the spin angle ζ. It is quickly recognized that for stable motion we must have $B > A$. Employing equations (4.7) and (4.8) as well, and making the assumptions that $\dot{\zeta} << \Omega$ and that $\zeta\eta \simeq 0$, there remains a system of two coupled linear differential equations:

$$A\ddot{\xi} - (A + B - C)\Omega\dot{\eta} = (B - C)\Omega^2\xi = 0 \tag{4.32a}$$

$$B\ddot{\eta} + (A + B - C)\Omega\dot{\xi} + 4(C - A)\Omega^2\eta = 0. \tag{4.32b}$$

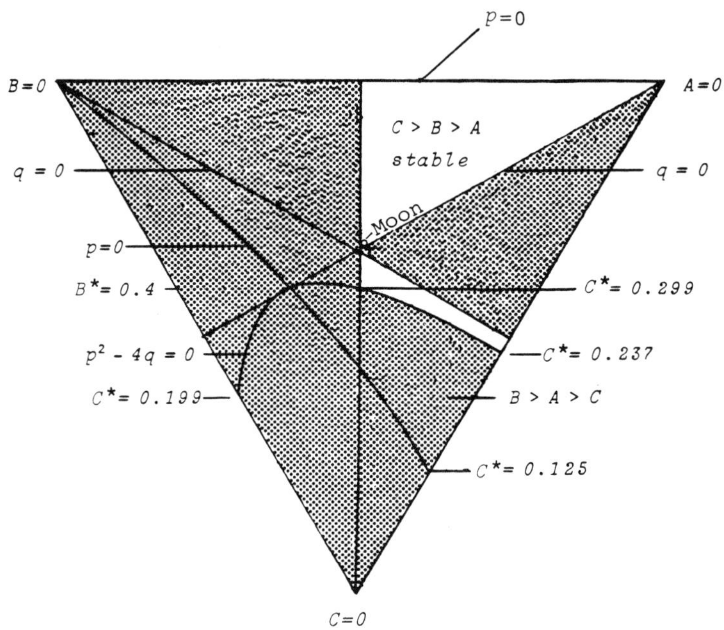

FIGURE 4.5. Magnus stability diagram for GG stabilized gyrosatellite configuration on a circular orbit (x-axis radially oriented, y-axis tangentially oriented, and z-axis perpendicular to orbital plane).

We use $e^{\lambda t}$ for a solution ansatz and obtain the following characteristic equation for λ

$$\begin{vmatrix} A\lambda^2 - (B-C)\Omega^2 & -(A+B-C)\Omega\lambda \\ (A+B-C)\Omega\lambda & B\lambda^2 + 4(C-A)\Omega^2 \end{vmatrix} = 0 \tag{4.33}$$

leading to

$$\left(\frac{\lambda}{\Omega}\right)^4 + p\left(\frac{\lambda}{\Omega}\right)^2 + q = 0 \tag{4.34}$$

with

$$p = 2 - 3\frac{A}{B} + 2\frac{C}{B} - \frac{C}{A} + \frac{C^2}{AB} \tag{4.35}$$

and

$$q = 4\left(\frac{C}{A} - 1\right)\left(\frac{C}{B} - 1\right). \tag{4.36}$$

The satellite will be stable if

$$B > A \tag{4.37a}$$

$$p > 0 \tag{4.37b}$$

$$q > 0 \tag{4.37c}$$

and

$$p^2 - 4q > 0. \tag{4.37d}$$

The Magnus shape triangle of Figure 4.5 shows the stable configurations. First there is a *static* stability region

$$C > B > A \tag{4.38}$$

and then there is a small *gyroscopic* stability island in the region $B > A > C$.

4.4 Arbitrary Satellite on Elliptical Orbit in Quasi-Locked Rotation

The mass center of a small gyrosatellite on an elliptic orbit obeys the Kepler equations

$$r = \frac{p}{1 + \epsilon\cos\theta} = \rho_C \tag{4.39}$$

and

$$\dot{\theta} = \sqrt{\frac{\mu}{p^3}}(1 + \epsilon\cos\theta)^2. \tag{4.40}$$

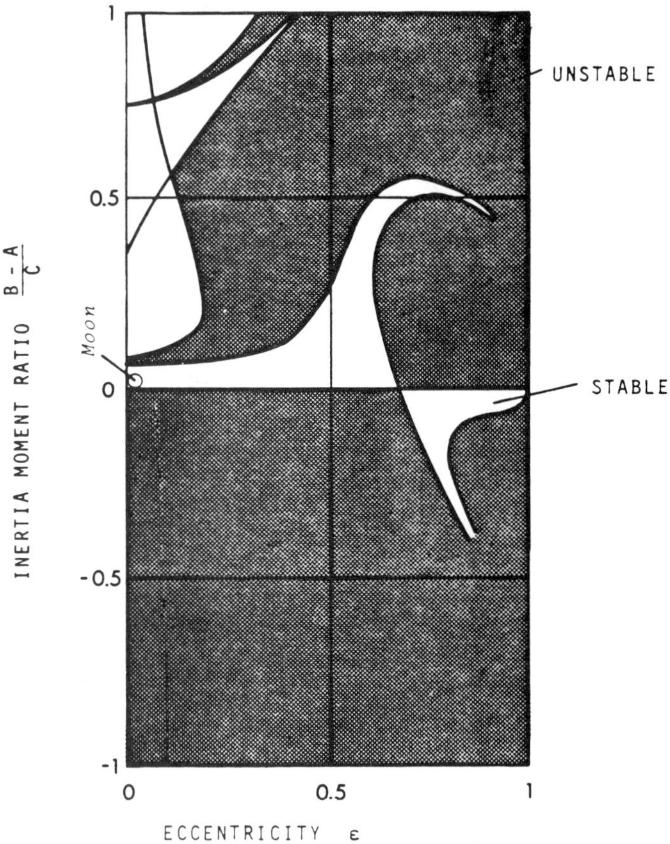

FIGURE 4.6. Zlatoustov stability diagram.

We restrict our study now to the case where the satellite's z-axis remains essentially normal to the orbital plane (Figure 4.4). The Cardan angle ξ and η thus remain small. The spin angle ζ, however, is allowed to assume any value. The angular velocity of the satellite can be approximated by

$$\boldsymbol{\omega} = [\mathbf{e}_x \ \mathbf{e}_y \ \mathbf{e}_z] \begin{bmatrix} 0 \\ 0 \\ \dot{\zeta} + \dot{\theta} \end{bmatrix}. \tag{4.41}$$

The direction cosines are, approximately, from equation (4.19),

$$\begin{bmatrix} \cos\alpha \\ \cos\beta \\ \cos\gamma \end{bmatrix} = \begin{bmatrix} \cos\zeta \\ -\sin\zeta \\ \eta \end{bmatrix}. \tag{4.42}$$

We now employ equation (4.9) and obtain

$$C(\ddot{\zeta} + \ddot{\theta}) + \frac{3\mu}{2\rho_C^3}(B - A)\sin 2\zeta = 0$$

or

$$C\ddot{\zeta} + \frac{3\mu}{2\rho_C^3}(B - A)\sin 2\zeta = -C\ddot{\theta}. \tag{4.43}$$

Introducing the true anomaly θ from equations (4.39) and (4.40) into equation (4.43) allows the elimination of time as the independent variable

$$\frac{d^2\zeta}{d\theta^2} - \frac{2\epsilon\sin\theta}{1 + \epsilon\cos\theta}\frac{d\zeta}{d\theta} + \frac{3}{2}\frac{B - A}{C(1 + \epsilon\cos\theta)}\sin 2\zeta = \frac{2\epsilon}{1 + \epsilon\cos\theta}\sin\theta. \tag{4.44}$$

Equation (4.44) is a nonhomogeneous nonlinear differential equation with parametric coefficients.

One can thus expect parametric vibrations in addition to the forced vibration caused by the nonuniform rotation of the satellite's position vector. Figure 4.6 shows the stability regions as determined by Zlatoustov et al. (1966). For an analysis of other cases see Magnus (1965).

Suggested Reading

1. Hughes, P.C. *Spacecraft Attitude Dynamics,* Wiley & Sons, 1985, 576 pp.

2. Kane, T.R.; E.L. Marsh; W.G. Wilson. "Letters to the Editor," *Journal of the Astronautical Sciences,* 108–109.

3. Magnus, K. "Die Stabilität partikulärer Drehbewegungen von Satelliten beliebiger Form auf einer Kreisbahn," *Ingenieur-Archiv,* 34, 1965, 129–138.

4. Thomson, W.T. "Spin Stabilization of Attitude against Gravity Torque," *Journal of the Astronautical Sciences,* 9, 1, 1962, 31–33.

5. Tyc, G. "A Method of Stability Analysis for Systems of Linear Differential Equations with Periodic Coefficients," M.A.S.c. Thesis, University of Toronto, 1987, 164 pp.

6. Wittenburg, J. *Dynamics of Systems of Rigid Bodies,* B.G. Teubner, 1977, 224 pp.

7. Zlatoustov, V.A.; D.E. Okhotsimsky; V.A. Sarychev; A.P. Torzhevsky. "Investigation of Satellite Oscillations in the Plane of an Elliptic Orbit," *Proceedings* (Ed. H. Görtler), 11th International Congress of Applied Mechanics, Munich, 1964, Springer-Verlag, 1966, 436–439.

Problems

4.1. Given is (a) $(3y - 1 - 2xy + x^2y^2)^2 = 4(4 - 3y + xy(3y - 5) + x^2y^2)$. Plot y versus x, for $-5 \leq x \leq 5$ and $0 \leq y \leq 2$. (b) Into the same diagram, plot $3y - 1 - 2xy + x^2y^2 = 0$. (c) Into the same diagram, plot $4 - 3y + xy(3y - 5) + x^2y^2 = 0$.

4.2. Given are:

$$p = 2 - 3\frac{A}{B} + 2\frac{C}{B} - \frac{C}{A} + \frac{C^2}{AB}$$

and

$$q = 4\left(\frac{C}{A} - 1\right)\left(\frac{C}{B} - 1\right).$$

Plot the curves (a) $p = 0$; (b) $q = 0$; and (c) $p^2 - 4q = 0$ into the Magnus shape triangle.

4.3. If the sole stability criterion for a satellite circling its master in locked rotation was that $x_G < 0$ in the equation:

$$x_G = \frac{3}{4m\rho_C}(2A - B - C).$$

(a) Establish the stable region in the Magnus shape triangle (let z be normal to the orbital plane.)

(b) Why is $x_G < 0$ not sufficient to represent attitude stability, even if the gyro satellite consisted of an absolutely rigid body?

4.4. Given is $\lambda^4 + p\lambda^2 + q = 0$. Show that $p > 0$, $q > 0$, and $p^2 - 4q > 0$ are required for stability (of the original equations).

4.5. The principal inertia moments of a spatial dumbbell gyro with concentrated tip masses are to satisfy $B > A > C$ and to fit into the small stability region (of the Magnus shape triangle) bounded by $C = A$ and $p^2 - 4q = 0$. If $a = 10$ m, what values for b and c satisfy the conditions?

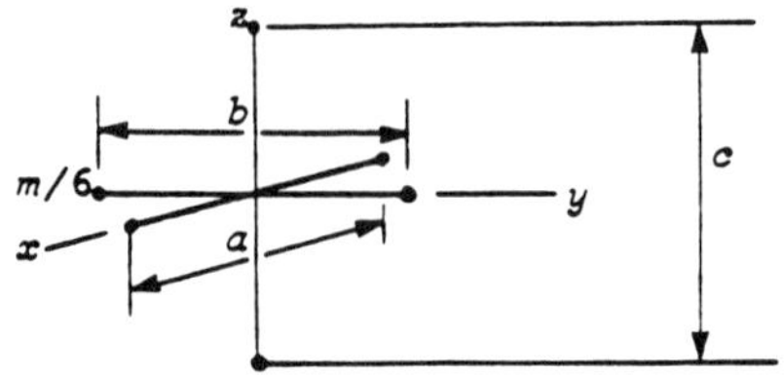

4.6. Depending upon the sequence of rotation, six kinds of Cardan angles can be defined.

Rotation Sequence	First Kind	Second Kind	Third Kind
1	ξ about x_0	η about y_0	ζ about z_0
2	η about y_1	ζ about z_1	ξ about x_1
3	ζ about z_2	ξ about x_2	η about y_2

Rotation Sequence	Fourth Kind	Fifth Kind	Sixth Kind
1	ξ about x_0	ζ about z_0	η about y_0
2	ζ about z_1	η about y_1	ξ about x_1
3	η about y_2	ξ about x_2	ζ about z_2

If the transformation equation for the angular velocity components in Cardan angles of the first kind is

$$\begin{bmatrix} \omega_x \\ \omega_y \\ \omega_z \end{bmatrix} = \begin{bmatrix} \cos\eta \quad \cos\zeta & \sin\zeta & 0 \\ -\cos\eta \quad \sin\zeta & \cos\zeta & 0 \\ \sin\eta & 0 & 1 \end{bmatrix} \begin{bmatrix} \dot{\xi} \\ \dot{\eta} \\ \dot{\zeta} \end{bmatrix}$$

establish the corresponding equation for (a) the second kind; (b) the third kind; (c) the fourth kind; (d) the fifth kind; and (e) the sixth kind.

4.7. A ship's attitude is expressed by roll ξ (about the longitudinal x-axis), pitch η (about the transverse y-axis), and yaw ζ (about the vertical z-axis). Which of the six kinds of Cardan angles is the most suitable to define roll, pitch, and yaw? Sketch. Explain.

4.8. By establishing the transformation matrix for each individual rotation separately, and then multiplying, establish the transformation matrix for the coordinates of a point $P(x_0, y_0, z_0)$ to the coordinates $P(x_3, y_3, z_3)$ for the same point. The $Ox_3y_3z_3$ coordinate system is obtained from $Ox_0y_0z_0$ coordinate system by rotation through Cardan angles of the first kind.

4.9. What is the direction cosine $\cos(x_3, y_0) =$? for the transformation of Problem 4.8?

4.10. On the spacecraft shown, a body-fixed coordinate system $Cxyz$ ($\equiv Cx_3y_3z_3$) is indicated. Sketch the systems $Cx_2y_2z_2$, $Cx_1y_1z_1$, and $Cx_0y_0z_0$, and (a) indicate which axes you would choose as roll axis, pitch axis, and yaw axis. The z_0-axis connects mass center C and Earth center (local vertical). The x_0-axis is tangent to the circular orbit. The y_0-axis is normal to the orbital plane. (b) If space engineers would choose y_0 as pitch axis, z_1 as yaw axis, and $x_2 = x_3 = x$ as roll axis, and their choice is different from yours, try and reason why they may have made that choice. [Hint: Is z_0 a space-fixed axis?]

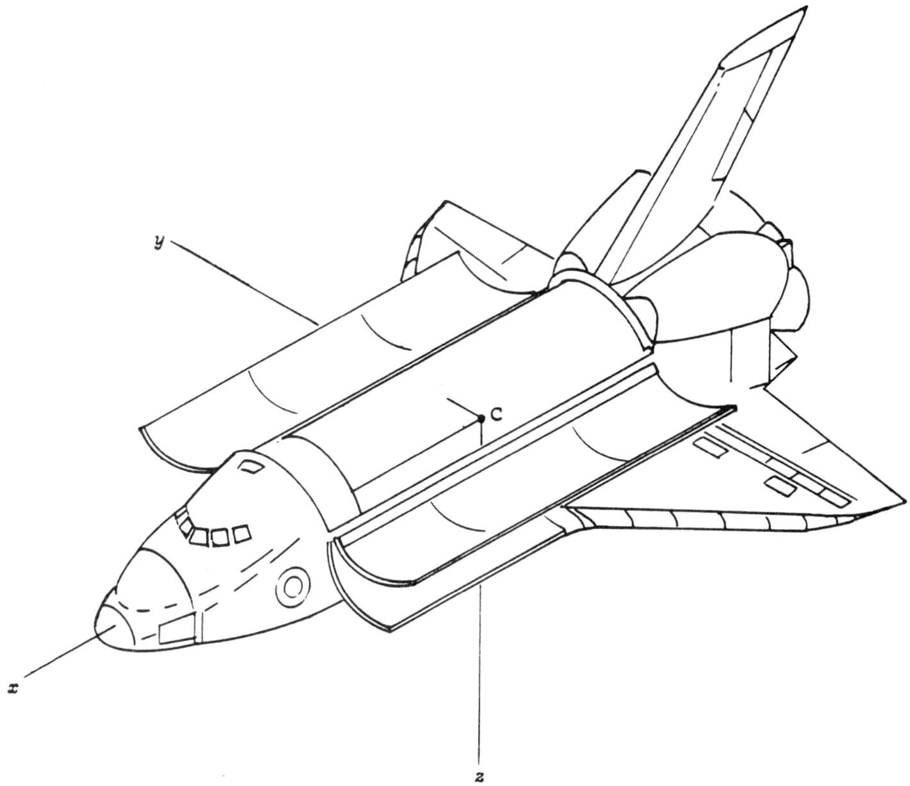

4.11. Use relations from spherical trigonometry and show that Cardan angles of the first kind and Euler angles are related by $\tan\xi = \cos\psi\tan\nu$; $\sin\eta = \sin\psi\sin\nu$; and $\tan(\zeta - \sigma) = \tan\psi\cos\nu$. Sketch the appropriate spherical triangles.

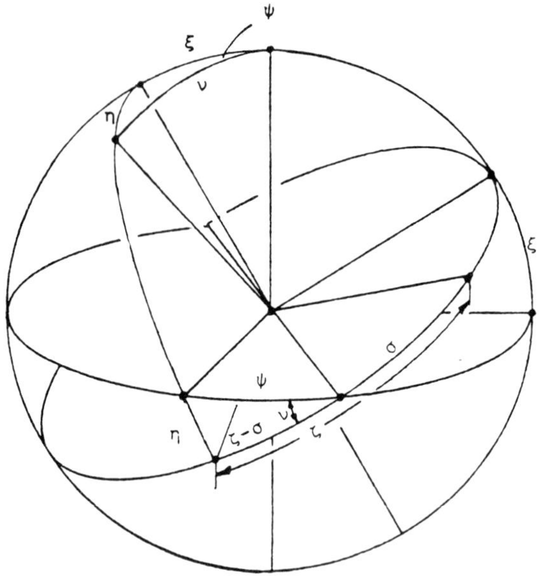

4.12. Show how

$$A\ddot{\xi} + (C\omega_0 - 2A\Omega)\dot{\eta} + (C\omega_0\Omega + A\Omega^2)\xi = 0$$
$$A\ddot{\eta} - (C\omega_0 - 2A\Omega)\dot{\xi} + (C\omega_0\Omega + 3C\Omega^2 - 4A\Omega^2)\eta = 0$$

are obtained.

4.13. An axisymmetric gyro has the shape of a solid circular cylinder and uniform mass distribution. It orbits the Earth on a circular path. If the absolute angular velocity ω_0 of the satellite is equal to three times the absolute angular velocity Ω of its orbital motion, what is the minimum ratio r/l that ensures attitude stability?

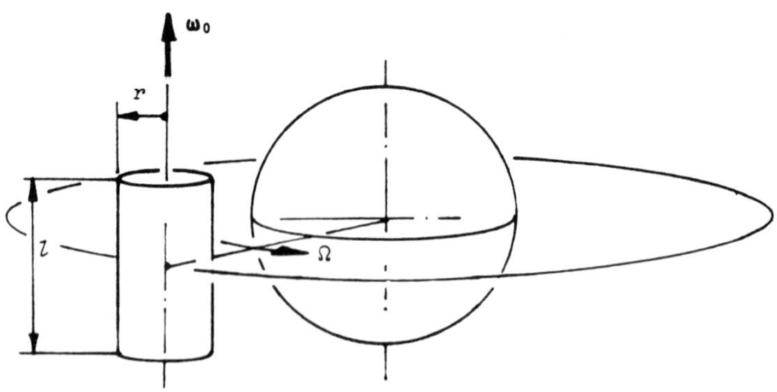

4.14. Sketch a Thomson diagram similar to Figure 4.3 for the motion of the Earth about the Sun and locate the Earth in it.

5

Torquefree Gyros

A torquefree gyro is a gyro upon which no external torque is acting

$$\mathbf{M} = 0. \tag{5.1}$$

According to the Euler law as expressed by equation (1.1), the time derivative $\dot{\mathbf{H}}$ of the angular momentum is consequently also zero. Thus its integral, the angular momentum itself, must remain constant.

$$\mathbf{H} = \text{constant}. \tag{5.2}$$

The assumption of zero torque is sometimes justified for satellites for short periods of time (e.g., for one orbit), or when they are very far away from their master.

The Euler equations (1.68 to 70) for a torquefree gyro, i.e., for $\mathbf{M} = \mathbf{0}$, become

$$\dot{\omega}_x = \frac{B - C}{A}\omega_y\omega_z \tag{5.3a}$$

$$\dot{\omega}_y = \frac{C - A}{B}\omega_z\omega_x \tag{5.3b}$$

$$\dot{\omega}_z = \frac{A - B}{C}\omega_x\omega_y. \tag{5.3c}$$

We have again assumed that the coordinate system chosen is such that the inertia tensor has the form

$$[I] = \begin{bmatrix} A & 0 & 0 \\ 0 & B & 0 \\ 0 & 0 & C \end{bmatrix}. \tag{5.4}$$

Furthermore the $Cxyz$ coordinate system is body-fixed and originates in the gyro's mass center.

5.1 Solution of Euler Equations for Torquefree Gyro

For the case that angular momentum

$$H = \sqrt{A^2\omega_x^2 + B^2\omega_y^2 + C^2\omega_z^2} \tag{5.5}$$

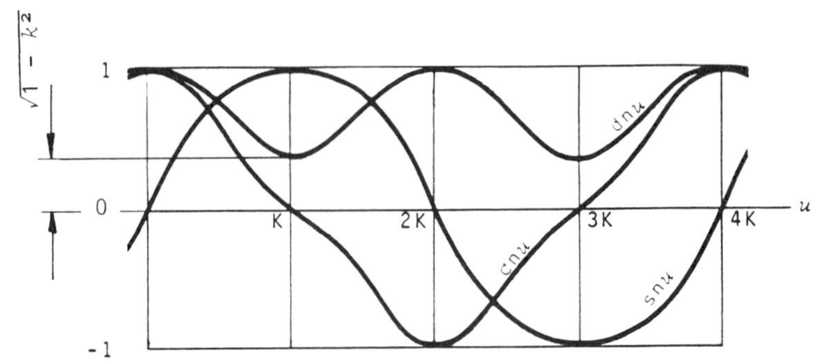

FIGURE 5.1. Typical shape of Jacobi elliptic functions.

and kinetic energy

$$T = \frac{1}{2}(A\omega_x^2 + B\omega_y^2 + C\omega_z^2) \tag{5.6}$$

are both constant, the Euler equations (5.3a–c) have solutions, which can be expressed in terms of the *Jacobi elliptic functions* sn, cn, and dn (Figure 5.1), as follows:

$$\omega_x = \sqrt{\frac{H^2 - 2TC}{A(A-C)}}\,\mathrm{cn}\, u \tag{5.7a}$$

$$\omega_y = \sqrt{\frac{H^2 - 2TC}{B(B-C)}}\,\mathrm{sn}\, u \tag{5.7b}$$

$$\omega_z = -\sqrt{\frac{2TA - H^2}{C(A-C)}}\,\mathrm{dn}\, u \tag{5.7c}$$

provided that

$$A > B > C \tag{5.8}$$

and that *epicycloidal* motion prevails, i.e., that

$$2TB > H^2. \tag{5.9}$$

See also Figure 5.2, where $(\omega_1)_y = (\omega_2)_x = 0$ and

$$\sin \nu_1 = \sqrt{\frac{1 - 2CT/H^2}{1 - C/A}}$$

$$\sin \nu_2 = \sqrt{\frac{1 - 2CT/H^2}{1 - C/B}}.$$

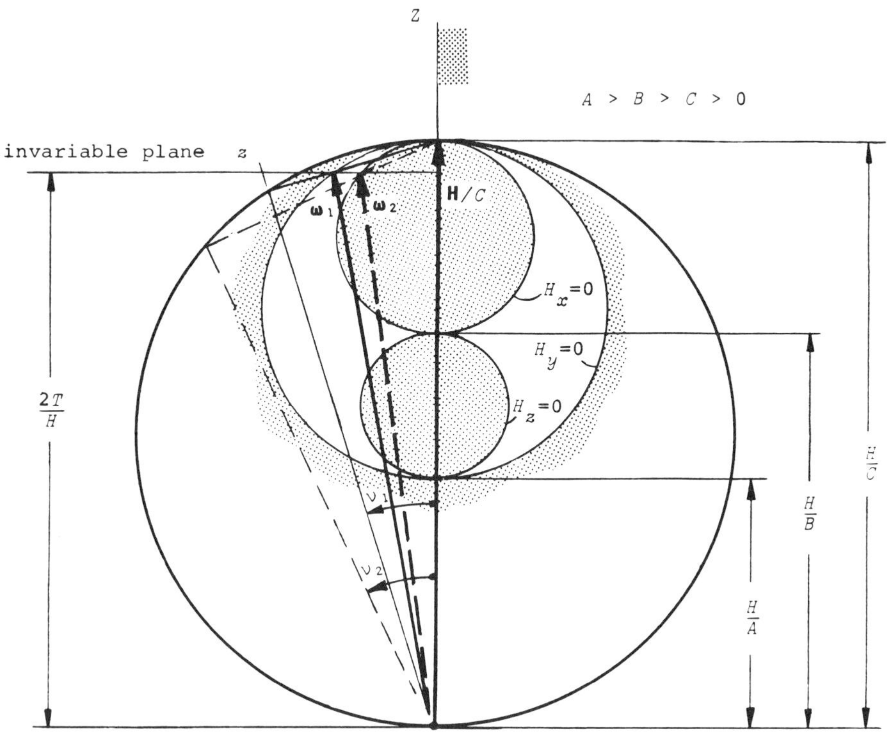

The diagram shown is for
H = 120 Js
A = 24 m²kg B = 15 m²kg C = 10 m²kg
T = 675 J

FIGURE 5.2. Attitude diagram.

The quantity u is dimensionless and related to the time by

$$u = \sqrt{\frac{(B-C)(2TA-H^2)}{ABC}}(t-t_0). \tag{5.10}$$

Jacobi elliptic functions are listed and plotted in tables of higher functions, such as Jahnke et al. (1960), where a *modulus* $k = \sin\alpha$ plays a role, which for our case is

$$k = \sin\alpha = \sqrt{\frac{H^2-2TC}{2TA-H^2}\frac{A-B}{B-C}}. \tag{5.11}$$

When $k = 0$, then cn $u = \cos u$, sn $u = \sin u$ and dn $u = 1$. When $k = 1$, then cn $u = \operatorname{dn} u = \operatorname{Sech} u$, and sn $u = \operatorname{Tanh} u$. The period of the Jacobi elliptical functions $\operatorname{sn} u$ and $\operatorname{cn} u$ (Figure 5.1) is $4K$, the period of the Jacobi elliptical function $\operatorname{dn} u$ is $2K$, where K is the complete elliptic integral of the first kind, given by

$$K = \int_0^1 \frac{d\xi}{\sqrt{(1-\xi^2)(1-k^2\xi^2)}} = \int_0^{\pi/2} \frac{d\psi}{\sqrt{1-\sin^2\alpha\sin^2\psi}}. \tag{5.12}$$

Complete elliptic integrals are also listed in Jahnke et al. (1960).

When *pericycloidal* motion prevails then

$$2TB < H^2 \tag{5.13}$$

and

$$\omega_x = +\sqrt{\frac{H^2-2TC}{A(A-C)}}\operatorname{dn} u \tag{5.14a}$$

$$\omega_y = +\sqrt{\frac{2TA-H^2}{B(A-B)}}\operatorname{sn} u \tag{5.14b}$$

$$\omega_z = -\sqrt{\frac{2TA-H^2}{C(A-C)}}\operatorname{cn} u \tag{5.14c}$$

with

$$u = \sqrt{\frac{(A-B)(H^2-2TC)}{ABC}}(t-t_0) \tag{5.15}$$

and a modulus

$$k = \sqrt{\frac{B-C}{A-B}\frac{2TA-H^2}{H^2-2TC}}. \tag{5.16}$$

For the borderline case

$$2TB = H^2 \tag{5.17}$$

we find

$$\omega_x = \sqrt{\frac{2T(B-C)}{A(A-C)}}\operatorname{Sech} u \tag{5.18a}$$

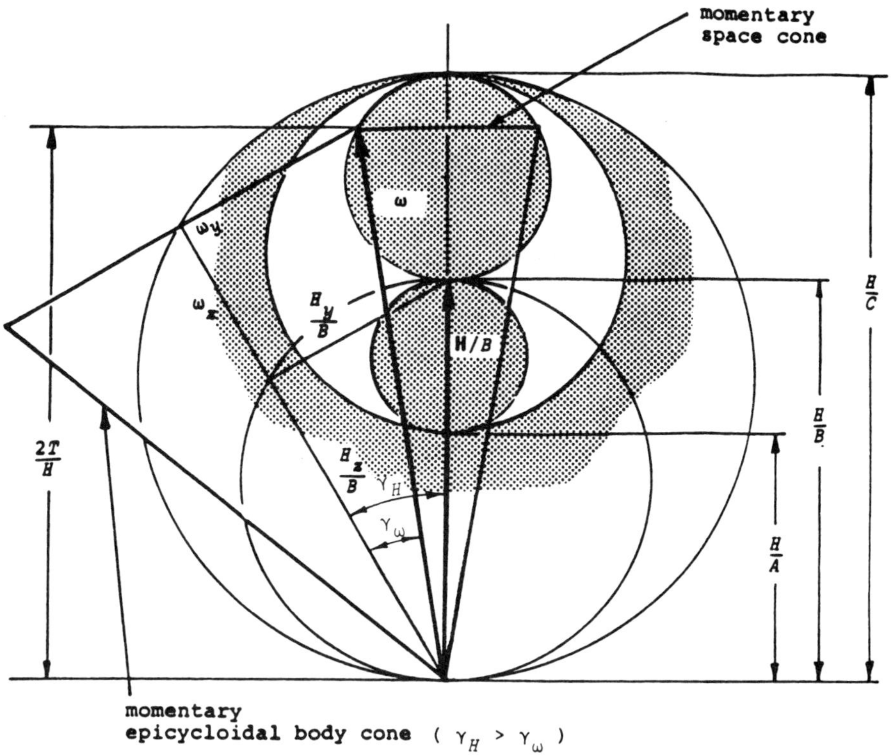

FIGURE 5.3. Epicyloidal motion.

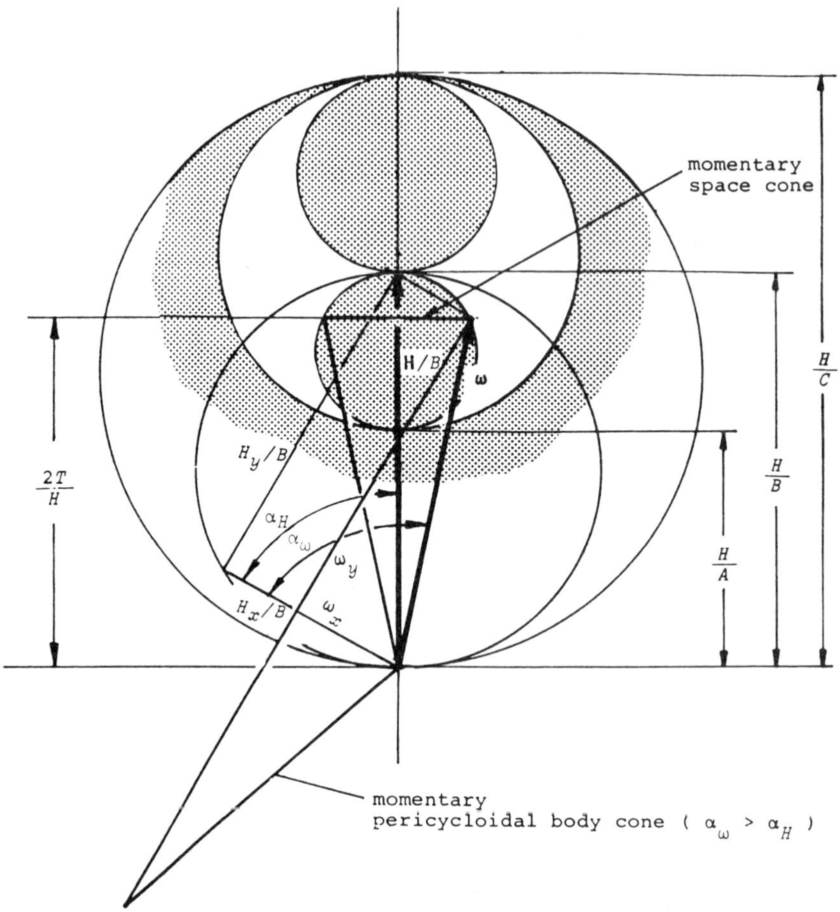

FIGURE 5.4. Pericycloidal motion.

$$\omega_y = \sqrt{\frac{2T}{B}}\ \mathrm{Tanh}\ u \tag{5.18b}$$

$$\omega_z = -\sqrt{\frac{2T(A-B)}{C(A-C)}}\mathrm{Sech}\ u \tag{5.18c}$$

with

$$u = \sqrt{\frac{2T(A-B)(B-C)}{ABC}}(t-t_0). \tag{5.19}$$

The signs appearing in equations (5.7), (5.14), and (5.18) supply but one possible solution out of four for each set, as one can readily establish by observing that

$$\frac{d}{du}(\mathrm{sn}\ u) = \mathrm{cn}\ u\ \mathrm{dn}\ u \tag{5.20a}$$

$$\frac{d}{du}(\mathrm{cn}\ u) = -\mathrm{sn}\ u\ \mathrm{dn}\ u \tag{5.20b}$$

$$\frac{d}{du}(\mathrm{dn}\ u) = -k^2\ \mathrm{sn}\ u\ \mathrm{cn}\ u \tag{5.20c}$$

and by entering the solutions into equations (5.3a–c). The following four sign combinations represent possible solutions

+	+	−	−
+	−	+	−
−	+	+	−

The terms *epicycloidal* and *pericycloidal* are readily interpreted by referring to the H/B level. When $\omega_x = \omega_z = 0$ and $2T/H = H/B$, then $\mathbf{H}/B$ and $\boldsymbol{\omega}$ coincide. If $2T/H > H/B$, then the attitude diagram of Figure 5.3 applies, where the momentary cone is seen to roll upon ("epi") the outside of the momentary space cone in an epicycloidal fashion. The axes of the two cones are on alternate sides of the $\boldsymbol{\omega}$ vector. If $2T/H < H/B$, then the attitude diagram of Figure 5.4 applies, where the momentary body cone is seen to roll about ("peri") the momentary space cone in a pericycloidal fashion. The axes of the two cones are on the same side of the $\boldsymbol{\omega}$ vector.

5.2 Phase Plane Representation

The nonlinear oscillations of the angular velocity components as given by equations (5.7a–c) can be shown to be due to linear and cubic restoring terms, and their behavior can be presented on $\omega\dot{\omega}$ phase planes.

Differentiation of equations (5.3a–c) of motion of a torquefree asymmetric gyro yields

$$\ddot{\omega}_x = \frac{B-C}{A}(\dot{\omega}_y\omega_z + \omega_y\dot{\omega}_z) \tag{5.21a}$$

$$\ddot{\omega}_y = \frac{C-A}{B}(\dot{\omega}_z\omega_x + \omega_z\dot{\omega}_x) \tag{5.21b}$$

$$\ddot{\omega}_z = \frac{A-B}{C}(\dot{\omega}_x\omega_y + \omega_x\dot{\omega}_y). \tag{5.21c}$$

From equations (5.3a–c), the time derivatives on the left-hand sides of equations (5.21a–c) can be substituted such that

$$\ddot{\omega}_x = \frac{B-C}{A}\left(\frac{A-B}{C}\omega_y^2 + \frac{C-A}{B}\omega_z^2\right)\omega_x \tag{5.22a}$$

$$\ddot{\omega}_y = \frac{C-A}{B}\left(\frac{A-B}{C}\omega_x^2 + \frac{B-C}{A}\omega_z^2\right)\omega_y \tag{5.22b}$$

$$\ddot{\omega}_z = \frac{A-B}{C}\left(\frac{C-A}{B}\omega_x^2 + \frac{B-C}{A}\omega_y^2\right)\omega_z. \tag{5.22c}$$

Equations (5.5) and (5.6) can be solved simultaneously for the squares of any two ω's as a function of the third, e.g.,

$$\omega_y^2 = \frac{2CT - H^2}{B(C-B)} - \frac{A(C-A)}{B(C-B)}\omega_x^2 \tag{5.23a}$$

$$\omega_z^2 = \frac{2BT - H^2}{C(B-C)} - \frac{A(B-A)}{C(B-C)}\omega_x^2. \tag{5.23b}$$

With the help of equations (5.23a,b), equation (5.22a) yields

$$\ddot{\omega}_x + \frac{(A-B)(2CT - H^2) + (C-A)(H^2 - 2BT)}{ABC}\omega_x + \frac{2(A-B)(A-C)}{BC}\omega_x^3 = 0 \tag{5.24a}$$

and in similar fashion we obtain

$$\ddot{\omega}_y + \frac{(B-C)(2AT - H^2) + (A-B)(H^2 - 2CT)}{ABC}\omega_y + \frac{2(B-A)(B-C)}{AC}\omega_y^3 = 0 \tag{5.24b}$$

$$\ddot{\omega}_z + \frac{(C-A)(2BT - H^2) + (B-C)(H^2 - 2AT)}{ABC}\omega_z + \frac{2(A-C)(B-C)}{AB}\omega_z^3 = 0. \tag{5.24c}$$

Each of the three equations (5.24a–c) is a homogeneous, undamped *Duffing* equation. The equations are *uncoupled,* but *interdependent* in the sense that the constants cannot be freely chosen. It is interesting to note that while the constants of the linear terms are functions of the inertia moments, of the angular momentum H, and of the kinetic energy T, the constants of the cubic terms are functions of the inertia moments only.

For notational convenience let

$$\ddot{\omega}_x + P_x\omega_x + Q_x\omega_x^3 = 0 \tag{5.25a}$$

$$\ddot{\omega}_y + P_y\omega_y + Q_y\omega_y^3 = 0 \tag{5.25b}$$

$$\ddot{\omega}_z + P_z\omega_z + Q_z\omega_z^3 = 0 \tag{5.25c}$$

where the coefficients P and Q are interdependent constants, given by the appropriate coefficients in equations (5.24a–c).

Equations (5.25a–c) have three immediate integrals of the form

$$\dot{\omega}_x^2 + P_x\omega_x^2 + \frac{1}{2}Q_x\omega_x^4 = K_x \tag{5.26a}$$

$$\dot{\omega}_y^2 + P_y\omega_y^2 + \frac{1}{2}Q_y\omega_y^4 = K_y \tag{5.26b}$$

$$\dot{\omega}_z^2 + P_z\omega_z^2 + \frac{1}{2}Q_z\omega_z^4 = K_z. \tag{5.26c}$$

The integration constants K can, e.g., be obtained by letting $\dot{\omega} = 0$ and then using equations (5.7a–c). The integration constants are

$$K_x = \frac{(2BT - H^2)(H^2 - 2CT)}{A^2BC} \tag{5.27a}$$

$$K_y = \frac{(2CT - H^2)(H^2 - 2AT)}{AB^2C} \tag{5.27b}$$

$$K_z = \frac{(2AT - H^2)(H^2 - 2BT)}{ABC^2}. \tag{5.27c}$$

An inspection of equations (5.26a–c), observing that $A > B > C > 0$, shows that the following possibilities exist:

$$P_x \gtrless 0$$

$$P_y > 0$$

$$P_z \gtrless 0$$

$$Q_x > 0$$

$$Q_y < 0$$

$$Q_z > 0.$$

The following qualitative observations can now be made:

Since the coefficients Q for the first and third equations are always positive it can immediately be hypothesized that all trajectories in the first and third phase planes must be closed.

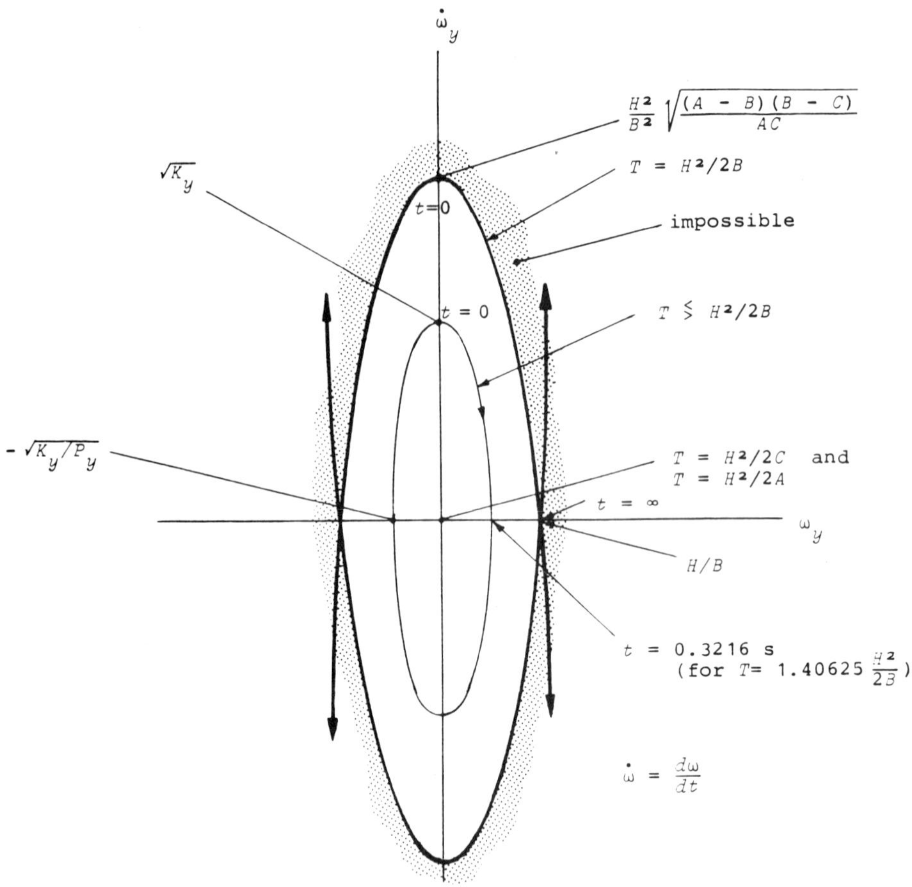

FIGURE 5.5. The $\omega_y \dot{\omega}_y$ phase plane.

Since the coefficient P_y is always positive and the coefficient Q_y always negative, and since the cubic term will eventually override the linear term (when ω_y is sufficiently large) the second equation could admit closed as well as open trajectories in the second phase plane.

The latter case is obviously the more interesting one and its phase plane trajectories will be discussed in the following.

For very small values of ω_y, the cubic term can be neglected. Then

$$\dot{\omega}_y^2 + P_y \omega_y^2 = K_y$$

which is the equation of an ellipse with semiaxes $\sqrt{K_y}$ and $\sqrt{K_y/P_y}$ and which represents the trajectory in the $\omega_y \dot{\omega}_y$ phase plane for very small values of ω_y. The other trajectories in the $\omega_y \dot{\omega}_y$ phase plane are distorted ellipses.

It can be shown (Junkins et al., 1973) that the closed trajectory region is bounded by the first pair of open trajectories associated with the kinetic energy level

$$T = \frac{H^2}{2B}$$

for which

$$P_y = \frac{4T(A-B)(B-C)}{ABC}$$

$$Q_y = -\frac{2(A-B)(B-C)}{AC}$$

$$K_y = \frac{4T^2(A-B)(B-C)}{AB^2C}$$

such that

$$P_y^2 + 2Q_yK_y = 0 \quad \left(\text{for } T = \frac{H^2}{2B}\right)$$

and the resulting curves are the two parabola-like open trajectories shown in Figure 5.5, intersecting the abscissa at $\pm H/B$, and the ordinate at $\pm H^2/B^2\sqrt{(A-B)(B-C)/AC}$. Trajectories further outside can be shown (Junkins et al., 1973) to be physically impossible.

5.3 Kinetic Energy Ellipsoid and Poinsot Ellipsoid

In order to facilitate an interpretation of the motion of a gyro, geometric representations are sometimes employed, in which the *(kinetic) energy ellipsoid*, the *Poinsot ellipsoid*, the *angular momentum ellipsoid*, and the *MacCullagh ellipsoid* play a role.

The rotational kinetic energy of a gyro is

$$T = \frac{1}{2}(A\omega_x^2 + B\omega_y^2 + C\omega_z^2). \tag{5.28}$$

It is constant, provided there is no internal energy loss. It is readily seen that equation (5.28) can be rewritten in the form

$$\frac{\omega_x^2}{2T/A} + \frac{\omega_y^2}{2T/B} + \frac{\omega_z^2}{2T/C} = 1 \tag{5.29}$$

which is the equation of the *energy ellipsoid* in a system whose coordinates are ω_x, ω_y, and ω_z. The semi-axes of the energy ellipsoid are

$$a = \sqrt{2T/A}$$

$$b = \sqrt{2T/B}$$

$$c = \sqrt{2T/C}.$$

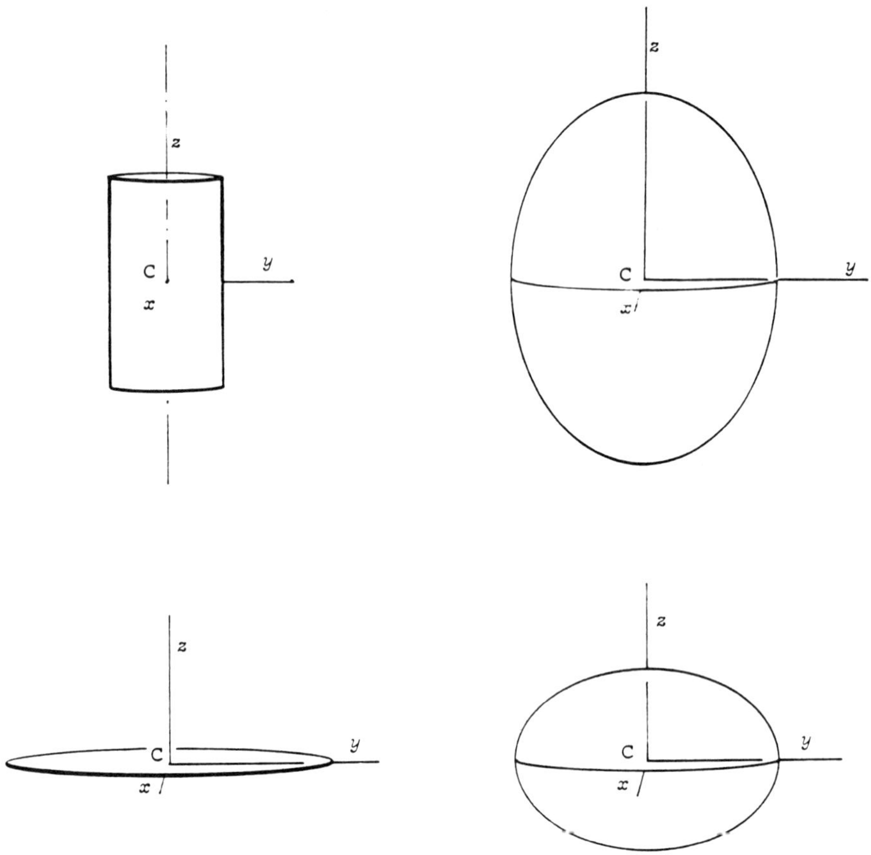

FIGURE 5.6. Axisymmetric bodies and associated inertia spheroids.

An arbitrary point on the energy ellipsoid has the components ω_x, ω_y, and ω_z, and thus represents one of the gyro's possible angular velocities, since

$$\boldsymbol{\omega} = [\mathbf{e}_x \ \mathbf{e}_y \ \mathbf{e}_z] \begin{bmatrix} \omega_x \\ \omega_y \\ \omega_z \end{bmatrix}.$$

The energy ellipsoid is typically less slim than the momentum ellipsoid, which will be introduced in the following section.

If ω_x is replaced by x, etc., and $2T = 1$, then the *Poinsot ellipsoid*

$$\frac{x^2}{1/A} + \frac{y^2}{1/B} + \frac{z^2}{1/C} = 1 \tag{5.30}$$

is obtained. Energy ellipsoid and Poinsot ellipsoid are consequently of the same shape.

The Poinsot ellipsoid of a given body is also called its *inertia ellipsoid.* The semi-axes of the Poinsot ellipsoid are

$$a = 1/\sqrt{A}$$
$$b = 1/\sqrt{B}$$
$$c = 1/\sqrt{C}.$$

For bodies with axisymmetric inertia moments the Poinsot ellipsoid becomes a spheroid, with

$$b = a = 1/\sqrt{A}$$
$$c = 1/\sqrt{C}.$$

Elongated bodies, such as the rod shown in Figure 5.4, are said to have a *prolate* inertia spheroid, while flattened bodies, such as the disk shown in Figure 5.6, have an *oblate* inertia spheroid.

The distance between the origin O and an arbitrary point $P(x, y, z)$ on the surface of the Poinsot ellipsoid is

$$\sqrt{x^2 + y^2 + z^2} = \frac{1}{\sqrt{I_{OP}}}.$$

Thus the distance is inversely proportional to the root of the body's inertia moment about the OP-axis.

5.4 Angular Momentum Ellipsoid

The square of the magnitude of the angular momentum of a gyro, rotating about its mass center, is

$$H^2 = A^2\omega_x^2 + B^2\omega_y^2 + C^2\omega_z^2. \tag{5.31}$$

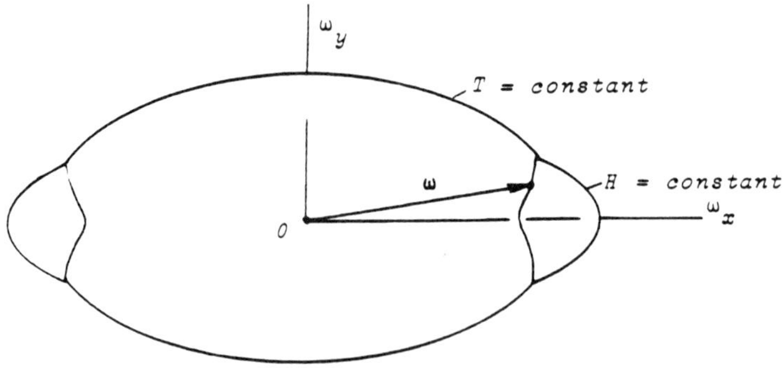

FIGURE 5.7. The spatial intersection curve between (plump) energy ellipsoid and (slim) momentum ellipsoid is the locus of the possible angular velocities.

It is constant, provided no external torque is applied to the gyro. It is readily seen that equation (5.31) can be rewritten to read

$$\frac{\omega_x^2}{(H/A)^2} + \frac{\omega_y^2}{(H/B)^2} + \frac{\omega_z^2}{(H/C)^2} = 1 \tag{5.32}$$

which is the equation of the *angular momentum ellipsoid* in a system whose coordinates are ω_x, ω_y, and ω_z. The semi-axes of the angular momentum ellipsoid are

$$a = \frac{H}{A}$$
$$b = \frac{H}{B}$$
$$c = \frac{H}{C}.$$

Any point on the angular momentum ellipsoid represents a possible angular velocity for the angular momentum given.

For a given angular momentum and a given kinetic energy, the angular momentum ellipsoid and the kinetic energy ellipsoid intersect along spatial curves, which represent the locus of the tip of all possible angular velocities $\boldsymbol{\omega}$ at that angular momentum and that kinetic energy (Figure 5.7).

5.5 MacCullagh Ellipsoid

The MacCullagh ellipsoid is defined by

$$\frac{H_x^2}{2TA} + \frac{H_y^2}{2TB} + \frac{H_z^2}{2TC} = 1 \tag{5.33}$$

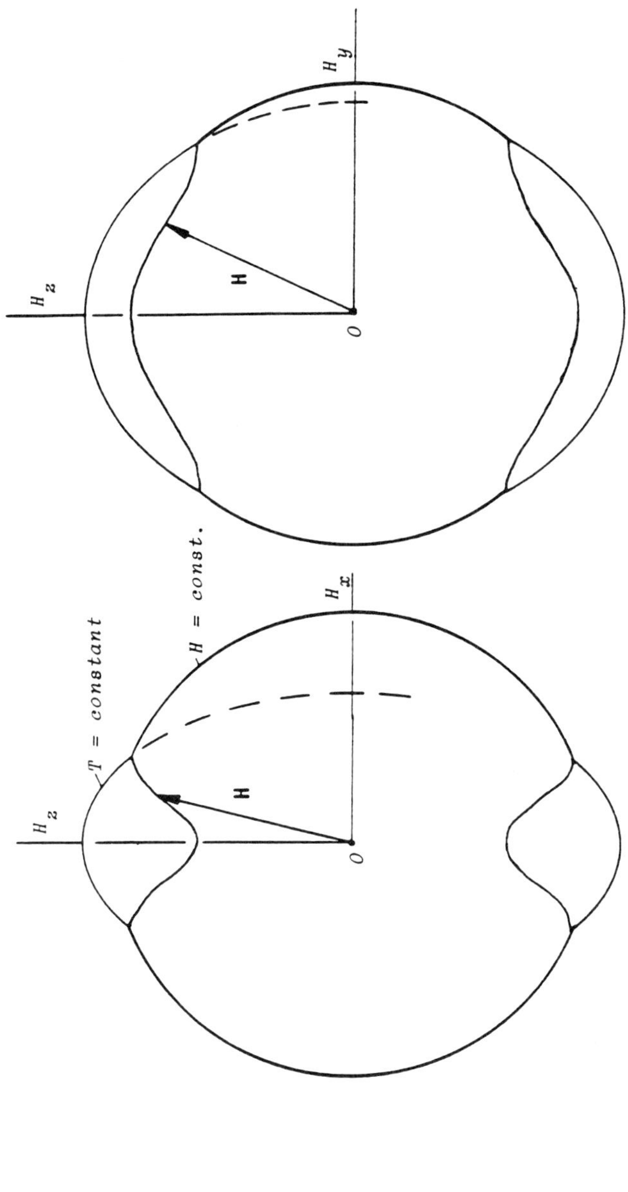

FIGURE 5.8. The spatial intersection curve between MacCullagh ellipsoid and angular momentum sphere is the locus of all possible angular momentum directions.

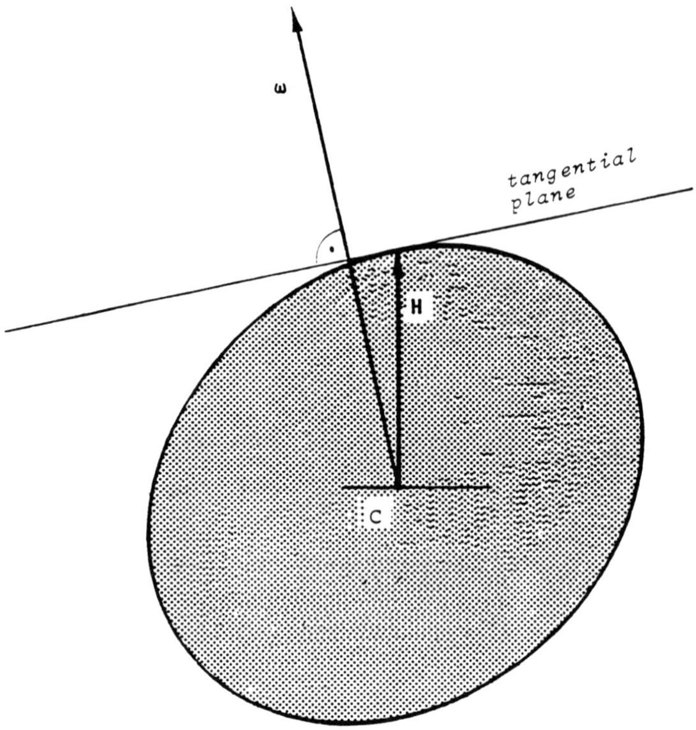

FIGURE 5.9. MacCullagh ellipsoid, angular momentum vector, and angular velocity vector.

where the $Cxyz$ coordinate system is a body-fixed principal coordinate system. The semi-axes of the MacCullagh ellipsoid are

$$a = \sqrt{2TA}$$
$$b = \sqrt{2TB}$$
$$c = \sqrt{2TC}.$$

The tip of the angular momentum vector **H** is located on the surface of the MacCullagh ellipsoid, and simultaneously on the surface of the angular momentum sphere

$$H_x^2 + H_y^2 + H_z^2 = H^2 \tag{5.34}$$

as shown in Figure 5.8. The angular velocity vector $\boldsymbol{\omega}$ is directed along the normal to the plane that passes through the tip of the angular momentum **H** and its tangent to the MacCullagh ellipsoid at that point (Figure 5.9).

5.6 Velocity and Acceleration of an Arbitrary Point within a Torquefree Gyro

An arbitrary point P_0 within a rigid gyro is subject to changing velocities and accelerations. Let point P_0 be located at

$$\mathbf{r} = [\mathbf{e}_x \; \mathbf{e}_y \; \mathbf{e}_z] \begin{bmatrix} x \\ y \\ z \end{bmatrix} \tag{5.35}$$

within the gyro, where the $Cxyz$ coordinate system is body-fixed and directed along the gyro's principal axes. The point's absolute velocity, then, is

$$\mathbf{v} = \dot{\mathbf{r}} = \boldsymbol{\omega} \times \mathbf{r} = [\mathbf{e}_x \; \mathbf{e}_y \; \mathbf{e}_z] \begin{bmatrix} \omega_y z - \omega_z y \\ \omega_z x - \omega_x z \\ \omega_x y - \omega_y x \end{bmatrix} \tag{5.36}$$

and its absolute acceleration is obtained from

$$\mathbf{a} = \ddot{\mathbf{r}} = \dot{\mathbf{r}}^{\circ} + \boldsymbol{\omega} \times \dot{\mathbf{r}} \tag{5.37}$$

and becomes

$$\mathbf{a} = [\mathbf{e}_x \; \mathbf{e}_y \; \mathbf{e}_z] \begin{bmatrix} z\dot{\omega}_y - y\dot{\omega}_z - x(\omega_y^2 + \omega_z^2) + (y\omega_y + z\omega_z)\omega_x \\ x\dot{\omega}_z - z\dot{\omega}_x - y(\omega_z^2 + \omega_x^2) + (z\omega_z + x\omega_x)\omega_y \\ y\dot{\omega}_x - x\dot{\omega}_y - z(\omega_x^2 + \omega_y^2) + (x\omega_x + y\omega_y)\omega_z \end{bmatrix}. \tag{5.38a}$$

Using equations (5.3a–c), equation (5.38a) can also be expressed as

$$\mathbf{a} = [\mathbf{e}_x \; \mathbf{e}_y \; \mathbf{e}_z] \begin{bmatrix} \frac{B+C-A}{BC}(B\omega_y y + C\omega_z z)\omega_x - (\omega_y^2 + \omega_z^2)x \\ \frac{C+A-B}{CA}(C\omega_z z + A\omega_x x)\omega_y - (\omega_z^2 + \omega_x^2)y \\ \frac{A+B-C}{AB}(A\omega_x x + B\omega_y y)\omega_z - (\omega_x^2 + \omega_y^2)z \end{bmatrix} \tag{5.38b}$$

or

$$\mathbf{a} = [\mathbf{e}_x\ \mathbf{e}_y\ \mathbf{e}_z] \begin{bmatrix} -(\omega_y^2+\omega_z^2)x+\frac{B+C-A}{C}\omega_x\omega_y y+\frac{B+C-A}{B}\omega_x\omega_z z \\ \frac{C+A-B}{C}\omega_y\omega_x x-(\omega_z^2+\omega_x^2)y+\frac{C+A-B}{A}\omega_y\omega_z z \\ \frac{A+B-C}{B}\omega_z\omega_x x+\frac{A+B-C}{A}\omega_z\omega_y y-(\omega_x^2+\omega_y^2)z \end{bmatrix} \tag{5.38c}$$

or

$$\mathbf{a} = [\mathbf{e}_x\ \mathbf{e}_y\ \mathbf{e}_z] \begin{bmatrix} -(\omega_y^2+\omega_z^2) & \frac{B+C-A}{C}\omega_x\omega_y & \frac{B+C-A}{B}\omega_x\omega_z \\ \frac{C+A-B}{C}\omega_x\omega_y & -(\omega_z^2+\omega_x^2) & \frac{C+A-B}{A}\omega_y\omega_z \\ \frac{A+B-C}{B}\omega_x\omega_z & \frac{A+B-C}{A}\omega_y\omega_z & -(\omega_x^2+\omega_y^2) \end{bmatrix} \begin{bmatrix} x \\ y \\ z \end{bmatrix}. \tag{5.38d}$$

5.7 Stability of Rotation about a Principal Axis for a Rigid Gyro

For an absolutely *rigid* gyro of inertia tensor

$$[I] = \begin{bmatrix} A & 0 & 0 \\ 0 & B & 0 \\ 0 & 0 & C \end{bmatrix} \tag{5.39}$$

the stability of its rotation can be investigated by assuming the gyro initially spinning about the z-axis at ω_0 with the other velocity components zero.

$$\boldsymbol{\omega} = [\mathbf{e}_x\ \mathbf{e}_y\ \mathbf{e}_z] \begin{bmatrix} 0 \\ 0 \\ \omega_0 \end{bmatrix}. \tag{5.40}$$

If we then permit a small deviation from this velocity state, i.e., superimpose ϵ, ω_x, and ω_y, keeping in mind that all three have magnitudes much smaller than ω_0, we have for the velocity vector

$$\boldsymbol{\omega} + \delta\boldsymbol{\omega} = [\mathbf{e}_x\ \mathbf{e}_y\ \mathbf{e}_z] \begin{bmatrix} \omega_x \\ \omega_y \\ \omega_0 + \epsilon \end{bmatrix}. \tag{5.41}$$

For a torquefree gyro, the angular momentum remains constant and thus equations (5.3a–c) apply. From equation (5.3c), after neglecting small terms of second order, we have

$$\dot{\omega}_z = 0 \tag{5.42}$$

leaving us with $\omega_z = \omega_0 + \epsilon =$ constant. From equations (5.3a) and (5.3b) we obtain, after neglecting small terms of second order

$$\dot{\omega}_x = \frac{B-C}{A}\omega_0\omega_y \tag{5.43}$$

and

$$\dot{\omega}_y = \frac{C-A}{B}\omega_0\omega_x. \tag{5.44}$$

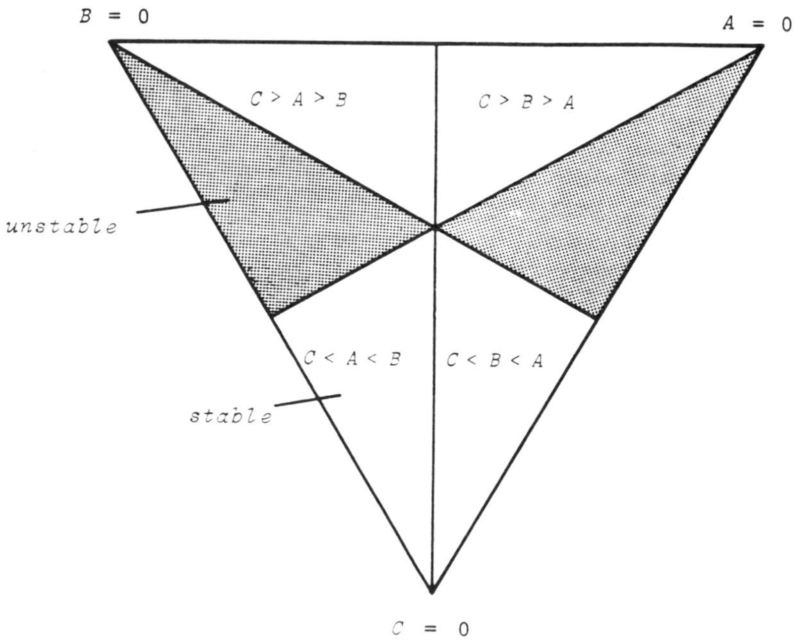

FIGURE 5.10. Magnus shape triangle for stable rotation about the z-axis for the theoretical case of an absolutely rigid gyroscope body.

Differentiating equation (5.12) again with respect to time and replacing $\dot{\omega}_x$, we have

$$\ddot{\omega}_y + \frac{(C-B)(C-A)}{AB}\omega_0^2\omega_y = 0. \tag{5.45}$$

Equation (5.13) will have a finite (stable) solution if the coefficient of ω_y is positive. This is the case if simultaneously

$$C > B \quad \text{and} \quad C > A \tag{5.46a}$$

and if simultaneously

$$C < B \quad \text{and} \quad C < A \tag{5.46b}$$

apply, i.e., *for a rotation about a principal axis to be stable, it must take place either about the axis of maximum inertia moment, or about the axis of minimum inertia moment* (Figure 5.10). This statement has been derived by assuming *absolute* rigidity. Thus, it is valid for absolutely rigid bodies only.

The assumption of absolute rigidity is apparently, acceptable in practice (Rimrott, 1984) for the determination of practical *stability in the short run* (e.g., rockets and bullets rotating about the axis with the smallest inertia moment are stable for the short duration of their flight), while it is not acceptable for the determination of ***stability in the long run*** (e.g., for spacecraft). Stability in the long run must take into consideration that real bodies do deform, at least slightly, under load; that there is consequently internal energy dissipation; and thus there is a decline of total mechanical energy.

If

$$A < C < B \tag{5.47}$$

or $B < C < A$, equation (5.45) has a solution where ω_y increases with time. Thus, the gyro's attitude is unstable when C is the intermediate inertia moment.

5.8 Stability of Rotation about a Principal Axis for a Deformable Gyro

Because of internal energy dissipation, the kinetic energy of a rotating gyro can only decrease with time. Thus, the stable attitude is the one for which the kinetic energy is the smallest possible, for a given angular momentum, which remains constant.

First consider a *rigid* gyro whose principal inertia tensor is

$$[I] = \begin{bmatrix} A & 0 & 0 \\ 0 & B & 0 \\ 0 & 0 & C \end{bmatrix}. \tag{5.48}$$

Initially (Figure 5.11), let this gyro have an arbitrary angular velocity

$$\boldsymbol{\omega}_0 = [\mathbf{e}_x \; \mathbf{e}_y \; \mathbf{e}_z \begin{bmatrix} \omega_{0x} \\ \omega_{0y} \\ \omega_{0z} \end{bmatrix}. \tag{5.49}$$

Since the inertia moments A, B, and C are readily determined for a rigid body, we assume that the gyro is *rigid* for $t \leq 0$. The gyro's angular momentum, thus, is

$$\mathbf{H}_0 = [\mathbf{e}_x \; \mathbf{e}_y \; \mathbf{e}_z] \begin{bmatrix} A\omega_{0x} \\ B\omega_{0y} \\ C\omega_{0z} \end{bmatrix} \tag{5.50a}$$

and has a magnitude

$$|\mathbf{H}_0| = H = \sqrt{A^2\omega_{0z}^2 + B^2\omega_{0y}^2 + C^2\omega_{0z}^2}. \tag{5.50b}$$

The rigid gyro's total mechanical energy is its kinetic energy, which for $t \leq 0$ is

$$T_0 = \frac{1}{2}(A\omega_{0x}^2 + B\omega_{0y}^2 + C\omega_{0z}^2). \tag{5.51}$$

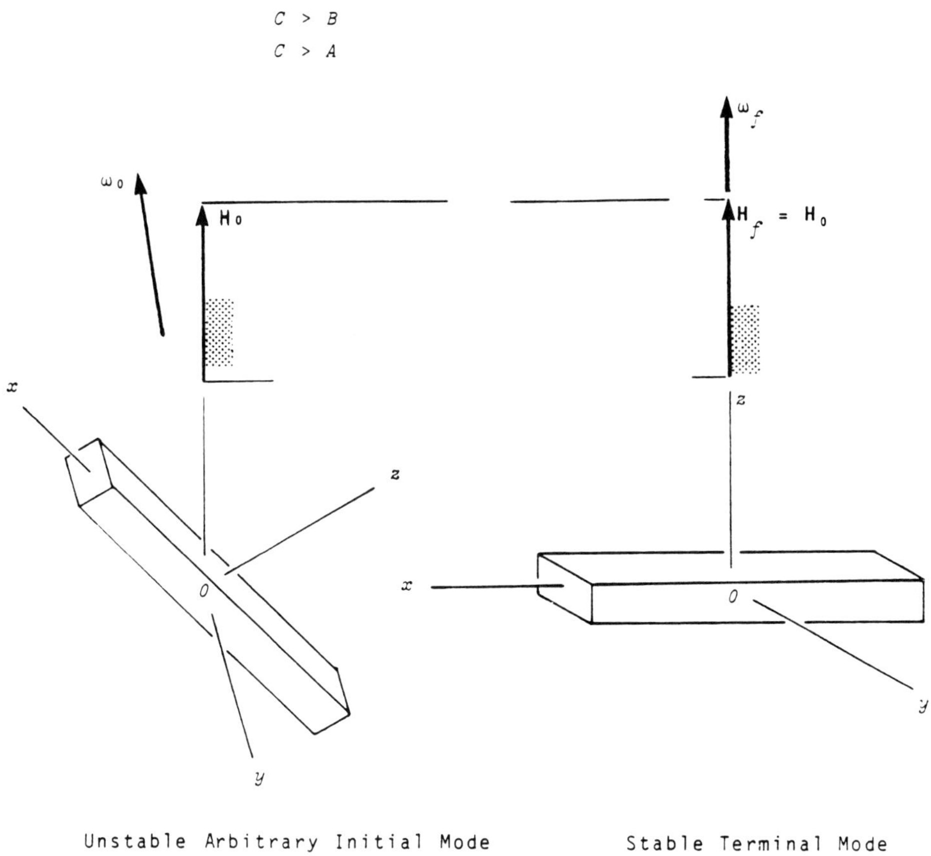

FIGURE 5.11. Unstable and stable attitude of a torquefree gyro.

Now let us suppose that for $0 < t < t_f$ the gyro assumes nonrigidity, i.e., since it is made of real materials, it will deform slightly and dissipate energy during the deformation process. As will be shown later, in chapter 7, the deforming gyro's inertia tensor can, e.g., be expressed by

$$[I] = \begin{bmatrix} A & I_{xy} & I_{xz} \\ I_{xy} & B & I_{yz} \\ I_{xz} & I_{yz} & C \end{bmatrix} \tag{5.52}$$

related to a $Cxyz$ system located close to the deforming gyro's principal axes system and chosen such, that the inertia products conveniently represent the deformation and are very small compared to the inertia moments, and changes ΔA, ΔB, and ΔC of the inertia moments can be neglected.

Let us further assume that the inertia products are harmonic functions of time and that they represent oscillating distortions during which energy is dissipated due to internal damping.

Eventually, at $t = t_f$, when the kinetic energy has reached the smallest possible value for a given (constant) angular momentum, no further energy dissipation and no further deformation is possible. Thus, the inertia tensor returns to its original form, as given by equation (5.48). The gyro has adjusted its attitude, has rigidized, and carries out a pure rotation about one of the body-fixed principal axes. Let this be the z-axis, which of necessity is collinear with the angular momentum, such that the final velocity is consequently

$$\boldsymbol{\omega}_f = [\mathbf{e}_x \ \mathbf{e}_y \ \mathbf{e}_z] \begin{bmatrix} 0 \\ 0 \\ \omega_f \end{bmatrix} \tag{5.53}$$

and the angular momentum can be expressed as

$$\mathbf{H}_f = [\mathbf{e}_x \ \mathbf{e}_y \ \mathbf{e}_z] \begin{bmatrix} 0 \\ 0 \\ C\omega_f \end{bmatrix} . \tag{5.54}$$

Since during the transition from state 0 to state f no external torque has acted on the gyro, the angular momentum must have remained constant (Figure 5.11), thus

$$\mathbf{H}_f = \mathbf{H}_0 . \tag{5.55}$$

From the condition of conservation of momentum, the magnitude of the final angular velocity can be determined and is

$$\omega_f = \sqrt{\frac{A^2}{B^2}\omega_{0x}^2 + \frac{B^2}{C^2}\omega_{0y}^2 + \omega_{0z}^2} . \tag{5.56}$$

The new kinetic energy is

$$T_f = \frac{1}{2} C\omega_f^2 \tag{5.57}$$

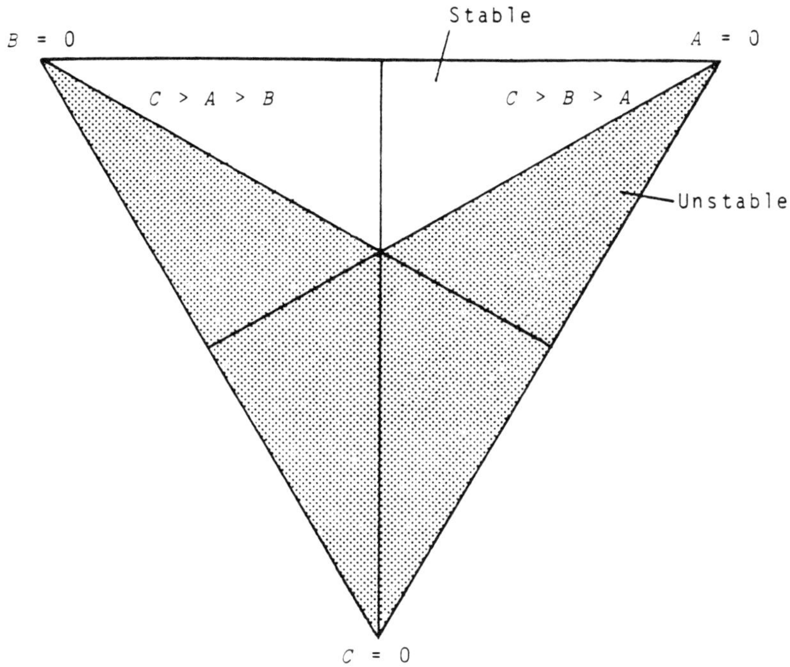

FIGURE 5.12. Magnus shape triangle for stable rotation about the z-axis.

or, upon use of equation (5.56),

$$T_f = \frac{1}{2}\left(\frac{A}{C}A\omega_{0x}^2 + \frac{B}{C}B\omega_{0y}^2 + C\omega_{0z}^2\right). \tag{5.58}$$

The new kinetic energy assumes the *smallest* possible magnitude if and only if simultaneously

$$C > B \quad \text{and} \quad C > A. \tag{5.59}$$

From rigid body stability criteria (5.46 and 5.47) we know that a gyro is unstable if C is the intermediate inertia moment. Thus, only pure rotation about the axis with the largest inertia moment is stable for a dissipative, i.e., real, solid gyro. Figure 5.12 shows the associated Magnus shape triangle.

In Figure 5.13, the constant angular momentum ellipsoid is shown, together with the energy ellipsoids for pure rotation about any one of the three principal axes. An inspection of Figure 5.13 shows that the *stable* principal axis is associated with a kinetic energy ellipsoid fully *inside* the angular momentum ellipsoid, while *unstable* principal axes are associated with kinetic energy ellipsoids fully or partially *outside* the angular momentum ellipsoid.

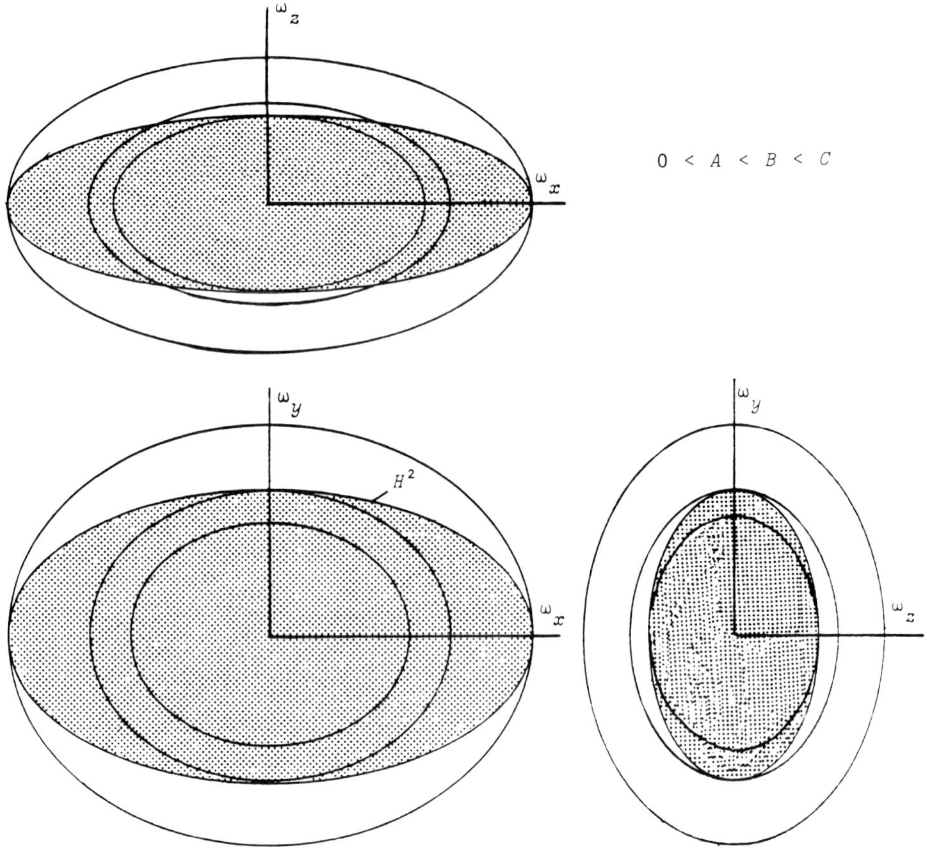

FIGURE 5.13. The three kinetic energy ellipsoids associated with pure rotation about each principal axis and the constant angular momentum ellipsoid.

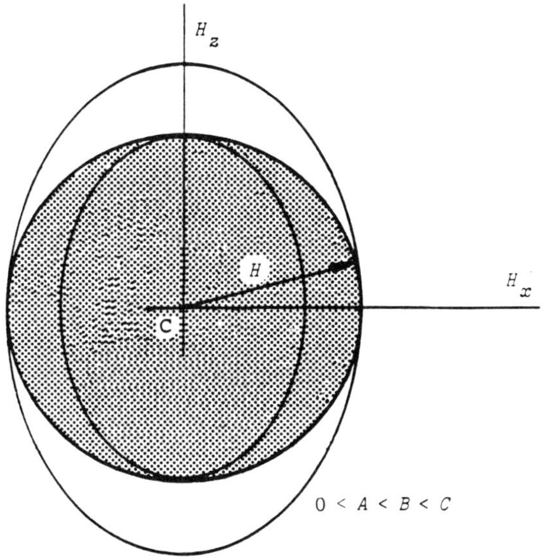

FIGURE 5.14. The MacCullagh ellipsoids associated with pure rotation about the gyro's x-axis and z-axis and the constant angular momentum sphere.

In Figure 5.14 the same situation is depicted by showing the (constant) angular momentum sphere, and the (large) MacCullagh ellipsoid associated with pure rotation about the axis with the minimum principal inertia moment, and the (small) MacCullagh ellipsoid associated with pure rotation about the axis with the maximum principal inertia moment. The smallest MacCullagh ellipsoid possible has the smallest kinetic energy. *Stability* in the long run is thus represented by the smallest MacCullagh ellipsoid located fully *inside* the angular momentum sphere, while unstable principal axes are associated with MacCullagh ellipsoids fully or partially *outside* the angular momentum sphere.

Note that in the derivation, changes in shape were expressed in terms of inertia products in equation (5.52). Changes in inertia moments were assumed negligible. Cases where large deflections occur, such as those described by Lohmeier (1975), are thus excluded.

We have shown that for a torquefree solid gyro, the kinetic energy is smallest when inequalities (5.59) apply. What remains to be shown is what effects, if any, the inertia products have that are inevitably present in a real solid gyro. As it turns out, these effects are minor as long as the gyro's inertia products indeed remain small (see Chapter 8).

In conclusion we now can describe the attitude stability of a real solid gyro in the following form:

Pure rotation of a torquefree real gyro about one of the principal axes is stable only if the rotation takes place about the axis with the largest inertia

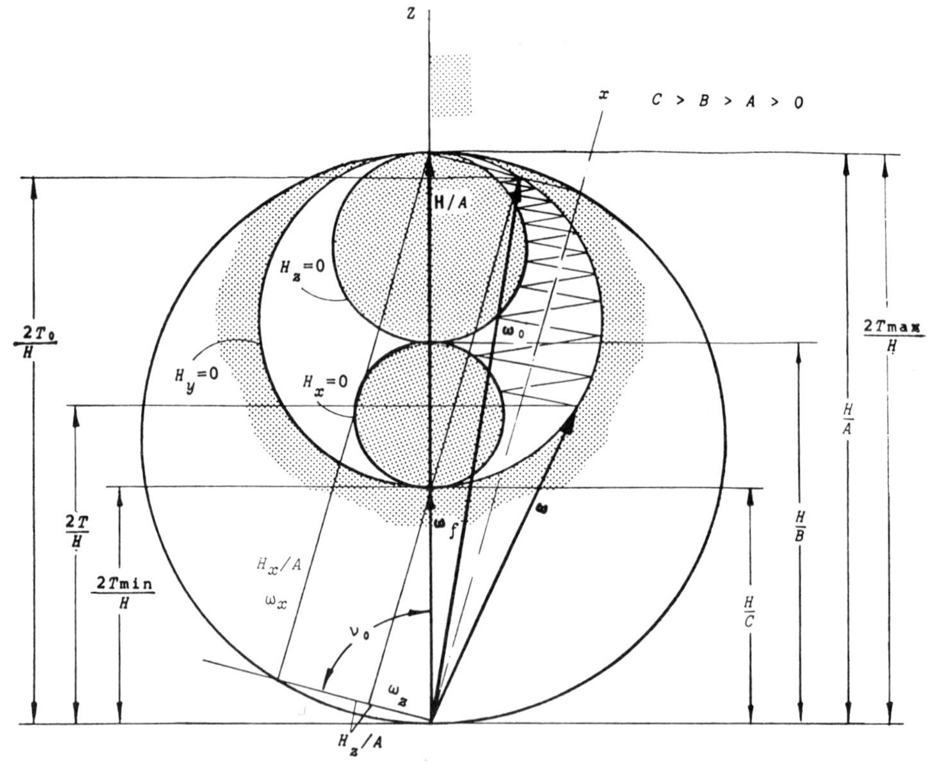

The diagram shown is for

H = 120 Js		
A = 10 m²kg	B = 15 m²kg	C = 24 m²kg
Tmax = 720 J	T_0 = 675 J	Tmin = 300 J

FIGURE 5.15. Attitude diagram for attitude drift from T_0 to $T_{\min}$.

moment. Rotation about the axis with the intermediate inertia moment is unstable, as is rotation about the axis with the smallest inertia moment.

All other forms of rotation are unstable as well and decay in the course of time towards the one stable terminal mode. The decrease of the rotational kinetic energy is brought about by internal friction (hysteresis losses) in the gyro's body. With the exception of a few special points, i.e., those located on the momentary axis of rotation, a tumbling gyroscope subjects any mass element within the gyro to periodic accelerations, which lead to cyclic deformation of the body and thus to a continuous loss of energy by internal friction.

This process stops when the gyro is in a state of pure rotation about one of the principal axes; then any arbitrary mass element of the gyro is subjected merely to a centripetal acceleration, and the motion causes a *rigidification* of the gyro's body. Small deviations from a state of pure rotation about a principal axis lead again to deformation and a loss of energy. And only in the case of rotation about the principal axis with the largest inertia moment will the small deviation decay.

In the two other cases, the small deviations will increase, eventually leading the gyro to reach the stable terminal mode of rotation about the axis with the largest inertia moment. The stable mode is reached more quickly in cases where inner friction losses are large, i.e., in the case of gyros with large internal damping, e.g., those with built-in energy absorbers.

The attitude drifting process is depicted in Figure 5.15, where the kinetic energy level is shown to decrease until it reaches $T_{\min}$, while the angular velocity cascades down from an initial $\boldsymbol{\omega}_0$ to a final $\boldsymbol{\omega}_f$.

5.9 Collinearity Theorems

The observations made in the preceding section and the fact that any solid body exhibits internal damping permit the establishment of three collinearity theorems involving the collinearity or noncollinearity of the body's angular velocity vector $\boldsymbol{\omega}$ and its angular momentum vector $\mathbf{H}$.

(1) *A torquefree gyro's attitude will drift towards the attitude associated with the lowest admissible kinetic energy for a given (constant) angular momentum.*

(2) *When angular momentum vector and angular velocity vector are collinear, there is no internal energy dissipation.*

(3) *When angular momentum vector and angular velocity vector are not collinear, there is internal energy dissipation, with the energy drawn from the gyro's kinetic energy and/or its elastic energy and/or work done by an externally applied torque.*

The angular velocity $\boldsymbol{\omega}$ referred to here of a gyro when it is deforming is the angular velocity of the phantom body, i.e., of the undeformed shape of the gyro.

Concerning energy rates, in general,

$$\dot{U} = \dot{T} + \dot{V} + \dot{D} \tag{5.60}$$

where $\dot{U}$ = rate of work done by external torque; $\dot{T}$ = kinetic energy rate; $\dot{V}$ = elastic energy rate; and $\dot{D}$ = energy dissipation rate. Theorems 2 and 3 are more general in nature and apply to nontorquefree gyros as well.

Suggested Reading

1. Jahnke, E.; F. Emde; F. Lösch. *Tables of Higher Functions,* McGraw-Hill, 1960, 318pp.

2. Junkins, J.L.; D. Jacobson; J.N. Blanton. "A Nonlinear Oscillator Analog of Rigid Body Dynamics," *Celestial Mechanics,* 7, 1973, 398–407.

3. Likins, P.W. "Effects of Energy Dissipation on the Free Body Motions of Spacecraft," *NASA,* TR 32-380, 1966.

4. Lohmeier, P. "On the Stability of Spinning Flexible Satellites," in *Satellite Dynamics,* Springer-Verlag, Berlin, 1975, 298–303.

5. Magnus, K. "Kreiselmechanik," *ZAMM,* 58, 1978, T56–T65.

6. Rimrott, F.P.J. "Practical Stability in the Short Run of a Torque-Free Axisymmetric Gyro," *ZAMM* 64, 4/5, 1984, T87–T88.

Problems

5.1. Given is

$$\omega_x = \sqrt{\frac{H^2 - 2TC}{A(A-C)}}\,\mathrm{cn}\, u \tag{P5.1}$$

$$\omega_y = \sqrt{\frac{H^2 - 2TC}{B(B-C)}}\,\mathrm{sn}\, u \tag{P5.2}$$

$$\omega_z = \sqrt{\frac{2TA - H^2}{C(A-C)}}\,\mathrm{dn}\, u. \tag{P5.3}$$

If

$$\frac{d}{du}(\mathrm{sn}\, u) = \mathrm{cn}\, u\, \mathrm{dn}\, u$$

$$\frac{d}{du}(\text{cn}\, u) = -\text{sn}\, u\, \text{dn}\, u$$

$$\frac{d}{du}(\text{dn}\, u) = -k^2\, \text{sn}\, u\, \text{cn}\, u$$

and if

$$u = \sqrt{\frac{(B-C)(2TA-H^2)}{ABC}}(t-t_0)$$

and

$$k = \sqrt{\frac{H^2-2TC}{2TA-H^2}\sqrt{A-B}B-C}$$

show that equations (P5.1)–(P5.3) satisfy the Euler equations for a torquefree gyro.

5.2. Given is a torquefree gyro with

$$\begin{bmatrix} A & 0 & 0 \\ 0 & B & 0 \\ 0 & 0 & C \end{bmatrix} = \begin{bmatrix} 3 & 0 & 0 \\ 0 & 2 & 0 \\ 0 & 0 & 1 \end{bmatrix} \text{Ws}^3$$

and initially, at $t = 0$,

$$\begin{bmatrix} \omega_x \\ \omega_y \\ \omega_z \end{bmatrix} = \begin{bmatrix} 5 \\ 10 \\ 150 \end{bmatrix} \text{rad/s.}$$

(a) Determine H^2, $2AT$, $2BT$, and $2CT$. (b) Find u_0, and from it t_0, both at time $t = 0$. (c) Find the components of $\boldsymbol{\omega}$ when $t = 0.01$ s. (d) Find and sketch the attitude with respect to the constant $\mathbf{H}$ vector (shown vertically) at $t = 0$ and at $t = 0.01$ s. [Hint: Assume initially that cn $u = \cos u$, sn $u = \sin u$, dn $u = 1$.]

5.3. Given is a torquefree gyro with

$$[I] = \begin{bmatrix} 24 & 0 & 0 \\ 0 & 15 & 0 \\ 0 & 0 & 10 \end{bmatrix} \text{m}^2\text{kg}$$

and $H = 13\,\text{Ws}^2$. For the condition $H^2 = 2BT$, plot the angular velocity components as a function of time, starting at $t = 0$ and ending at $t = 40$ s, if $t_0 = 20$ s. The initial conditions (at $t = 0$) are $\omega_x = \omega_z = +0$, $\omega_y = 0.8$ rad/s.

5.4. Given are the differential equations

$$\dot{\omega}_x - \frac{B-C}{A}\omega_y\omega_z = 0$$

$$\dot{\omega}_y - \frac{C-A}{B}\omega_z\omega_x = 0$$

$$\dot{\omega}_z - \frac{A-B}{C}\omega_x\omega_y = 0.$$

With H^2 constant and $T =$ constant, show that

$$\omega_x = \sqrt{\frac{H^2 - 2TC}{A(A-C)}}\,\mathrm{dn}\, u$$

$$\omega_y = \sqrt{\frac{2TA - H^2}{B(A-B)}}\,\mathrm{sn}\, u$$

$$\omega_z = -\sqrt{\frac{2TA - H^2}{C(A-C)}}\,\mathrm{cn}\, u$$

for the case of pericycloidal motion, i.e., when $H^2 > 2TB$.

5.5. Given are the *Jacobi Elliptic Functions*

$$\mathrm{sn}\, u = \frac{1}{\sqrt{k}}\frac{\theta_1(v)}{\theta_4(v)}$$

$$\mathrm{cn}\, u = \sqrt{\frac{k'}{k}}\frac{\theta_2(v)}{\theta_4(v)}$$

$$\mathrm{dn}\, u = \sqrt{k'}\frac{\theta_3(v)}{\theta_4(v)}$$

where $k' = \sqrt{1-k^2}$; $k = \sin\alpha$; $k' = \cos\alpha$; $v = u/2K$; and $\theta_1(v)$, $\theta_2(v)$, $\theta_3(v)$, and $\theta_4(v)$ are *theta functions,* whose values are shown on the accompanying table. Plot altogether six curves, three for $\alpha = 63°$, and three for $\alpha = 81°$, for the range from 0 to $8v$ on the abscissa.

Theta Functions

$2v$	$\alpha = 0°$	9°	18°	27°	36°	45°	54°	63°	72°	81°	$2v_1$
0.0	1.0000	0.9970	0.9874	0.9712	0.9471	0.9135	0.8680	0.8052	0.7152	0.5694	1.0
1	1.0000	0.9970	0.9881	0.9725	0.9497	0.9178	0.8744	0.8147	0.7190	0.5898	0.9
2	1.0000	0.9975	0.9899	0.9766	0.9572	0.9300	0.8931	0.8424	0.7691	0.6494	0.8
3	1.0000	0.9982	0.9927	0.9831	0.9689	0.9493	0.9223	0.8853	0.8318	0.7429	0.7
4	1.0000	0.9991	0.9961	0.9911	0.9836	0.9732	0.9592	0.9397	0.9110	0.8419	0.6
0.5	1.0000	1.0000	1.0000	1.0000	1.0000	1.0000	1.0000	0.9999	0.9992	0.9956	0.5
6	1.0000	1.001	1.004	1.009	1.016	1.027	1.041	1.060	1.088	1.131	0.4
7	1.0000	1.002	1.007	1.017	1.031	1.051	1.078	1.115	1.168	1.254	0.3
8	1.0000	1.003	1.010	1.023	1.043	1.070	1.107	1.158	1.231	1.353	0.2
9	1.0000	1.003	1.012	1.028	1.050	1.082	1.126	1.186	1.272	1.417	0.1
1.0	1.0000	1.003	1.013	1.029	1.053	1.086	1.132	1.195	1.286	1.439	0.0
0.0	0.0000	0.0000	0.0000	0.0000	0.0000	0.0000	0.0000	0.0000	0.0000	0.0000	1.0
1	0.0000	0.06206	0.08804	0.1084	0.1260	0.1419	0.1566	0.1700	0.1810	0.1843	0.9
2	0.0000	0.1226	0.1739	0.2140	0.2488	0.2804	0.3098	0.3368	0.3597	0.3698	0.8
3	0.0000	0.1801	0.2555	0.3145	0.3657	0.4123	0.4560	0.4968	0.5335	0.5563	0.7
4	0.0000	0.2332	0.3308	0.4073	0.4736	0.5343	0.5917	0.6467	0.6989	0.7413	0.6
0.5	0.0000	0.2805	0.3980	0.4900	0.5700	0.6436	0.7139	0.7827	0.8517	0.9200	0.5
6	0.0000	0.3210	0.4553	0.5607	0.6524	0.7372	0.8188	0.9007	0.9870	1.085	0.4
7	0.0000	0.3535	0.5015	0.6176	0.7188	0.8129	0.9041	0.9972	1.100	1.227	0.3
8	0.0000	0.3773	0.5353	0.6592	0.7675	0.8682	0.9669	1.069	1.184	1.337	0.2
9	0.0000	0.3918	0.5559	0.6847	0.7973	0.9022	1.005	1.113	1.237	1.407	0.1
1.0	0.0000	0.3967	0.5629	0.6933	0.8074	0.9135	1.018	1.128	1.255	1.431	0.0

From Jahnke et al. (1960).

5.6. Draw the trajectories in all three phase planes for the case $H^2/2BT = 1.01$, if $A = 5000\,\mathrm{m^2kg}$, $B = 4000\,\mathrm{m^2kg}$, $C = 2500\,\mathrm{m^2kg}$, and $T = 1250$ J.

5.7. What is (a) the velocity and (b) the acceleration of point 1 of the gyroscope shown, at the moment when

$$\boldsymbol{\omega} = [\mathbf{e}_x\ \mathbf{e}_y\ \mathbf{e}_z] \begin{bmatrix} 3 \\ 4 \\ 12 \end{bmatrix} \text{rad/s}$$

where $Cxyz$ is a body-fixed coordinate system and $a = 3$ m, $b = 4$ m, and $c = 5$ m. Each of the six point masses has a mass of 2 kg. The rods are rigid and massless.

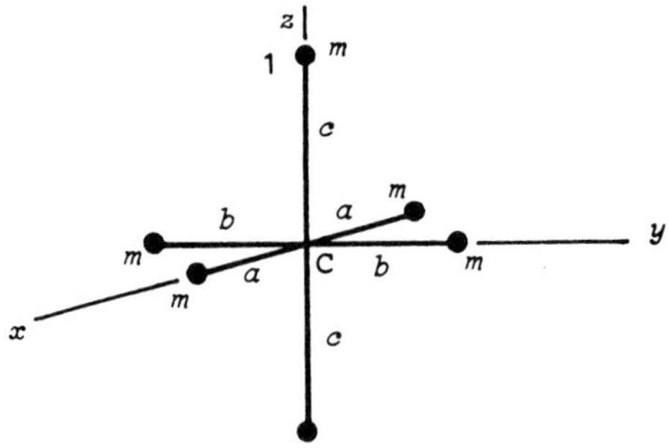

5.8. Given is a rigid gyro with

$$[I] = \begin{bmatrix} 18 & 0 & 0 \\ 0 & 32 & 0 \\ 0 & 0 & 50 \end{bmatrix} \mathrm{m^2kg}.$$

Draw the angular momentum ellipsoid ($H^2 = 25\,600\,\mathrm{W^2s^4}$):

$$A^2\omega_x^2 + B^2\omega_y^2 + C^2\omega_z^2 = H^2.$$

5.9. Given is a rigid gyro with

$$[I] = \begin{bmatrix} 18 & 0 & 0 \\ 0 & 32 & 0 \\ 0 & 0 & 50 \end{bmatrix} \mathrm{m^2kg}.$$

Draw the kinetic energy ellipsoid (make $2T = 32\,\mathrm{Ws}$).

5.10. Given is a torquefree gyro, with

$$[I] = \begin{bmatrix} 18 & 0 & 0 \\ 0 & 32 & 0 \\ 0 & 0 & 50 \end{bmatrix} \mathrm{m^2kg}$$

and

$$\boldsymbol{\omega} = [\mathbf{e}_x\ \mathbf{e}_y\ \mathbf{e}_z] \begin{bmatrix} 10 \\ 20 \\ 30 \end{bmatrix} \mathrm{rad/s}.$$

Determine (a) the angular momentum $\mathbf{H}$; (b) the magnitude H of the angular momentum; and (c) the kinetic energy T. (d) Make a sketch, showing the gyro's shape (parallelepiped), the angular velocity vector, and the angular momentum vector.

5.11. Given is a torquefree gyro of

$$[I] = \begin{bmatrix} 3 & 0 & 0 \\ 0 & 2 & 0 \\ 0 & 0 & 1 \end{bmatrix} \mathrm{m^2kg}$$

and an angular velocity of

$$\boldsymbol{\omega} = [\mathbf{e}_x \ \mathbf{e}_y \ \mathbf{e}_z] \begin{bmatrix} 5 \\ 10 \\ 0 \end{bmatrix} \text{rad/s}.$$

(a) Compute the magnitude of the angular momentum. (b) Plot the angular momentum ellipsoid. (c) Compute the kinetic energy. (d) Plot the kinetic energy ellipsoid into the same figure as the angular momentum ellipsoid. (e) Show $\boldsymbol{\omega}$ in the figure. (f) Why is the angular momentum ellipsoid slimmer (less spherelike) than the kinetic energy ellipsoid?

5.12. A thin steel plate has a mass of 24 kg. It is rectangular and has side lengths of $a = 4$ m and $b = 3$ m. (a) Determine the inertia tensor. If the steel plate has an angular velocity of $\boldsymbol{\omega}_0 = 10\mathbf{e}_x + 20\mathbf{e}_y + 30\mathbf{e}_z$ rad/s, determine (b) its angular momentum and (c) its kinetic energy. If no external torque is applied to the steel plate, (d) how much kinetic energy will be dissipated due to internal damping in the course of time? (e) Make a sketch of the initial and final attitude of the steel plate (with respect to the **H** vector).

5.13. A torquefree axisymmetric gyro of inertia tensor

$$\begin{bmatrix} 200 & 0 & 0 \\ 0 & 200 & 0 \\ 0 & 0 & 300 \end{bmatrix} \text{m}^2\text{kg}$$

has, at a given point in time, an angular velocity of

$$\boldsymbol{\omega} = [\mathbf{e}_x \ \mathbf{e}_y \ \mathbf{e}_z] \begin{bmatrix} 10 \\ 0 \\ 30 \end{bmatrix} \text{rad/s}.$$

The $Cxyz$ coordinate system is body-fixed and along the principal axes of the gyro. Determine (a) the angular momentum, (b) the kinetic energy, and (c) the MacCullagh ellipsoid. (d) Make a sketch showing the angular momentum, the MacCullagh ellipsoid, and the angular velocity.

5.14. An inertia ellipsoid has the following semi-axes:

$$b = a = \frac{1}{\sqrt{2500}}$$

$$c = \frac{1}{\sqrt{4000}}.$$

(a) Sketch the inertia ellipsoid. Then determine the dimensions and make sketches of the following bodies: (b) a hollow thin-walled circular cylinder, (c) a solid circular cylinder, (d) a spatial cross, and (e) an ellipsoid; each of uniform mass distribution with $m = 600$ kg mass, and of an inertia tensor as represented by the inertia ellipsoid (a).

6

Torquefree Axisymmetric Gyros

Satellites often have rotational symmetry. The properties and characteristics of torquefree gyros with rotational symmetry are the subject of the subsequent discussion, where it shall be understood that symmetry refers to *massive* symmetry as expressed by equation (6.1), which does not presuppose geometric symmetry. The equations derived in this chapter apply when the origin of the xyz-coordinate system is either the gyro's mass center C or its fixed point 0 (Figure 1.2). If mass center C and fixed point 0 coincide, then the gyroscopic motion is not only torquefree but also *forcefree;* if they do not coincide, an external force is present to bring about the mass center's linear accelerations associated with the gyroscopic motion, among which the centripetal acceleration is most common.

6.1 Angular Velocity Components

We shall select the body-fixed z-axis as the axis of symmetry. The inertia tensor thus assumes the form

$$[I] = \begin{bmatrix} A & 0 & 0 \\ 0 & A & 0 \\ 0 & 0 & C \end{bmatrix}. \tag{6.1}$$

The angular velocity of the satellite is

$$\boldsymbol{\omega} = [\mathbf{e}_x \ \mathbf{e}_y \ \mathbf{e}_z] \begin{bmatrix} \omega_x \\ \omega_y \\ \omega_z \end{bmatrix}. \tag{6.2}$$

The angular momentum, thus, is

$$\mathbf{H} = [\mathbf{e}_x \ \mathbf{e}_y \ \mathbf{e}] \begin{bmatrix} A\omega_x \\ A\omega_y \\ C\omega_z \end{bmatrix}. \tag{6.3}$$

The torque applied is zero. Consequently, the angular momentum remains constant. We apply the Euler equations (1.68–1.70) set $B = A$ and $M_x = M_y = M_z = 0$ and get

$$0 = A\dot{\omega}_x - (A - C)\omega_y\omega_z \tag{6.4}$$

$$0 = A\dot{\omega}_y - (C - A)\omega_z\omega_x \tag{6.5}$$

$$0 = C\dot{\omega}_z. \tag{6.6}$$

From equation (6.6) we conclude right away that the z-component of the angular velocity is constant

$$\omega_z = \text{ constant.} \tag{6.7}$$

We now introduce an auxiliary frequence ω'_z, defined by

$$\omega'_z = \frac{C - A}{A}\omega_z \tag{6.8}$$

which is also constant. With the aid of this frequency, equations (6.4) and (6.5) assume the form

$$\dot{\omega}_x + \omega'_z\omega_y = 0 \tag{6.9}$$

$$\dot{\omega}_y - \omega'_z\omega_x = 0. \tag{6.10}$$

Multiplying equation (6.9) with ω_x and equation (6.10) with ω_y and then adding the two equations results in

$$\omega_x\dot{\omega}_x + \omega_y\dot{\omega}_y = 0 \tag{6.11}$$

and upon integration

$$\sqrt{\omega_x^2 + \omega_y^2} = \text{constant.} \tag{6.12}$$

The constant of equation (6.12) is obviously an angular velocity, and we shall designate it ω_v.

$$\omega_v = \sqrt{\omega_x^2 + \omega_y^2}. \tag{6.13}$$

After introducing a floating $Cuvz$ coordinate system (Figure 6.1) we can now express the angular velocity $\boldsymbol{\omega}$ by using either the $Cxyz$ or the $Cuvz$ coordinate system

$$\boldsymbol{\omega} = [\mathbf{e}_x \ \mathbf{e}_y \ \mathbf{e}_z] \begin{bmatrix} \omega_x \\ \omega_y \\ \omega_z \end{bmatrix} = [\mathbf{e}_x \ \mathbf{e}_y \ \mathbf{e}_z] \begin{bmatrix} 0 \\ \omega_v \\ \omega_z \end{bmatrix}. \tag{6.14}$$

The ω_v-component of the angular velocity lies in the xy-plane. We now pose the question as to whether, and if so how fast, the ω_v-vector rotates in the xy-plane. We form the derivative of equation (6.9) with respect to time

$$\ddot{\omega}_x + \omega'_z\dot{\omega}_y = 0 \tag{6.15}$$

and replace $\dot{\omega}_y$ from equiation (6.10)

$$\ddot{\omega}_x + \omega_z'^2\omega_x = 0. \tag{6.16}$$

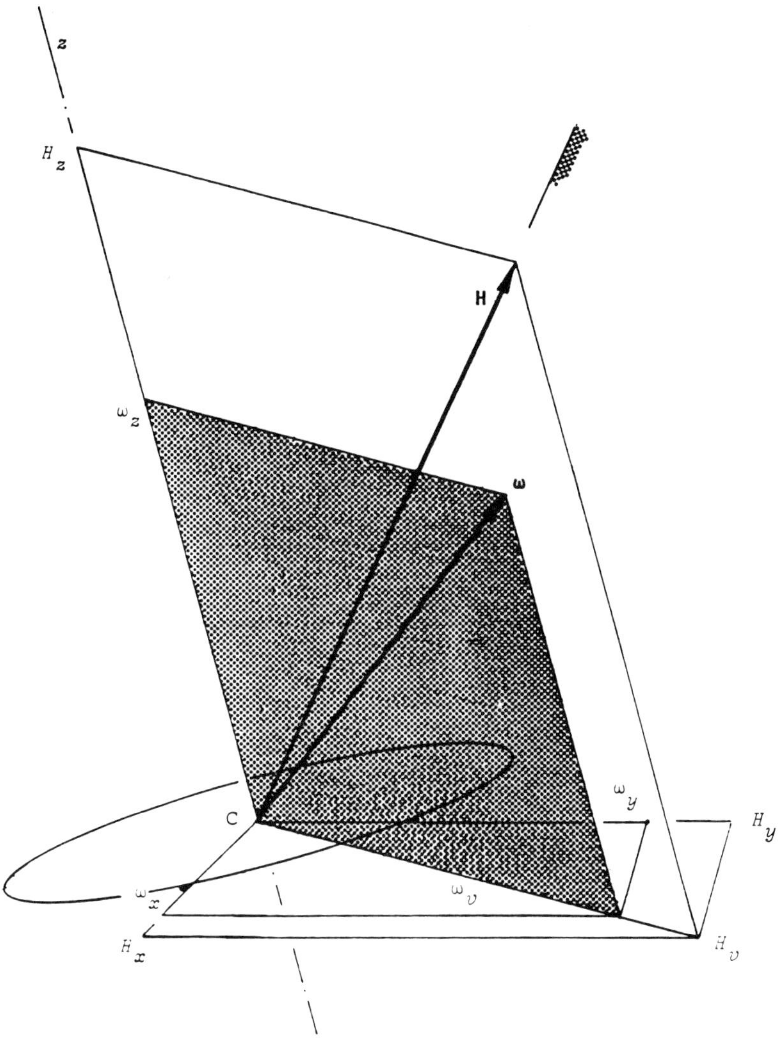

FIGURE 6.1. For a torquefree axisymmetric gyro, angular velocity $\boldsymbol{\omega}$, angular momentum $\mathbf{H}$, and the gyro's symmetry axis lie in one and the same plane.

We are now confronted by a linear differential equation of the second order, whose solution can be written

$$\omega_x = C_1 \cos(\omega_z' t + C_2). \tag{6.17}$$

The ω_y-component can be found by applying equation (6.9):

$$\omega_y = C_1 \sin(\omega_z' t + C_2). \tag{6.18}$$

In order to simplify further discussion, we can without loss of generality set

$$C_2 = 0. \tag{6.19}$$

In order to find the value of integration constant C_1 we employ equation (6.13), and obtain

$$C_1 = \omega_v . \tag{6.20}$$

The final result is

$$\omega_x = \omega_v \cos \omega_z' t \tag{6.21a}$$

$$\omega_y = \omega_v \sin \omega_z' t. \tag{6.21b}$$

These two equations inform us that the component ω_v and, consequently, the v-axis and the u-axis, the latter being normal to v and z, rotate at a relative angular velocity (equation (6.8))

$$\omega_z' = \frac{C - A}{A} \omega_z$$

in the xy-plane of the body-fixed xyz-coordinate system (Figure 6.2). The z-component of the absolute angular velocity of the v-axis is

$$\omega_z + \omega_z' = \frac{C}{A} \omega_z . \tag{6.22}$$

For a flattened gyroscope ($C > A$) we find the v-axis rotating *faster* than the x-axis, for an elongated gyroscope ($A > C$) the v-axis rotates more *slowly* than the x-axis.

The $Cuvz$ coordinate system thus moves at an angular velocity of

$$\boldsymbol{\Omega} = [\mathbf{e}_u \ \mathbf{e}_v \ \mathbf{e}_z] \begin{bmatrix} 0 \\ \omega_v \\ \frac{C}{A}\omega_z \end{bmatrix} . \tag{6.23}$$

The angular velocity $\boldsymbol{\omega}$ of the gyro body and the angular velocity $\boldsymbol{\Omega}$ of the floating $Cuvz$ coordinate system are related by

$$\boldsymbol{\omega} = \boldsymbol{\Omega} + [\mathbf{e}_u \ \mathbf{e}_v \ \mathbf{e}_z] \begin{bmatrix} 0 \\ 0 \\ \dot{\sigma} \end{bmatrix} \tag{6.24}$$

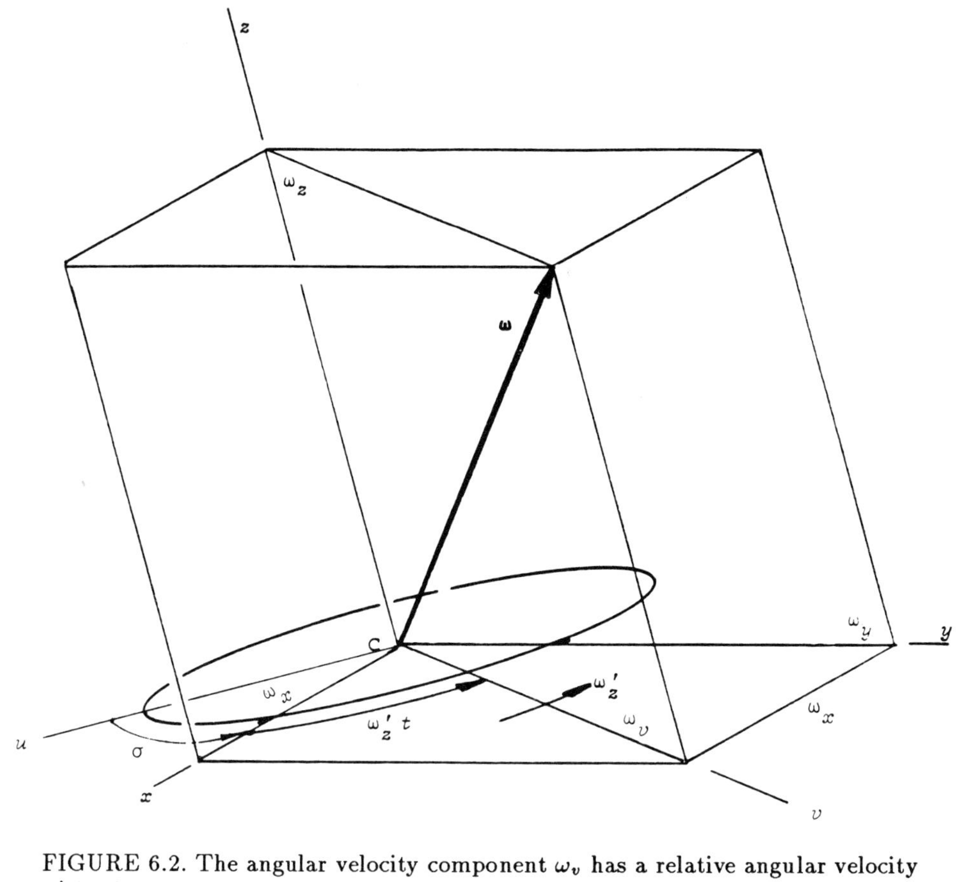

FIGURE 6.2. The angular velocity component ω_v has a relative angular velocity ω_z' in the xy-plane.

where $\dot{\sigma}$ is the spin of the gyro, a quantity that is to be derived in the next section.

The gyro's kinetic energy of rotation is

$$T = \frac{1}{2}\mathbf{H} \cdot \boldsymbol{\omega} = \frac{1}{2}(A\omega_x^2 + A\omega_y^2 + C\omega_z^2) = \frac{1}{2}(A\omega_v^2 + C\omega_z^2). \qquad (6.25)$$

6.2 Euler Frequencies

We locate the origin 0 of the coordinate system in the mass center of the gyroscope. Since there is no external torque applied to the gyroscope, the angular momentum remains constant.

As a consequence, the direction of the angular momentum vector may be used to define a space-fixed coordinate axis, which we will designate the Z-axis. The Z-axis thus also becomes the precession axis.

The z-axis of the rotating $Ouvz$-coordinate system is the symmetry axis of the gyroscope; the v-axis lies in a plane formed by the Z- and z-axis. We write

$$\mathbf{H} = [\mathbf{e}_X \ \mathbf{e}_Y \ \mathbf{e}_Z] \begin{bmatrix} 0 \\ 0 \\ H \end{bmatrix} = [\mathbf{e}_v \ \mathbf{e}_u \ \mathbf{e}_z] \begin{bmatrix} 0 \\ A\omega_v \\ C\omega_z \end{bmatrix}. \qquad (6.26)$$

The nutation angle ν is the angle between the z- and Z-axes, and since Z and $\mathbf{H}$ coincide, we can write

$$\omega_u = 0 \qquad (6.27a)$$

$$\omega_v = \frac{H}{A}\sin\nu \qquad (6.27b)$$

$$\omega_z = \frac{H}{C}\cos\nu. \qquad (6.27c)$$

The nutation angle remains constant and the gyroscope is carrying out a steady precession about the angular momentum vector. Since $\omega_v = \dot{\psi}\sin\nu$ (Figure 6.3), we obtain

$$\dot{\psi} = \frac{C}{A\cos\nu}\omega_z = \frac{H}{A} \qquad (6.28)$$

$$\dot{\sigma} = \frac{A-C}{A}\omega_z = -\omega_z' = \frac{A-C}{AC}H\cos\nu. \qquad (6.29)$$

Note that (Figure 6.2)

$$\sigma = 90^\circ - \omega_z' t$$

and for the angle γ between the z-axis and angular velocity $\boldsymbol{\omega}$ (Figure 6.3), we have $\tan\gamma = \omega_v/\omega_z$, which can be converted into

$$\tan\gamma = \frac{C}{A}\tan\nu. \qquad (6.30)$$

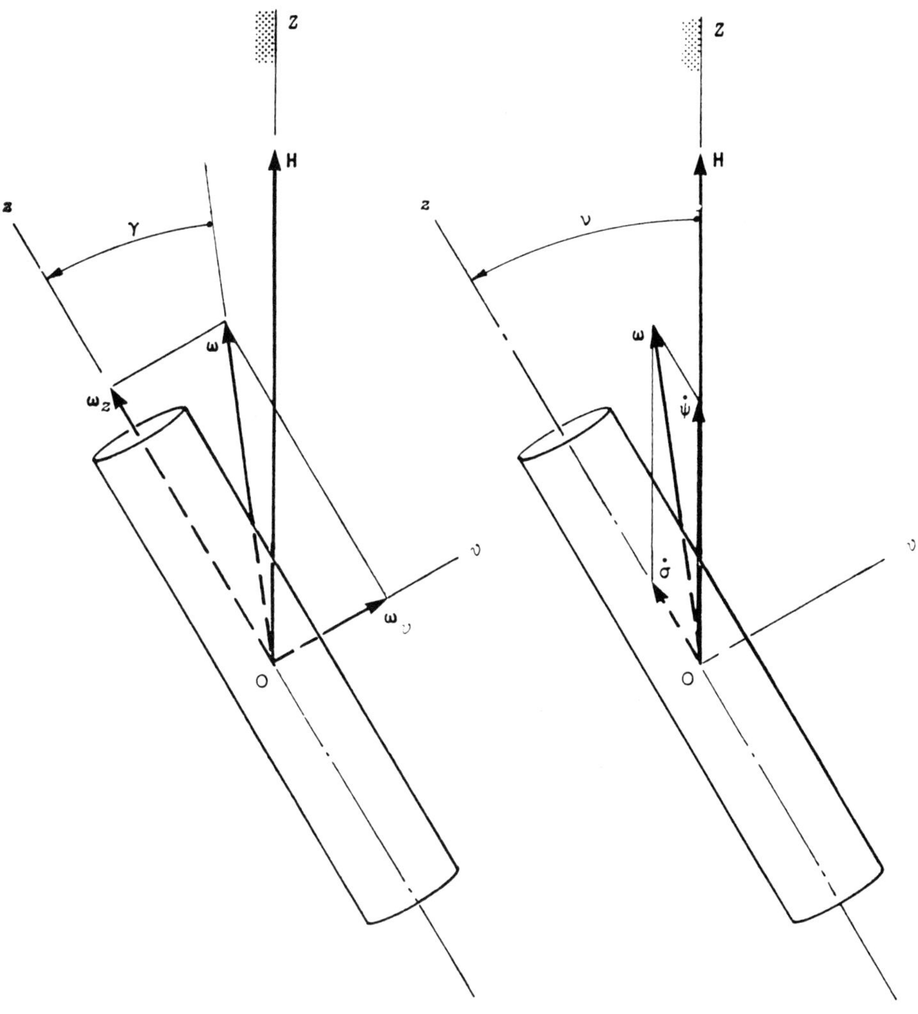

FIGURE 6.3. The components of the angular velocity for an elongated axisymmetric gyro.

The motion of a torquefree gyro with rotational symmetry can be visualized by imagining a *space cone* and a *body cone.* The space cone is generated by successive positions of the angular velocity $\boldsymbol{\omega}$ in the space-fixed coordinate system. The body cone is generated by $\boldsymbol{\omega}$ in the body-fixed coordinate system. Since the body cone and space cone have, at a given instant, a common generator (i.e., the $\boldsymbol{\omega}$-vector), the body cone rolls at an angular velocity $\boldsymbol{\omega}$ on the fixed space cone about the $\boldsymbol{\omega}$-vector, which represents their instantaneous line of contact.

The axis of the space cone is the space-fixed Z-axis. The axis of the body cone is the body-fixed z-axis of symmetry of the gyroscope.

For an elongated gyroscope, characterized by $A > C$, we find that $\gamma < \nu$, the precession is *direct,* and the outside of the body cone rolls on the outside of the space cone (Figures 6.4 and 6.5) in an *epicycloidal* fashion. The term *direct* refers to the fact that $\dot{\sigma}$ and $\dot{\psi}$ have the same sign:

$$\dot{\psi} = \frac{C}{(A - C)\cos\nu}\dot{\sigma}. \tag{6.31}$$

For a flattened gyroscope, characterized by $A < C$, we find that $\gamma > \nu$, the precession is *retrograde,* and the inside of the body cone rolls on the outside of the space cone (Figure 6.6) in a *pericycloidal* fashion. The term *retrograde* refers to the fact that $\dot{\sigma}$ and $\dot{\psi}$ have opposite signs:

$$\dot{\psi} = -\frac{C}{(C - A)\cos\nu}\dot{\sigma}. \tag{6.32}$$

The $\boldsymbol{\omega}$-vector moves along the space cone at an angular velocity $\dot{\psi}$. An observer standing on the gyro would notice the $\boldsymbol{\omega}$-vector moving along the body cone at an angular velocity of $\omega'_z = -\dot{\sigma}$.

With the help of equations (6.27b and c) and Figure 6.3, we can express ω_v as a function of ω_z, giving

$$\omega_v = \frac{C\tan\nu}{A}\omega_z. \tag{6.33}$$

Figures 6.7 and 6.8 show *attitude diagrams,* in which all quantities have the same dimensional units, i.e., rad/s.

Expressed in Euler angles, the gyro's kinetic energy of rotation (6.25) is

$$T = \frac{H^2}{2A}\left(1 + \left(\frac{A}{C} - 1\right)\cos^2\nu\right) \tag{6.34a}$$

or

$$T = \frac{H^2}{2C}\left(1 + \left(\frac{C}{A} - 1\right)\sin^2\nu\right). \tag{6.34b}$$

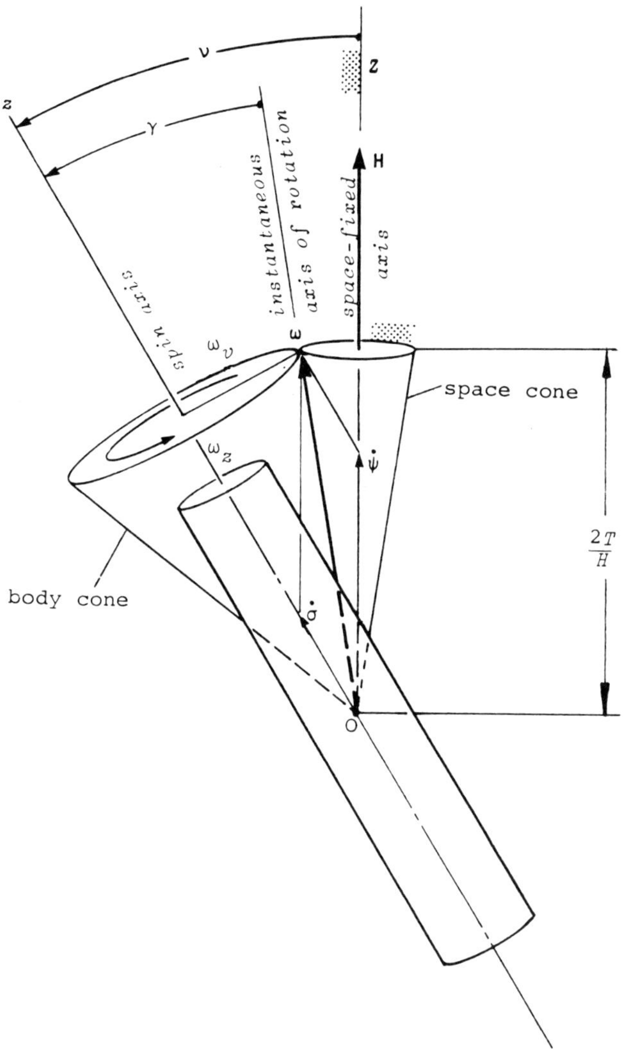

FIGURE 6.4. Space cone and body cone for an elongated gyro (epicycloidal case).

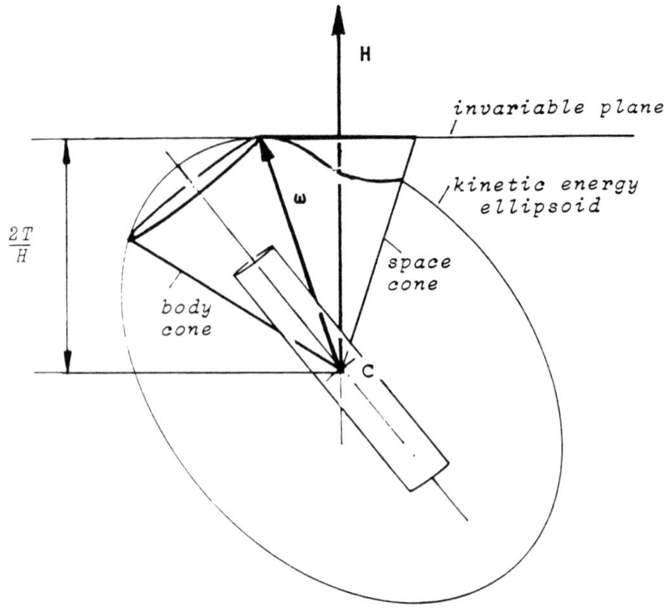

FIGURE 6.5. While the body cone rolls off the space cone, the kinetic energy ellipsoid rolls on the invariable plane.

6.3 Velocity and Acceleration of a Point

An arbitrary body-fixed point P within a rigid gyro is located at

$$\mathbf{r} = [\mathbf{e}_x \ \mathbf{e}_y \ \mathbf{e}_z] \begin{bmatrix} x \\ y \\ z \end{bmatrix}. \tag{6.35}$$

The velocity of point P is

$$\mathbf{v} = \boldsymbol{\omega} \times \mathbf{r}. \tag{6.36}$$

With $\boldsymbol{\omega}$ as given by equation (6.2), the point's velocity becomes

$$\mathbf{v} = [\mathbf{e}_x \ \mathbf{e}_y \ \mathbf{e}_z] \begin{bmatrix} \omega_y z - \omega_z y \\ \omega_z x - \omega_x z \\ \omega_x y - \omega_y x \end{bmatrix}. \tag{6.37}$$

The acceleration of point P is

$$\mathbf{a} = \dot{\mathbf{v}} = \overset{\circ}{\mathbf{v}} + \boldsymbol{\omega} \times \mathbf{v}. \tag{6.38}$$

Considering that $\dot{\omega}_z = 0$ for an axisymmetric gyro, the relative derivative

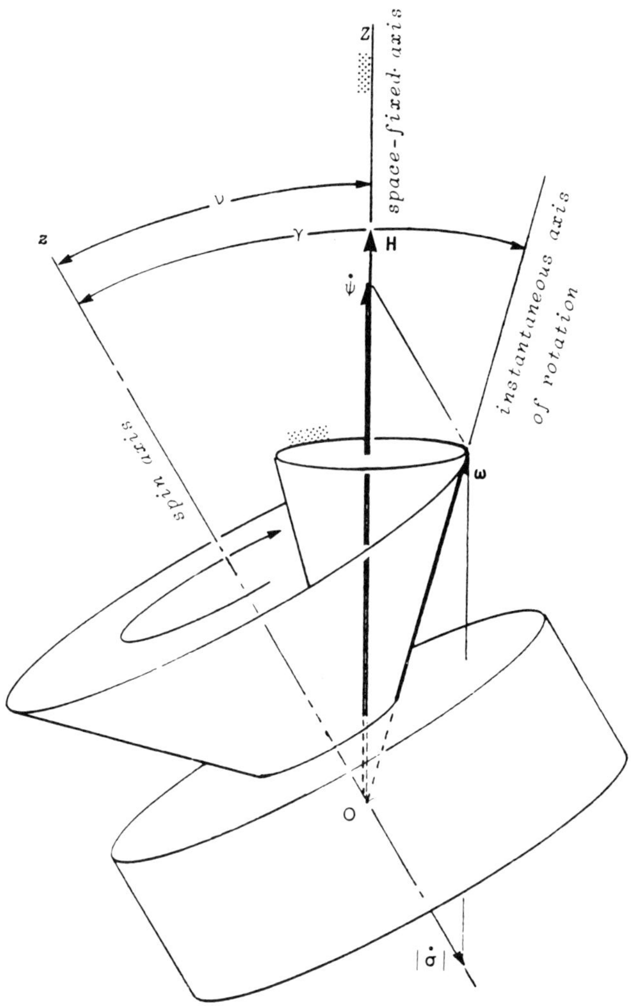

FIGURE 6.6. Space cone and body cone for a flattened gyro (pericycloidal case).

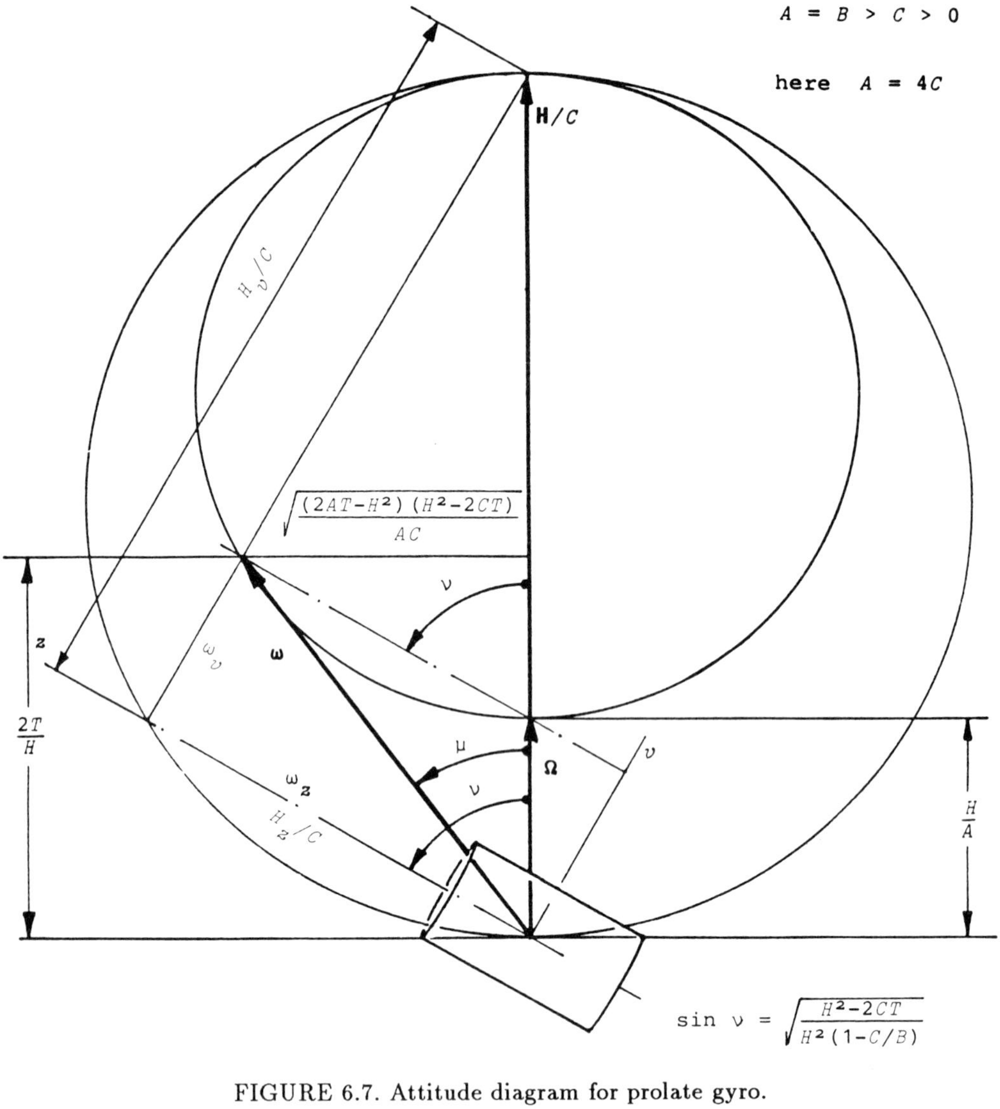

FIGURE 6.7. Attitude diagram for prolate gyro.

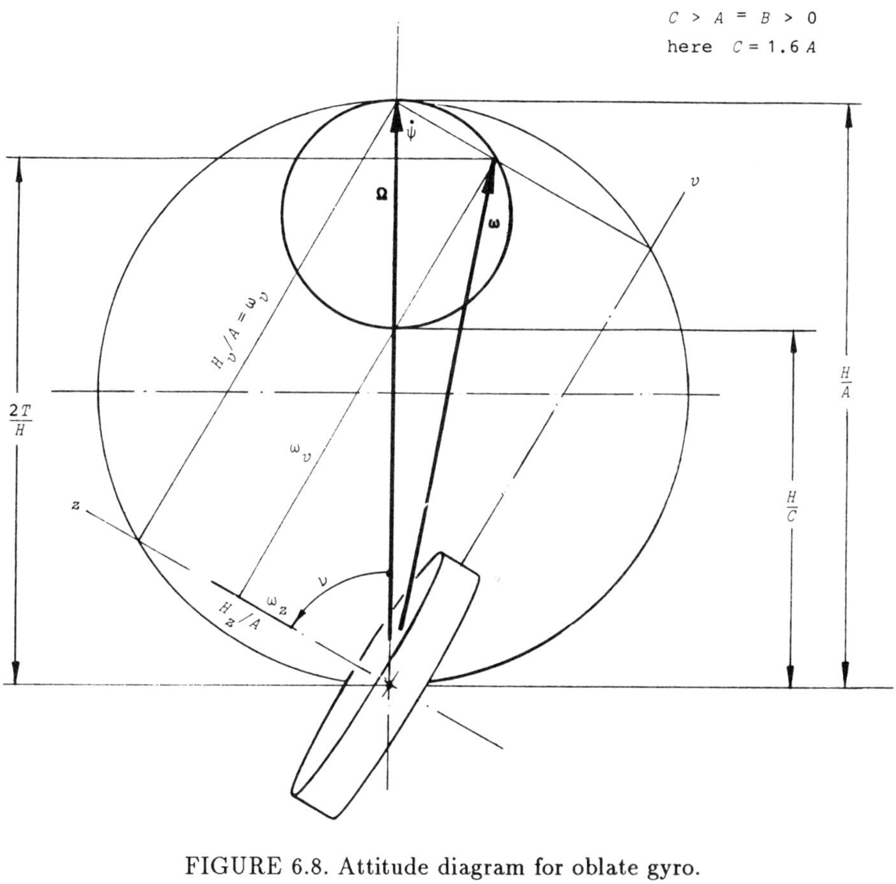

FIGURE 6.8. Attitude diagram for oblate gyro.

becomes

$$\overset{\circ}{\mathbf{v}} = [\mathbf{e}_x \ \mathbf{e}_y \ \mathbf{e}_z] \begin{bmatrix} \dot{\omega}_y z \\ -\dot{\omega}_x z \\ \dot{\omega}_x y - \dot{\omega}_y x \end{bmatrix} \tag{6.39}$$

while $\boldsymbol{\omega} \times \mathbf{v}$ is obtained from equations (6.2) and (6.37). Together they give for the acceleration

$$\mathbf{a} = [\mathbf{e}_x \ \mathbf{e}_y \ \mathbf{e}_z] \begin{bmatrix} z\dot{\omega}_y - x(\omega_y^2 + \omega_z^2) + (y\omega_y + z\omega_z)\omega_x \\ -z\dot{\omega}_x - y(\omega_z^2 + \omega_x^2) + (z\omega_z + x\omega_x)\omega_y \\ y\dot{\omega}_x - x\dot{\omega}_y - z(\omega_x^2 + \omega_y^2) + (x\omega_x + y\omega_y)\omega_z \end{bmatrix} . \tag{6.40}$$

The acceleration of a body-fixed point P of a torquefree rigid gyro may, in view of the Euler equations (6.4) and (6.5), also be expressed as

$$\mathbf{a} = [\mathbf{e}_x \ \mathbf{e}_y \ \mathbf{e}_z] \begin{bmatrix} -(\omega_y^2 + \omega_z^2) & \omega_x\omega_y & \frac{C}{A}\omega_x\omega_z \\ \omega_x\omega_y & -(\omega_z^2 + \omega_x^2) & \frac{C}{A}\omega_y\omega_z \\ (2 - \frac{C}{A})\omega_x\omega_z & (2 - \frac{C}{A})\omega_y\omega_z & -(\omega_x^2 + \omega_y^2) \end{bmatrix} \begin{bmatrix} x \\ y \\ z \end{bmatrix} . \tag{6.41}$$

Transformation from one coordinate system, i.e., from the $Cxyz$ system, to another, i.e., the $Cuvz$ system, is accomplished by writing

$$\begin{bmatrix} a_u \\ a_v \\ a_z \end{bmatrix} = [T] \begin{bmatrix} a_x \\ a_y \\ a_z \end{bmatrix} \tag{6.42}$$

with

$$[T] = \begin{bmatrix} \cos\sigma & -\sin\sigma & 0 \\ \sin\sigma & \cos\sigma & 0 \\ 0 & 0 & 1 \end{bmatrix} \tag{6.43}$$

where $\sigma = ((A - C)/A)\omega_z t$ is the angle measured from the floating u-coordinate to the body-fixed x-coordinate. Since $\omega_x = \omega_v \sin\sigma$ and $\omega_y = \omega_v \cos\sigma$, the acceleration expressed in floating $Cuvz$ coordinates becomes

$$\mathbf{a} = [\mathbf{e}_u \ \mathbf{e}_v \ \mathbf{e}_z] \begin{bmatrix} -\omega^2\cos\sigma & \omega^2\sin\sigma & 0 \\ -\omega_z^2\sin\sigma & -\omega_z^2\cos\sigma & \frac{C}{A}\omega_v\omega_z \\ (2 - \frac{C}{A})\omega_v\omega_z\sin\sigma & (2 - \frac{C}{A})\omega_v\omega_z\cos\sigma & -\omega_v^2 \end{bmatrix} \begin{bmatrix} x \\ y \\ z \end{bmatrix} . \tag{6.44}$$

With the exception of σ, which is time-dependent, all other terms appearing in equation (6.44) are constant. One can thus conclude that any arbitrary body-fixed point P of a rigid axisymmetric gyro is subject to an acceleration that has constant terms and sinusoidal terms of frequency $\dot{\sigma}$.

It is interesting to note that for the limiting case of a fully oblate gyro with $C = 2A$, all mass elements are located in the Cxy plane and consequently $z = 0$. The acceleration of an arbitrary body-fixed point $P(x, y, 0)$ is thus

$$\mathbf{a} = [\mathbf{e}_u \ \mathbf{e}_v \ \mathbf{e}_z] \begin{bmatrix} -x\omega^2\cos\sigma + y\omega^2\sin\sigma \\ -x\omega_z^2\sin\sigma - y\omega_z^2\cos\sigma \\ 0 \end{bmatrix} \tag{6.45}$$

and it can be concluded that for this special case, only accelerations within the body plane, i.e., the xy-plane, occur.

6.4 The Inertial Force Field

Any arbitrary body-fixed point P within a rigid gyro is subjected to an acceleration as given by equation (6.38). For a mass element dm located at point P, the inertial force (i.e., centrifugal and/or reversed Coriolis force) is $d\mathbf{f} = -\mathbf{a}dm$, or

$$d\mathbf{f} = [\mathbf{e}_u \ \mathbf{e}_v \ \mathbf{e}_z] \begin{bmatrix} \omega^2(x\cos\sigma - y\sin\sigma) \\ \omega_z^2(x\sin\sigma + y\cos\sigma) - \frac{C}{A}\omega_v\omega_z z \\ (\frac{C}{A} - 2)\omega_v\omega_z(x\sin\sigma + y\cos\sigma) + \omega_v^2 z \end{bmatrix} dm. \tag{6.46}$$

Since, from Figure 6.2, $u = x\cos\sigma - y\sin\sigma$ and $v = x\sin\sigma + y\cos\sigma$, equation (6.39) may also be written

$$d\mathbf{f} = [\mathbf{e}_u \ \mathbf{e}_v \ \mathbf{e}_z] \begin{bmatrix} u\omega^2 dm \\ v\omega_z^2 dm - \frac{C}{A} z\omega_v\omega_z dm \\ z\omega_v^2 dm - (2 - \frac{C}{A})v\omega_v\omega_z dm \end{bmatrix}. \tag{6.47a}$$

Because $C/A = \Omega_z/\omega_z$, equation (6.47a) may also be written

$$d\mathbf{f} = [\mathbf{e}_u \ \mathbf{e}_v \ \mathbf{e}_z] \begin{bmatrix} u\omega^2 dm \\ v\omega_z^2 dm - z\omega_v\Omega_z dm \\ z\omega_v^2 dm - v\omega_v(2\omega_z - \Omega_z)dm \end{bmatrix}. \tag{6.47b}$$

Since $\int_m u dm = \int_m v dm = \int_m z dm = 0$, because the gyro's mass center C and the origin of the $Cuvz$ coordinate system coincide,

$$\int_m d\mathbf{f} = \mathbf{0}. \tag{6.48}$$

In words, the resultant of the inertial force system vanishes, as is to be expected.

With

$$\mathbf{r} = [\mathbf{e}_u \ \mathbf{e}_v \ \mathbf{e}_z] \begin{bmatrix} u \\ v \\ z \end{bmatrix} \tag{6.49}$$

and since $\int uv dm = \int vz dm = \int uz dm = 0$, because the $Cuvz$ system is a principal axes system, and since $\int v^2 dm = C/2$ because the gyro is axisymmetric, it can be shown that

$$\int_m \mathbf{r} \times d\mathbf{f} = \mathbf{0}. \tag{6.50}$$

In words, the resultant torque of the inertial force system vanishes, as is to be expected.

An inspection of equation (6.40) indicates that the components of the inertial force field when expressed in $Cuvz$ coordinates are all constant with respect to time, leading to the conclusion that *the inertial force field rotates at the angular velocity* $\boldsymbol{\Omega}$ *of the floating* $Cuvz$ *system,* while the gyro body itself rotates at the angular velocity $\boldsymbol{\omega}$ of the body-fixed $Cxyz$ system.

6.5 Changing Inertia Moments

If the inertia tensor changes with time, but such that the inertia products remain zero, i.e., if

$$[\dot{I}] = \begin{bmatrix} \dot{A} & 0 & 0 \\ 0 & \dot{A} & 0 \\ 0 & 0 & \dot{C} \end{bmatrix} \tag{6.51}$$

then the Euler equations (1.60) become

$$A\dot{\omega}_x + \dot{A}\omega_x + (C - A)\omega_z\omega_y = 0 \tag{6.52a}$$

$$A\dot{\omega}_y + \dot{A}\omega_y - (C - A)\omega_z\omega_x = 0 \tag{6.52b}$$

$$C\dot{\omega}_z + \dot{C}\omega_z = 0. \tag{6.52c}$$

Equation (6.52c) can be integrated immediately to give

$$C\omega_z = \text{constant}. \tag{6.53}$$

Multiplication of equation (6.52a) with ω_x, and of equation (6.52b) with ω_y, addition of the two, and introduction of

$$\omega_v = \sqrt{\omega_x^2 + \omega_y^2} \tag{6.54}$$

leads to

$$A\dot{\omega}_v + \dot{A}\omega_v = 0 \tag{6.55}$$

which can immediately be integrated to give

$$A\omega_v = \text{constant}. \tag{6.56}$$

Since equations (6.53) and (6.56) represent the angular momentum components it can be deduced that the nutation angle ν remains constant when the inertia moments change according to equation (6.51).

By introducing temporarily

$$d\alpha = \frac{C - A}{A}\omega_z dt \tag{6.57}$$

equations (6.52a) and (6.52b) can be combined to yield

$$\frac{d^2(A\omega_x)}{d\alpha^2} + (A\omega_x) = 0 \tag{6.58}$$

and solved to give

$$\omega_x = \omega_v \cos\alpha \tag{6.59a}$$

$$\omega_y = \omega_v \sin\alpha \tag{6.59b}$$

with

$$\alpha = \int_0^t \frac{C-A}{A}\omega_z dt \tag{6.60}$$

and

$$\omega_v = \frac{A_0\omega_{v0}}{A} \tag{6.61}$$

where the subscript o refers to values at time $t = 0$. Many satellites have long appendages, such as antennas. These cannot be carried, in an extended state, inside the fairing of a rocket or the loading bay of a space shuttle. Consequently these appendages are extended in orbit (MacNaughton, 1963). The attitude dynamics of this extension process is discussed, e.g., by Hughes (1972), while a more general case of slowly varying inertia moments is given by Sherman and Graham (1966).

Inertia moments of satellites change, when booms, such as antennas, or when solar power collector arrays, etc., are extented. Sometimes it is still permissible to model core satellite and appendages as a rigid body of changing inertia tensor, in other cases the appendages can no longer be assumed rigid, but their flexure and torsion must be taken into account (Cherchas, 1971; Cherchas and Gossain, 1974).

As an example, consider an axisymmetric satellite with $A_0 = 200\,\mathrm{m^2kg}$, and $C_0 = 100\,\mathrm{m^2kg}$. Four coilable antennas of uniform mass distribution and of $m = 6$ kg mass each are stored so close to the satellite's mass center that their contribution to the inertia moments can be ignored. Then they are slowly and uniformly extended radially (with motors turning in opposing directions to ensure zero effect on the satellite) until they reach a length of 50 m each from the satellite center. The satellite has an angular momentum of $H = 1000$ Ws^2, and precesses at a nutation angle of $\nu = 5°$. The new inertia moments, with the antennas assumed rigid, are

$$A = A_0 + 2\frac{1}{3}ml^2 = 200 + 2\frac{1}{3}6(50^2) = 10\,200\,\mathrm{m^2kg}$$

$$C = C_0 + 4\frac{1}{3}ml^2 = 100 + 4\frac{1}{3}6(50^2) = 20\,100\,\mathrm{m^2kg}.$$

The new angular velocity components are

$$\omega_v = \frac{H\sin\nu}{A} = \frac{1000\sin 5°}{10\,200} = 0.008\,544\,\mathrm{rad/s}$$

$$\omega_z = \frac{H\cos\nu}{C} = \frac{1000\cos 5^\circ}{21\,000} = 0.049\,562\,\mathrm{rad/s}.$$

The old kinetic energy was

$$T_0 = \frac{H^2}{2}\left(\frac{\sin^2\nu}{A_0} + \frac{\cos^2\nu}{C_0}\right) = 4981\,\mathrm{J}.$$

The new kinetic energy is

$$T = \frac{H^2}{2}\left(\frac{\sin^2\nu}{A} + \frac{\cos^2\nu}{C}\right) = 25.059\,\mathrm{J}.$$

The difference is provided by the braking action of the motors and friction losses during the extension process.

6.6 Capture

If a satellite, tumbling freely about its mass center C, is to be retrieved by another spacecraft, such as the STS (space transportation system) *space shuttle* ($\simeq$ 95 000 kg mass), the retriever will extend an RMS (remote manipulator system) cantilever boom with an *end effector* at the free end (Doetsch, 1978; Wagner–Bartak, 1981). The retrieval process consists of three phases, viz.

> approach

> capture

> rigidization.

Let the *approach* be so gradual that the relative velocity between retrievee and retriever can be neglected for purposes of the present analysis. Let us also assume that mass and inertia tensor of the retriever are so much larger than those of the retrievee, that the retriever can be considered fixed in space. We shall also assume the RMS boom to be rigid, and the end effector closing suddenly, i.e., in a very short time interval Δt, as it captures the satellite.

If the *grapple fixture* on the satellite, which mates with the end effector, is located at point O (Figure 6.9) on the satellite's axis of symmetry, then the angular velocity of the satellite after *capture* can be readily calculated. Before capture, the angular momentum with respect to point O is, according to equation (1.9),

$$\mathbf{H}_0 = \mathbf{H}_C + m\mathbf{r}_C \times \mathbf{v}_C.$$

For purposes of the capture study we set $\mathbf{v}_C = 0$, and obtain

$$\mathbf{H}_0 = [\mathbf{e}_u\ \mathbf{e}_v\ \mathbf{e}_z]\begin{bmatrix} 0 \\ A\omega_v \\ C\omega_z \end{bmatrix} \tag{6.62}$$

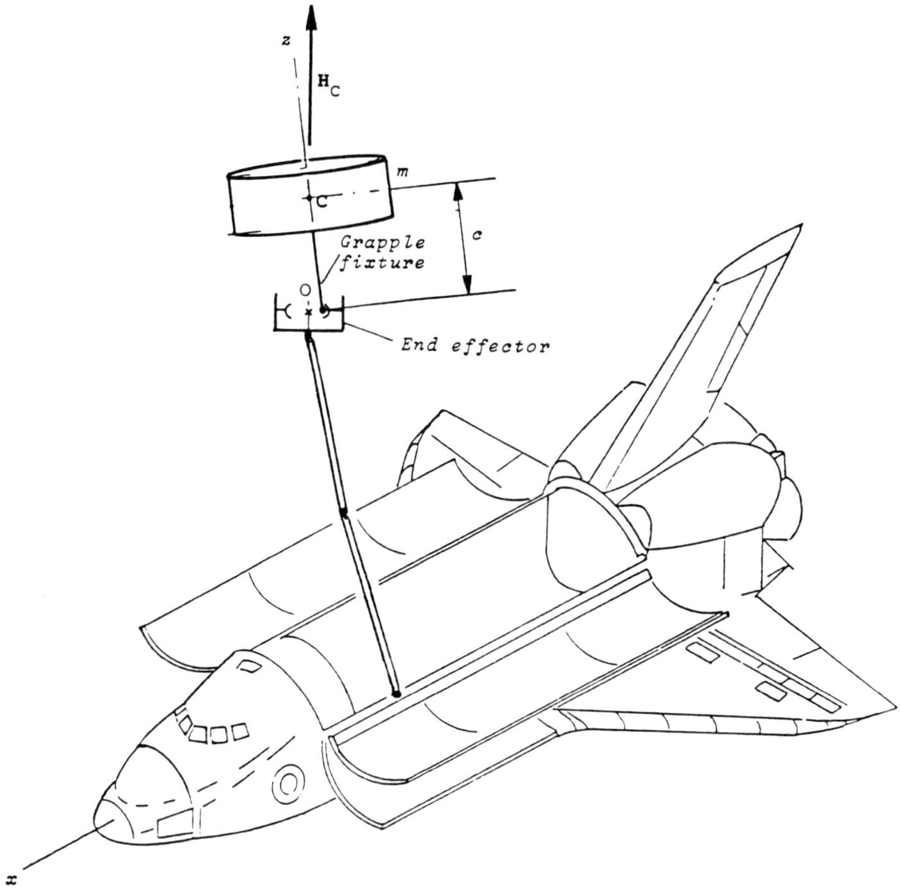

FIGURE 6.9. Capture of a tumbling satellite.

where A and C are principal inertia moments with respect to the satellite's mass center.

If there is no torque transmitted through the end effector O, then there will be no change of angular momentum with respect to point O. Using primes for the angular velocity after capture, we obtain

$$\mathbf{H}_{\mathrm{O}} = [\mathbf{e}_u \ \mathbf{e}_v \ \mathbf{e}_z] \begin{bmatrix} 0 \\ (A + mc^2)\omega_v' \\ C\omega_z' \end{bmatrix} \tag{6.63}$$

and from equations (6.62) and (6.63),

$$\boldsymbol{\omega}' = [\mathbf{e}_u \ \mathbf{e}_v \ \mathbf{e}_z] \begin{bmatrix} 0 \\ \frac{A}{A+mc^2}\omega_v \\ \omega_z \end{bmatrix}. \tag{6.64}$$

The linear impulse transmitted through the end effector O can be obtained by using the change in the angular momentum with respect to the satellite's mass center C. If we furthermore assume that the force components are constant throughout the capture time interval, then the linear impulse $\mathbf{S} = \mathbf{F}\Delta t$, and

$$\mathbf{r} \times \mathbf{F}\Delta t = \mathbf{H}_{\mathrm{C}}' - \mathbf{H}_{\mathrm{C}} = [\mathbf{e}_u \ \mathbf{e}_v \ \mathbf{e}_z] \begin{bmatrix} 0 \\ \frac{A^2}{A+mc^2}\omega_v - A\omega_v \\ C\omega_z - C\omega_z \end{bmatrix}. \tag{6.65}$$

Since $\mathbf{r} = -c\mathbf{e}_z$, the individual force components are

$$F_u = \frac{mc}{A + mc^2}\frac{A\omega_v}{\Delta t} \tag{6.66a}$$

$$F_v = 0 \tag{6.66b}$$

$$F_z = 0. \tag{6.66c}$$

It is readily seen that a satellite with pure rotation about the z-axis, i.e., with $\omega_v = 0$, is captured without causing any forces in the end effector.

The kinetic energy of the satellite was originally

$$T_0 = \frac{1}{2}(A\omega_v^2 + C\omega_z^2). \tag{6.67}$$

After capture it is

$$T' = \frac{1}{2}((A + mc^2)\omega_v'^2 + C\omega_z^2). \tag{6.68}$$

The capture process thus changes the kinetic energy by

$$\Delta T = T' - T = -\frac{1}{2}\frac{mc^2}{A + mc^2}A\omega_v^2 \tag{6.69}$$

i.e., there is an energy loss.

The third phase is *rigidization,* where any relative motion between retrievee and end effector is eliminated. The kinetic energy of a captured satellite, tumbling at $\boldsymbol{\omega}'$ about point O in the end effector is given by equation (6.68). It must be dissipated during the rigidization phase.

Take, for example, a satellite of $m = 100$ kg and

$$[I] = \begin{bmatrix} 180 & 0 & 0 \\ 0 & 180 & 0 \\ 0 & 0 & 200 \end{bmatrix} \text{m}^2\text{kg}.$$

Just prior to capture, it has an angular velocity of

$$\boldsymbol{\omega} = [\mathbf{e}_u\ \mathbf{e}_v\ \mathbf{e}_z] \begin{bmatrix} 0 \\ 0.5 \\ 1.2 \end{bmatrix} \text{rad/s}.$$

The grapple fixture O is located along the z-axis at $z = -2$ m. The capture time interval is $\Delta t = 0.05$ s. What is the angular velocity immediately after capture? What is the force applied to the satellite at the grappling fixture during capture? What is the kinetic energy loss?

With $mc^2 = 100(4) = 400\text{m}^2\text{kg}$

$$\boldsymbol{\omega}' = [\mathbf{e}_u\ \mathbf{e}_v\ \mathbf{e}_z] \begin{bmatrix} 0 \\ \frac{A}{A+mc^2}\omega_v \\ \omega_z \end{bmatrix} = [\mathbf{e}_u\ \mathbf{e}_v\ \mathbf{e}_z] \begin{bmatrix} 0 \\ 0.115 \\ 1.2 \end{bmatrix} \text{rad/s}$$

$$\mathbf{F} = [\mathbf{e}_u\ \mathbf{e}_v\ \mathbf{e}_z] \begin{bmatrix} \frac{mc}{A+mc^2}\frac{A\omega_v}{\Delta t} \\ 0 \\ 0 \end{bmatrix} = [\mathbf{e}_u\ \mathbf{e}_v\ \mathbf{e}_z] \begin{bmatrix} 621 \\ 0 \\ 0 \end{bmatrix} \text{N}.$$

The kinetic energy of the captured satellite is

$$T = \frac{1}{2}\left((180 + 400)0.155^2 + (200)1.2^2\right) = 151 \text{ J}.$$

The satellite's original kinetic energy was 166 J, of which 15 J were dissipated during capture, and the remaining 151 J must be dissipated during rigidization.

Suggested Reading

1. Cherchas, D.B. "Dynamics of Spin Stabilized Satellites during Extension of Long Flexible Booms," *Journal of Spacecraft and Rockets,* 8, 7, 1971, 802–804.

2. Cherchas, D.B.; D.M. Gossain. "Dynamics of a Flexible Solar Array during Deployment from a Spinning Spacecraft," *CASI Transactions,* 7, 1, 1974, 10–18.

3. Doetsch, K.-H., Jr. "The Remote Manipulator System for the Space Shuttle Orbiter," *Zeitschrift für Flugwissenschaften and Weltraumforschung,* 2, 4, 1978, 261–269.

4. Hughes, P.C. "Dynamics of a Spin Stabilized Satellite during Extension of Rigid Booms," *CASI Transactions,* 5, 1, 1972, 11–14.

5. MacNaughton, J.D. "Unfurlable Metal Structures for Spacecraft," *Canadian Aeronautics and Space Journal,* 9, 4, 1963, 103–116.

6. Sherman, B.C.; J.D. Graham. "Coning Motion of a Spinning Body with Slowly Varying Inertias," *AIAA Journal,* 4, 8, 1966, 1467–1469.

7. Wagner-Bartak, C.G. "Future Applications of Remote Manipulators in Space," *Canadian Aeronautics and Space Journal,* 27, 1, 1981, 17–25.

Problems

6.1. A steel disk is 1 cm thick, and has a radius of 1 m. (a) Determine its principal inertia tensor $[I]$. (b) Determine the angular momentum $\mathbf{H}$, if $\boldsymbol{\omega} = 30\mathbf{e}_v + 40\mathbf{e}_z$. (c) Calculate the kinetic energy. (d) How fast does ω_v rotate within the xy-plane? (e) What is the absolute velocity of the ω_v-vector?

6.2. The space station shown has a mass of 1200 kg and is known to precess about the fixed direction OZ at the rate of one revolution per hour. Assuming that the station is dynamically equivalent to a homogeneous cylinder of length 40 m and radius 4 m, determine (a) the inertia matrix of the space station, (b) its spin rate, (c) its angular velocity, (d) its angular momentum, (e) its kinetic energy, and (f) the linear velocity of point C_1 located at $z = 10$ m. The space station is to be separated into two equal portions.

There is no interference between the two portions during or after separation. Each portion of the station is dynamically equivalent to a cylinder of length 20 m and radius 4 m. Knowing that when the link is severed, the station is oriented as shown, determine for portion 1 alone (g) the inertia matrix, (h) the axis of precession, (i) the precession rate [Hint: The angular velocity after separation is equal to the angular velocity before separation], (j) the spin rate, (k) the angular momentum; and show that (l) the total angular momentum and (m) the total kinetic energy of the two portions have not changed as a result of the separation.

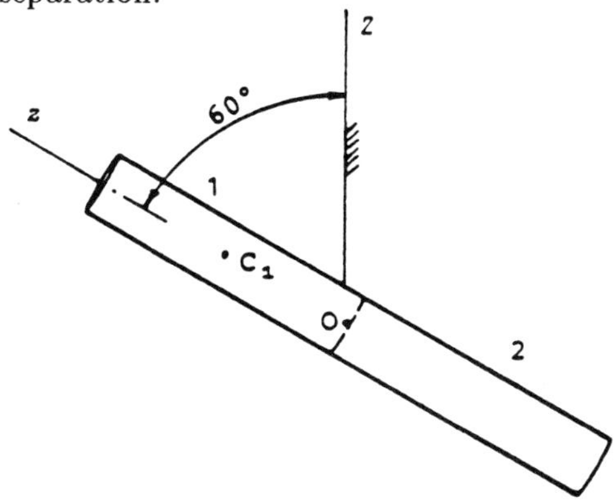

6.3. A disk gyro has $C = 2A$. (a) Sketch the gyro. Its angular velocity is

$$\boldsymbol{\omega} = [\mathbf{e}_x\ \mathbf{e}_y\ \mathbf{e}_z] \begin{bmatrix} 3 \\ 4 \\ 12 \end{bmatrix} \text{rad/s}.$$

(b) Determine the angular momentum, if $A = 2000\,\text{m}^2\text{kg}$. (c) Determine the auxiliary frequency ω_z'. (d) Determine the frequency component ω_v. (e) Determine the z-component of the absolute frequency of the v-axis.

6.4. A very flat disk gyro has a radius of 3 m and a mass of 20 kg. (a) Determine its inertia tensor. At a certain point in time, the following velocity vector is measured

$$\boldsymbol{\omega} = [\mathbf{e}_u\ \mathbf{e}_v\ \mathbf{e}_z] \begin{bmatrix} 0 \\ 20 \\ 10 \end{bmatrix} \text{rad/s}.$$

(b) Determine the angular momentum $\mathbf{H}$ and make a sketch. (c) Determine the nutation angle ν. (d) Determine the angle γ between $\boldsymbol{\omega}$-vector and z-axis. (e) What is the precession frequency $\dot{\psi}$? (f) What is the spin frequency $\dot{\sigma}$? (g) How fast does the $\mathbf{e}_v$-vector rotate with respect to a body-fixed vector $\mathbf{e}_x$? (h) What is the velocity of the body point located at

$$\mathbf{r} = [\mathbf{e}_x \ \mathbf{e}_y \ \mathbf{e}_z] \begin{bmatrix} 1.5 \\ 0 \\ 0 \end{bmatrix} \text{m?}$$

(i) Sketch body and space cones.

6.5. A very thin plate of square shape, with sides $a = b = 2$ m, is free of any external torque and located such that

$$\psi = 90°$$

$$\nu = ?$$

$$\sigma = 15°.$$

At that moment, the plate's angular velocity is

$$\boldsymbol{\omega} = [\mathbf{e}_x \ \mathbf{e}_y \ \mathbf{e}_z] \begin{bmatrix} 30 \\ 40 \\ 5 \end{bmatrix} \text{rad/s.}$$

Determine (a) the Euler frequencies, and (b) make a sketch, showing the x- and y-axes.

6.6. Given is a spatial dumbbell gyro with concentrated masses at the tips. The total mass of the gyro is $m = 60$ kg. The rods are rigid and massless. The gyro is torquefree. The nutation angle is 30°. Determine the force acting on the mass $m/6$ located at $z = c/2$, at the moment where $\omega_z' t = 2\pi$, for $a = b = 12$ m, $c = 4$ m, and $\omega_z = 30$ rad/s.

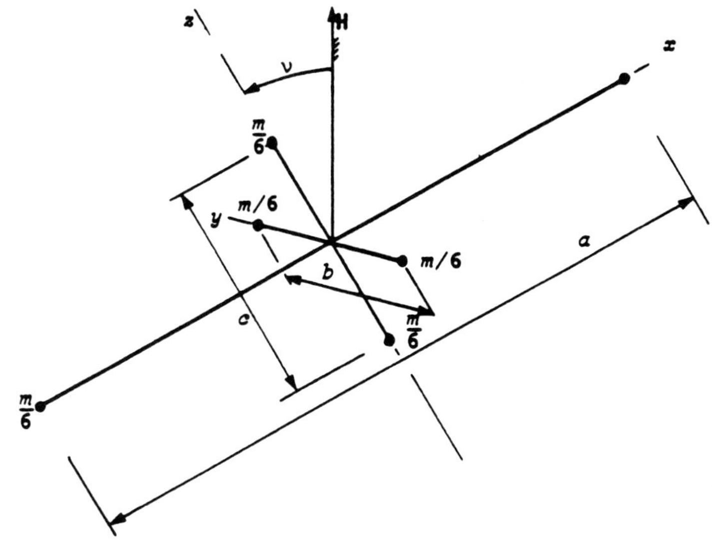

6.7. Given is a satellite, which consists of four thin, long, rigid rods forming a cross. Each rod weights 6 kg and is 100 m long (total mass of satellite is 24 kg). The body-fixed x- and y-axes run along the rods. The torquefree satellite has an angular velocity of constant magnitude, which has, at time $t = 0$, the following value

$$\boldsymbol{\omega} = [\mathbf{e}_x \ \mathbf{e}_y \ \mathbf{e}_z] \begin{bmatrix} 0 \\ 8 \\ 10 \end{bmatrix} 10^{-4} \text{ rad/s}.$$

(a) Find the inertia tensor $[I]$. (b) Determine the angular momentum $\mathbf{H}$. (c) Sketch space cone and body cone for time $t = 0$. (d) Determine the relative angular velocity ω_z'. (e) Enter a floating $Cuvz$ coordinate system into your sketch.

Introduce a space-fixed $CXYZ$ coordinate system, such that the Z-axis coincides with $\mathbf{H}$, and the X-axis and the x-axis coincide at $t = 0$. Subsequently (f) determine the spin frequency $\dot{\sigma}$, (g) determine the nutation angle ν, (h) determine the precession frequency $\dot{\psi}$, and (i) sketch space cone and body cone at $t = 16\,043$ s.

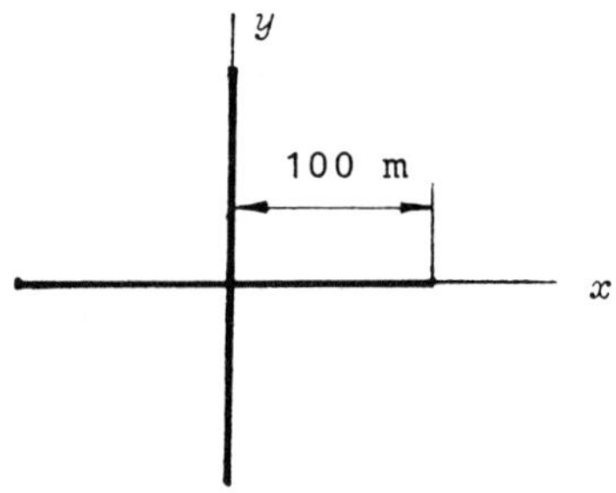

6.8. Given is a torquefree rigid spatial dumbbell satellite as shown, with $a = b = 3$ m, and $c_o = 4$ m, a total mass of 300 kg, and an angular velocity of

$$\boldsymbol{\omega}_o = [\mathbf{e}_u\ \mathbf{e}_v\ \mathbf{e}_z] \begin{bmatrix} 0 \\ 5 \\ 10 \end{bmatrix} \text{rad/s.}$$

(a) Determine the angular momentum, (b) sketch the satellite's attitude with respect to it, and (c) enter the angular velocity into the sketch.

At $t = 0$ the rigid massless z-booms are extended according to $c = c_0 + 0.05t$, with t in seconds, until $c_f = 8$ m is reached. Determine the angular velocity (d) when $c = 6$ m, (e) when $c = c_f = 8$ m. (f) Make a sketch.

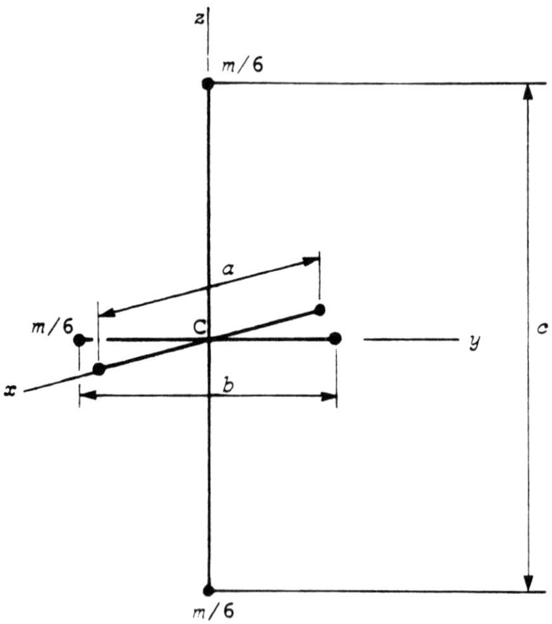

6.9. A torquefree axisymmetric gyro with an angular velocity of

$$\boldsymbol{\omega} = [\mathbf{e}_x \ \mathbf{e}_y \ \mathbf{e}_z] \begin{bmatrix} 5\cos 8t \\ -5\sin 8t \\ 12 \end{bmatrix} \text{rad/s}$$

has $\psi = 0$ and $\sigma = 0$ at $t = 0$. Determine (a) the spin angle σ (in revolutions), and (b) the precession angle ψ (in revolutions) when $t = 90$ s. (c) What was the nutation angle ν throughout the motion? (d) Make a sketch of the z-H-plane, and show $\boldsymbol{\omega}$, $\dot{\sigma}$, and $\dot{\psi}$. (e) Is the gyro prolate or oblate?

6.10. Given is the body-fixed point P of a torquefree rigid axisymmetric gyro. The gyro's angular velocity is

$$\boldsymbol{\omega} = [\mathbf{e}_u \ \mathbf{e}_v \ \mathbf{e}_z] \begin{bmatrix} 0 \\ \omega_v \\ \omega_z \end{bmatrix}.$$

Use the floating $Cuvz$ coordinate system and derive the acceleration of point P.

6.11. A point mass of 2 kg is located in a rigid axisymmetric satellite at $x = 2$ m, $y = 0$, and $z = 3$ m. The attitude angle is $\nu = 10°$. The inertia moments are $A = B = 1000\,\mathrm{Ws}^3$, and $C = 200\,\mathrm{Ws}^3$. The angular momentum is $H = 4000\,\mathrm{Ws}^2$. The x-axis of the satellite is just passing the v-axis of the floating $Cuvz$ coordinate system. (a) What is the inertial loading the mass is experiencing (expressed in $Cuvz$ coordinates)? (b) What is the magnitude of the load?

6.12. A thin disk, with $A = 600\,\mathrm{m}^2\mathrm{kg}$ and a radius of 1.5 m, rotates at $\nu = 20°$, with an angular momentum of 2400 Ws^2. What is the acceleration of a point on the disk's circumference at the moment when the point passes through the u-axis?

7

Deformable Axisymmetric Gyros

In the present chapter a near-rigid solid is defined. The concept is then introduced into gyrodynamics. After defining the properties of a suitable floating coordinate system, the latter is used to obtain suitable equations of gyrodynamics for deformable bodies. The concept of a body of constant configuration in a floating coordinate system is introduced. The theory is then demonstrated on typical examples.

A typical single solid gyro, e.g., a spacecraft, is made of materials that exhibit elastic-hysteretic behavior. Thus, deformations due to inertial forces are inevitable. Such deformations are usually small. We *impose* the condition that deformations be small and refer to such a solid body from now on as a *near-rigid body.*

In the present chapter it is shown in which fashion a typical single solid body gyro deviates from its rigid body configuration.

7.1 The Single Solid Body

Let us proceed by assuming that a typical solid of principal inertia tensor

$$[I]_{\text{undef}} = \begin{bmatrix} A & 0 & 0 \\ 0 & B & 0 \\ 0 & 0 & C \end{bmatrix} \tag{7.1}$$

before deformation will change to

$$[I]_{\text{def}} = \begin{bmatrix} A + \Delta A & I_{xy} & I_{xz} \\ I_{xy} & B + \Delta B & I_{yz} \\ I_{xz} & I_{yz} & C + \Delta C \end{bmatrix} \tag{7.2}$$

after deformation, when one and the same Cxyz coordinate system is used to describe both tensors. The orientation of the Cxyz system is defined by the principal axes of the undeformed solid. Its origin is the body's mass center C. If the small principal inertia moment changes ΔA, ΔB, and ΔC are neglected, then

$$[I]_{\text{def}} = \begin{bmatrix} A & I_{xy} & I_{xz} \\ I_{xy} & B & I_{yz} \\ I_{xz} & I_{yz} & C \end{bmatrix} . \tag{7.3}$$

Now let the principal coordinate system of the deformed gyro be $Cx_1y_1z_1$, then

$$[I_1]_{\text{def}} = \begin{bmatrix} A_1 & 0 & 0 \\ 0 & B_1 & 0 \\ 0 & 0 & C_1 \end{bmatrix}. \tag{7.4}$$

Mathematically, these principal inertia moments are the eigenvalues of the matrix of equation (7.2).

With the principal axes system $Cx_1y_1z_1$ of the deformed gyro assumed known, the inertia tensor of the deformed gyro in the principal axes system $Cxyz$ of the undeformed gyro can be found from

$$[I]_{\text{def}} = [T]^T[I_1]_{\text{def}}[T]. \tag{7.5}$$

For very small Cardan angles ξ, η, and ζ (Appendix C), measured from the principal axes system of the undeformed state to the principal axes system of the deformed state,

$$[T] = \begin{bmatrix} 1 & \zeta & -\eta \\ -\zeta & 1 & \xi \\ \eta & -\xi & 1 \end{bmatrix}. \tag{7.6}$$

Neglecting products and squares of the Cardan angles,

$$[I]_{\text{def}} = \begin{bmatrix} A_1 & (A_1 - B_1)\zeta & (C_1 - A_1)\eta \\ (A_1 - B_1)\zeta & B_1 & (B_1 - C_1)\xi \\ (C_1 - A_1)\eta & (B_1 - C_1)\xi & C_1 \end{bmatrix}. \tag{7.7a}$$

Equation (7.7a) corresponds to equation (7.2). If ΔA, ΔB, and ΔC are neglected, then

$$[I]_{\text{def}} = \begin{bmatrix} A & (A - B)\zeta & (C - A)\eta \\ (A - B)\zeta & B & (B - C)\xi \\ (C - A)\eta & (B - C)\xi & C \end{bmatrix}. \tag{7.7b}$$

Equation (7.7b) corresponds to equation (7.3). Summing up, it may be concluded that a near-rigid solid can consequently be characterized by

$$\frac{|\Delta A|}{A} << 1 \quad \frac{|\Delta B|}{B} << 1 \quad \frac{|\Delta C|}{C} << 1 \tag{7.8a}$$

$$|\xi| << 1 \quad |\eta| << 1 \quad |\zeta| << 1. \tag{7.8b}$$

7.2 The Axisymmetric Solid Body

We shall define as *axisymmetric* a gyro which in its undeformed state satisfies the condition

$$B = A. \tag{7.9}$$

From equations (7.7b) and (7.9) we then deduce that the inertia product I_{xy} vanishes, and that in its deformed state a near-rigid axisymmetric gyro has an inertia tensor of

$$[I]_{\text{def}} = \begin{bmatrix} A & 0 & (C-A)\eta \\ 0 & A & (A-C)\xi \\ (C-A)\eta & (A-C)\xi & C \end{bmatrix}. \tag{7.10}$$

While ξ and η are small, the angle ζ can now assume any arbitrary value.

7.3 Observations

We conclude that a near-rigid solid whose inertia tensor in the undeformed state is

$$[I]_{\text{undef}} = \begin{bmatrix} A & 0 & 0 \\ 0 & B & 0 \\ 0 & 0 & C \end{bmatrix} \tag{7.11}$$

will typically have an inertia tensor in the deformed state of

$$[I]_{\text{def}} = \begin{bmatrix} A & I_{xy} & I_{xz} \\ I_{xy} & B & I_{yz} \\ I_{xz} & I_{yz} & C \end{bmatrix} \tag{7.12}$$

with

$$I_{xy} = (A-B)\zeta \tag{7.13a}$$

$$I_{yz} = (B-C)\xi \tag{7.13b}$$

$$I_{zx} = (C-A)\eta \tag{7.13c}$$

and all inertia moments and inertia products expressed in terms of the $Cxyz$ principal coordinate system of the near-rigid solid in its *undeformed* state. The orientation of the $Cxyz$ system within the deforming body is arbitrary as long as it remains close to (or coincides with) the principal axes system of the deforming body. It is typically chosen such that it allows a convenient description of the gyro's deformation. It is worth noting that for the practically important class of axisymmetric gyros, $B = A$, and consequently $I_{xy} = 0$ in their deformed state.

A near-rigid solid in its deformed state is characterized by small deformations leading to (a) *small inertia moment changes and* (b) *small inertia products in comparison with the principal inertia moments of the solid in its undeformed state. The small inertia moment changes ΔA, ΔB, and ΔC are insignificant (in comparison with the inertia moments A, B, and C) and can be neglected. The small inertia products I_{xy}, I_{xz}, and I_{yz} are significant (in comparison with zero) and must be retained.*

To avoid unnecessary complications we specifically exclude from consideration the special case $B = A$, $C = 0$ (i.e., a rod of zero cross section), which is of no practical interest.

7.4 Floating Coordinates

Considerations similar to those used to derive equation (7.10) can also be followed to derive the inertia tensor of a deformed gyro, which was axisymmetric in its undeformed state, using a floating $Cuvz$ coordinate system (Canavin and Likins, 1977) instead of a body-fixed $Cxyz$ coordinate system. Starting with a deformed gyro and its inertia tensor in a principal coordinate system $Cx_1y_1z_1$, as given by equation (7.4), and a transition matrix $[T]$ for Cardan angles of the first kind (Appendix C) with small angles ξ and η, and an arbitrary angle ζ,

$$[T] = \begin{bmatrix} \cos\zeta & \sin\zeta & \xi\sin\zeta - \eta\cos\zeta \\ -\sin\zeta & \cos\zeta & \xi\cos\zeta + \eta\sin\zeta \\ \eta & -\xi & 1 \end{bmatrix}. \tag{7.14}$$

The inertia tensor $[T]$ in the floating $Cuvz$ coordinate system is obtained by forming

$$[I]_{\text{def}} = [T]^T [I_1]_{\text{def}} [T]. \tag{7.15}$$

Carrying out the matrix multiplication, and neglecting products and squares of the two small Cardan angles, results in

$$[I]_{\text{def}} =$$

$$\begin{bmatrix} A_1\cos^2\zeta + B_1\sin^2\zeta & (A_1 - B_1)\sin\zeta\cos\zeta & \begin{matrix} A_1(\xi\sin\zeta - \eta\cos\zeta)\cos\zeta - \\ B_1(\xi\cos\zeta + \eta\sin\zeta)\sin\zeta + C_1\eta \end{matrix} \\ (A_1 - B_1)\sin\zeta\cos\zeta & A_1\sin^2\zeta + B_1\cos^2\zeta & \begin{matrix} A_1(\xi\sin\zeta - \eta\cos\zeta)\sin\zeta + \\ B_1(\xi\cos\zeta + \eta\sin\zeta)\cos\zeta - C_1 xi \end{matrix} \\ \begin{matrix} A_1(\xi\sin\zeta - \eta\cos\zeta)\cos\zeta - \\ B_1(\xi\cos\zeta + \eta\sin\zeta)\sin\zeta + C_1\eta \end{matrix} & \begin{matrix} A_1(\xi\sin\zeta - \eta\cos\zeta)\sin\zeta + \\ B_1(\xi\cos\zeta + \eta\sin\zeta)\cos\zeta - C_1\xi \end{matrix} & \begin{matrix} A_1(\xi\sin\zeta - \eta\cos\zeta)^2 + \\ B_1(\xi\cos\zeta + \eta\sin\zeta)^2 + C_1 \end{matrix} \end{bmatrix}. \tag{7.16}$$

If one now sets $B_1 = A_1 = A$ and $C_1 = C$, the inertia tensor of the deformed axisymmetric gyro, expressed in floating $Cuvz$ coordinates of the undeformed gyro, becomes

$$[I] = \begin{bmatrix} A & 0 & I_{uz} \\ 0 & A & I_{vz} \\ I_{uz} & I_{vz} & C \end{bmatrix} \tag{7.17}$$

with

$$I_{uz} = (C - A)\eta \tag{7.18a}$$

$$I_{vz} = (A - C)\xi \tag{7.18b}$$

where this time ξ and η are Cardan angles measured about the u-axis and the (carried) v-axis, respectively.

If the inertia products $I_{uz} = (C - A)\eta$ and $I_{vz} = (A - C)\xi$ of equations (7.18a and b) are constant, then the body retains its configuration expressed in $Cuvz$ coordinates. Such a body shall be referred to as a *body of constant configuration in a floating coordinate system.* For I_{uz} and I_{vz} to be constant, the body must not only be dynamically symmetrical (i.e., $B = A$) but must also be structurally symmetrical. Thus, e.g., flexible beams with rectangular cross sections are generally not permissible.

7.5 Auxiliary Inertia Products and Moments

The development of the gyrodynamics of deforming axisymmetric solids, which is to follow, requires the introduction of auxiliary inertia products.

As an example let us use as typical an inertia product I_{uz}, which is defined as

$$I_{uz} = -\int_m uz\, dm. \tag{7.19}$$

The position coordinates of the mass element dm are

$$u = \rho_u + \delta_u \tag{7.20}$$

$$z = \rho_z + \delta_z \tag{7.21}$$

where ρ_u and ρ_z are the position of the mass element dm in an undeformed *phantom* gyro and δ_u and δ_z are the displacements of the mass element due to the gyro's deformation. The axes u and z are principal axes of the phantom gyro originating at the mass center. Now from equations (7.19), (7.20), and (7.21),

$$I_{uz} = -\int \rho_u \rho_z\, dm - \int \rho_u \delta_z\, dm - \int \rho_z \delta_u\, dm - \int \delta_u \delta_z\, dm. \tag{7.22}$$

For a typical solid the displacements δ_u and δ_z are small compared to ρ_u and ρ_z, such that the product $\delta_u \delta_z$ becomes negligible. Furthermore $\int_m \rho_u \rho_z\, dm = 0$ because u and v are principal axes of the phantom body. That leaves

$$I_{uz} = -\int_m \rho_u \delta_z\, dm - \int_m \rho_z \delta_u\, dm. \tag{7.23a}$$

By the same token (Figures 7.1 and 7.2)

$$I_{vz} = -\int_m \rho_v \delta_z\, dm - \int_m \rho_z \delta_v\, dm \tag{7.23b}$$

while

$$I_{uv} = 0 \tag{7.23c}$$

because of axisymmetry.

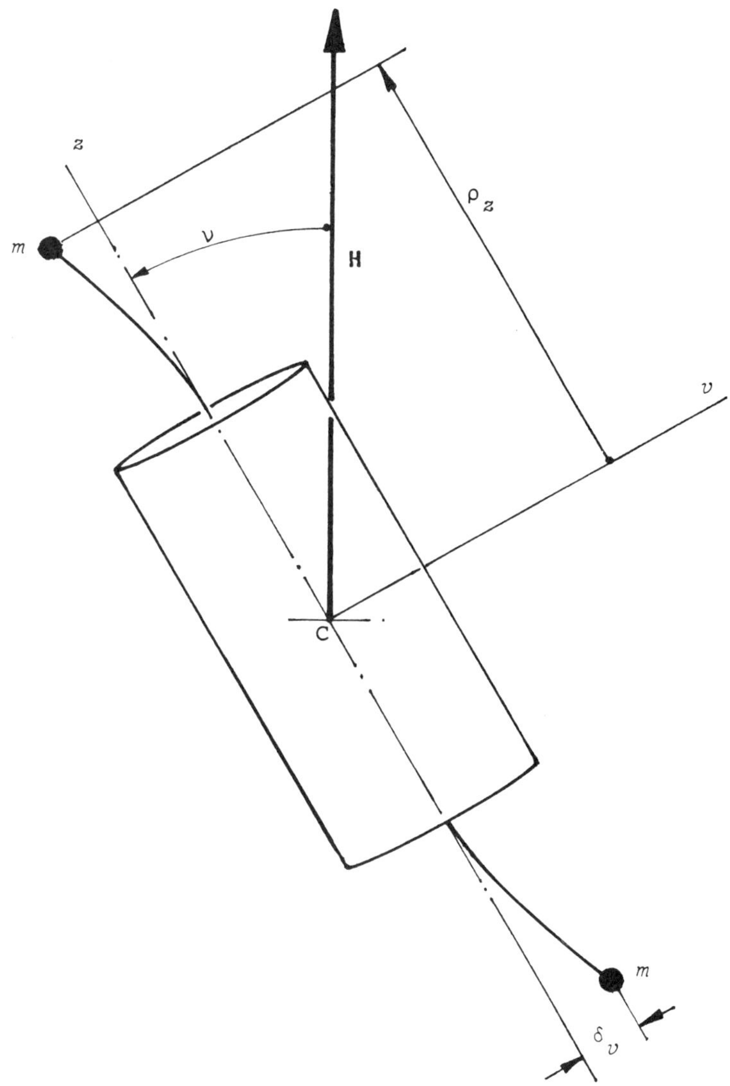

FIGURE 7.1. Model for a gyro exhibiting a partial inertia product J_{zv}.

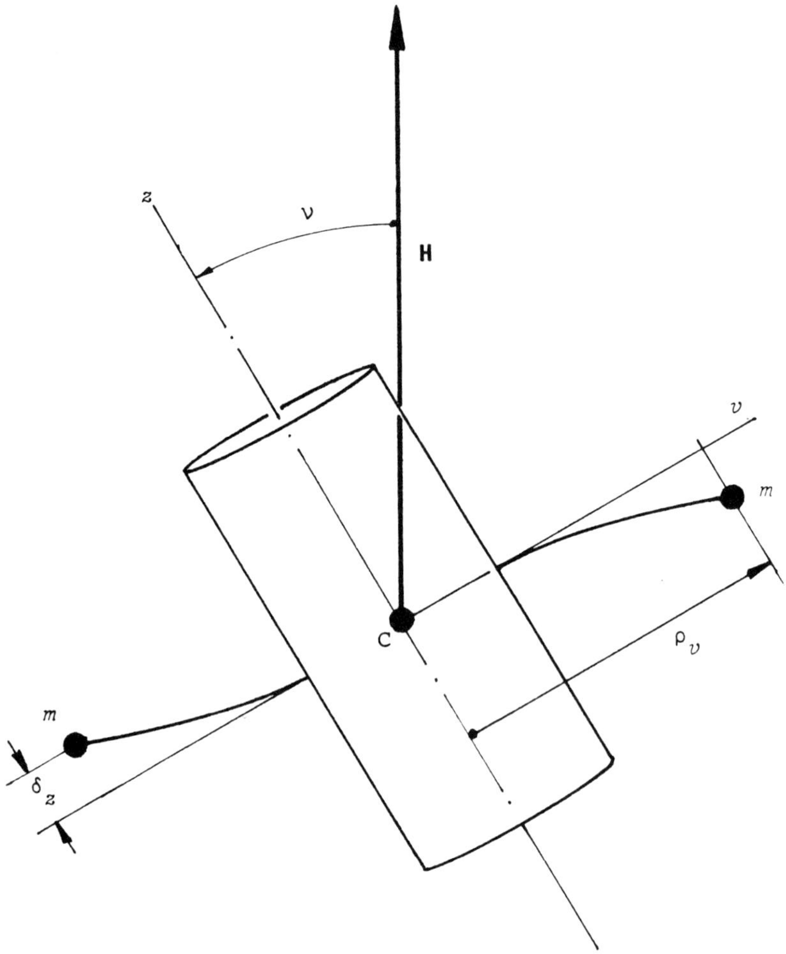

FIGURE 7.2. Model for a gyro exhibiting a partial inertia product J_{vz}.

For subsequent convenience we now introduce auxiliary expressions, which we shall call *partial inertial products*

$$J_{uz} = -\int_m \rho_u \delta_z dm \tag{7.24a}$$

and

$$J_{zu} = -\int_m \rho_z \delta_u dm \tag{7.24b}$$

as well as the *modified inertia product*

$$Y_{uz} = \left(1 + \frac{\dot{\sigma}}{\omega_z}\right) J_{uz} + \left(1 - \frac{\dot{\sigma}}{\omega_z}\right) J_{zu}. \tag{7.24c}$$

The inertia product I_{uz} is consequently composed of two partial inertia products such that

$$I_{uz} = J_{uz} + J_{zu}. \tag{7.24d}$$

Correspondingly there are (Figure 7.3)

$$J_{vz} = -\int_m \rho_v \delta_z dm \tag{7.25a}$$

$$J_{zv} = -\int_m \rho_z \delta_v dm \tag{7.25b}$$

$$Y_{vz} = \left(1 + \frac{\dot{\sigma}}{\omega_z}\right) J_{vz} + \left(1 - \frac{\dot{\sigma}}{\omega_z}\right) J_{zv} \tag{7.25c}$$

$$I_{vz} = J_{vz} + J_{zv}. \tag{7.25d}$$

7.6 Angular Momentum of a Deforming Body

The angular momentum $\mathbf{H}$ of a body of mass m is defined by

$$\mathbf{H} = \int_m \mathbf{r} \times \dot{\mathbf{r}}\, dm. \tag{7.26}$$

Any mass elements dm that is located at a position $\boldsymbol{\rho}$ in a solid body in its undeformed state will have been displaced by a (small) distance $\boldsymbol{\delta}$ in the deformed state, such that

$$\mathbf{r} = \boldsymbol{\rho} + \boldsymbol{\delta}. \tag{7.27}$$

The angular velocity of the body (more accurately, of the $Cxyz$ principal coordinate system fixed to the phantom (or shadow) body, i.e., the undeformed state of the solid) is $\boldsymbol{\omega}$. The angular velocity of the floating

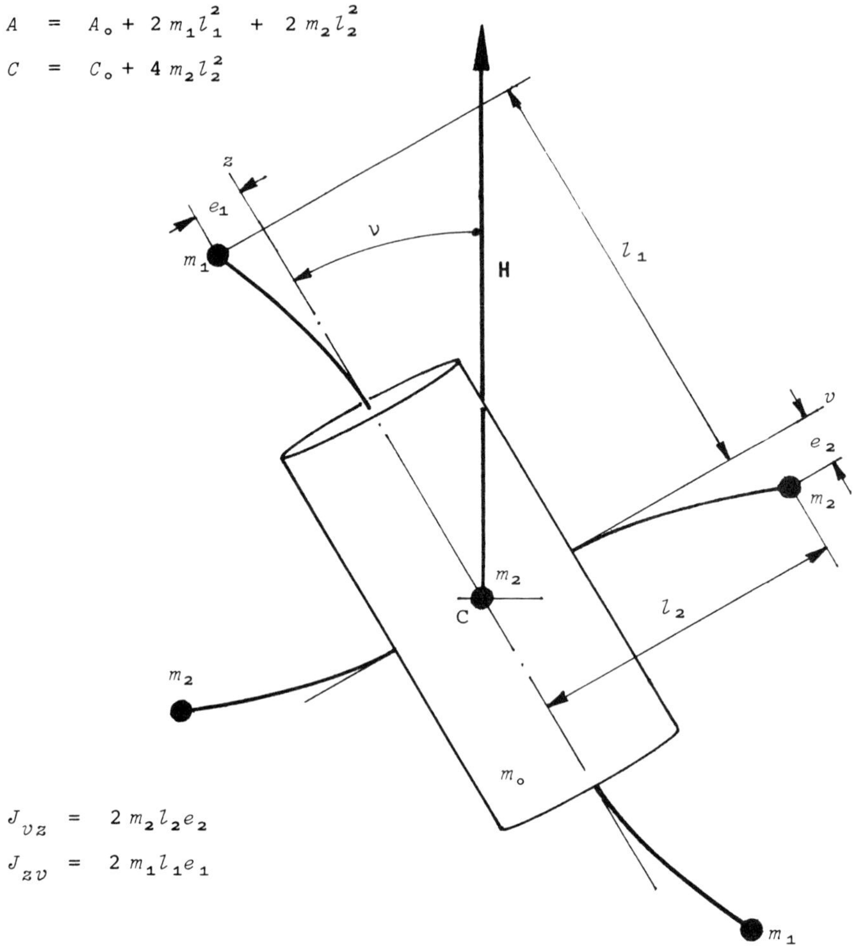

FIGURE 7.3. Model of an originally axisymmetric gyro with deformations leading to partial inertia products.

$Cuvz$ principal coordinate system of the phantom body is $\boldsymbol{\Omega}$. Thus for the velocity

$$\dot{\mathbf{r}} = \overset{\circ}{\mathbf{r}} + \boldsymbol{\Omega} \times \mathbf{r} \tag{7.28}$$

$$\dot{\mathbf{r}} = \overset{\circ}{\boldsymbol{\rho}} + \overset{\circ}{\boldsymbol{\delta}} + \boldsymbol{\Omega} \times (\boldsymbol{\rho} + \boldsymbol{\delta}). \tag{7.29}$$

When an axisymmetric phantom body is involved, then

$$\overset{\circ}{\boldsymbol{\rho}} = \dot{\sigma}\mathbf{e}_z \times \boldsymbol{\rho} \tag{7.30}$$

and the angular momentum can be written

$$\begin{aligned}\mathbf{H} = &\int \boldsymbol{\rho} \times (\boldsymbol{\Omega} \times \boldsymbol{\rho})dm + \int \boldsymbol{\rho} \times (\boldsymbol{\Omega} \times \boldsymbol{\delta})dm \\ &+ \int \boldsymbol{\delta} \times (\boldsymbol{\Omega} \times \boldsymbol{\rho})dm + \int \boldsymbol{\delta} \times (\boldsymbol{\Omega} \times \boldsymbol{\delta})dm \\ &+ \int \boldsymbol{\rho} \times (\dot{\sigma}\mathbf{e}_z \times \boldsymbol{\rho})dm + \int \boldsymbol{\delta} \times (\dot{\sigma}\mathbf{e}_z \times \boldsymbol{\rho})dm \\ &+ \int \boldsymbol{\rho} \times \overset{\circ}{\boldsymbol{\delta}}\, dm + \int \boldsymbol{\delta} \times \overset{\circ}{\boldsymbol{\delta}}\, dm.\end{aligned} \tag{7.31a}$$

The fourth and eighth terms are negligible. The seventh term $\int \boldsymbol{\rho} \times \overset{\circ}{\boldsymbol{\delta}}\, dm = \int \boldsymbol{\delta} \times \overset{\circ}{\boldsymbol{\rho}}\, dm$ for bodies of constant configuration, i.e., $\int \boldsymbol{\rho} \times \boldsymbol{\delta} dm =$ constant, in the floating $Cuvz$ coordinate system selected, which is thus neither a Tisserand frame nor a Buckens frame (Canavin and Likins, 1977). Since $\boldsymbol{\omega} = \boldsymbol{\Omega} + \dot{\sigma}\mathbf{e}_z$, the first and fifth terms can be combined, and so can the third, sixth, and seventh terms, giving eventually

$$\mathbf{H} = \int \boldsymbol{\rho} \times (\boldsymbol{\omega} \times \boldsymbol{\rho})dm + \int \boldsymbol{\delta} \times ((2\boldsymbol{\omega} - \boldsymbol{\Omega}) \times \boldsymbol{\rho})dm + \int \boldsymbol{\rho} \times (\boldsymbol{\Omega} \times \boldsymbol{\delta})dm. \tag{7.31b}$$

Equation (7.31b) contains the phantom body's angular momentum plus two correction terms associated with the phantom body's angular velocity $\boldsymbol{\omega}$ and the floating coordinate system's angular velocity $\boldsymbol{\Omega}$.

Equation (7.31b) can be brought into the following form

$$\begin{aligned}\mathbf{H} = &[\mathbf{e}_u\ \mathbf{e}_v\ \mathbf{e}_z]\begin{bmatrix} A & 0 & 0 \\ 0 & A & 0 \\ 0 & 0 & C \end{bmatrix}\begin{bmatrix} \omega_u \\ \omega_v \\ \omega_z \end{bmatrix} \\ &+ [\mathbf{e}_u\ \mathbf{e}_v\ \mathbf{e}_z]\begin{bmatrix} J_{uu} & J_{uv} & J_{uz} \\ J_{vu} & J_{vv} & J_{vz} \\ J_{zu} & J_{zv} & J_{zz} \end{bmatrix}\begin{bmatrix} \omega_u \\ \omega_v \\ \omega_z + \dot{\sigma} \end{bmatrix} \\ &+ [\mathbf{e}_u\ \mathbf{e}_v\ \mathbf{e}_z]\begin{bmatrix} J_{uu} & J_{vu} & J_{zu} \\ J_{uv} & J_{vv} & J_{zv} \\ J_{uz} & J_{vz} & J_{zz} \end{bmatrix}\begin{bmatrix} \omega_u \\ \omega_v \\ \omega_z - \dot{\sigma} \end{bmatrix}\end{aligned} \tag{7.31c}$$

or, more compactly,

$$\mathbf{H} = [\mathbf{e}_u\ \mathbf{e}_v\ \mathbf{e}_z]([I]\{\omega\} + [J]\{2\omega - \Omega\} + [J]^T\{\Omega\}). \tag{7.31d}$$

Note that the matrix $[J]$ is not symmetrical.

The definitions (7.24) and (7.25) apply, as well as

$$J_{uv} = -\int \rho_u \delta_u dm \tag{7.32a}$$

$$J_{vu} = -\int \rho_v \delta_u dm \tag{7.32b}$$

$$I_{uv} = -\int (\rho_u \delta_v + \rho_v \delta_u) dm = J_{uv} + J_{vu} \tag{7.32c}$$

$$J_{uu} = \int (\rho_v \delta_v + \rho_z \delta_z) dm = \frac{1}{2}\Delta A \tag{7.33a}$$

$$J_{vv} = \int (\rho_z \delta_z + \rho_u \delta_u) dm = \frac{1}{2}\Delta B \tag{7.33b}$$

$$J_{zz} = \int (\rho_u \delta_u + \rho_v \delta_v) dm = \frac{1}{2}\Delta C. \tag{7.33c}$$

Because of their insignificance with respect to A and C, the quantities ΔA, ΔB, and ΔC can typically be neglected.

Since $\Omega_u = \omega_u$, $\Omega_v = \omega_v$, and since $I_{uv} = J_{uv} + J_{vu} = 0$ because of axisymmetry, equation (7.31) can also be written

$$\mathbf{H} = [\mathbf{e}_u\ \mathbf{e}_v\ \mathbf{e}_z] \begin{bmatrix} A & 0 & Y_{uz} \\ 0 & A & Y_{vz} \\ I_{uz} & I_{vz} & C \end{bmatrix} \begin{bmatrix} \omega_u \\ \omega_v \\ \omega_z \end{bmatrix} \tag{7.34}$$

with Y_{uz} and Y_{vz} as given by equations (7.24c) and (7.25c).

For the special case that $\delta_z = 0$, there results $J_{uz} = J_{vz} = 0$ and consequently

$$\mathbf{H} = [\mathbf{e}_u\ \mathbf{e}_v\ \mathbf{e}_z] \begin{bmatrix} A & 0 & \frac{\Omega_z}{\omega_z} I_{uz} \\ 0 & A & \frac{\Omega_z}{\omega_z} I_{vz} \\ I_{uz} & I_{vz} & C \end{bmatrix} \begin{bmatrix} \omega_u \\ \omega_v \\ \omega_z \end{bmatrix}. \tag{7.35}$$

Equation (7.35) would, for instance, be applicable for a gyro consisting of a rigid bus body, with flexible longitudinal antennas such as shown in Figure 7.1.

For the special case that $\delta_u = \delta_v = 0$, there results $J_{zu} = J_{zv} = 0$, and equation (7.31) can be brought into the form

$$\mathbf{H} = [\mathbf{e}_u\ \mathbf{e}_v\ \mathbf{e}_z] \begin{bmatrix} A & 0 & \left(1 + \frac{\dot{\sigma}}{\omega_z}\right) I_{uz} \\ A & 0 & \left(1 + \frac{\dot{\sigma}}{\omega_z}\right) I_{vz} \\ I_{uz} & I_{vz} & C \end{bmatrix} \begin{bmatrix} \omega_u \\ \omega_v \\ \omega_z \end{bmatrix}. \tag{7.36}$$

Equation (7.34), rewritten, gives for the angular momentum of a deforming body

$$\mathbf{H} = [\mathbf{e}_u \ \mathbf{e}_v \ \mathbf{e}_z] \begin{bmatrix} A\omega_u + Y_{uz}\omega_z \\ A\omega_v + Y_{vz}\omega_z \\ C\omega_z + I_{uz}\omega_u + I_{vz}\omega_v \end{bmatrix}. \tag{7.37}$$

In practical applications the term $I_{uz}\omega_u$ is small of second order, and negligible. Furthermore the terms with Y_{vz} are small of first order, and lead to minor adjustments only. They can typically also be neglected. This leaves in effect

$$\mathbf{H} = [\mathbf{e}_u \ \mathbf{e}_v \ \mathbf{e}_z] \begin{bmatrix} A\omega_u + Y_{uz}\omega_z \\ A\omega_v \\ C\omega_z \end{bmatrix} \tag{7.38}$$

for the angular momentum of a deforming axisymmetric gyro for secular attitude drift studies.

As an example consider an oblate gyro with $C = 300$ m^{2}kg and $A = 200$ m^{2}kg, $J_{uz} = 0.01$ m^{2}kg, $J_{zu} = 0.02$ m^{2}kg, $\omega_u = -0.007$ rad/s, $\omega_v = 30$ rad/s, and $\omega_z = 40$ rad/s. (a) Assume ΔA, ΔB, ΔC, J_{vz}, and J_{zv} are negligible. (b) Repeat the problem with $\Delta A = \Delta B = \Delta C = J_{zv} = 0.01$ m^{2}kg, and $J_{vz} = 0.005$ m^{2}kg.

Since $\Omega_u = \omega_u$, $\Omega_v = \omega_v$, $\Omega_z = (C/A)\omega_z$, and $\dot{\sigma} = ((A - C)/A)\omega_z$ the angular momentum, from equation (7.31), is for case (a)

$$\mathbf{H} = [\mathbf{e}_u \ \mathbf{e}_v \ \mathbf{e}_z] \begin{bmatrix} -200(0.007) + 0.01(20) + 0.02\frac{300}{200}40 \\ 200(30) \\ 300(40) \end{bmatrix}$$

$$= [\mathbf{e}_u \ \mathbf{e}_v \ \mathbf{e}_z] \begin{bmatrix} 0 \\ 6000 \\ 12000 \end{bmatrix} \mathrm{Ws}^2$$

and for case (b)

$$\mathbf{H} = [\mathbf{e}_u \ \mathbf{e}_v \ \mathbf{e}_z] \begin{bmatrix} -0.00007 \\ 6001 \\ 12000.85 \end{bmatrix} \mathrm{Ws}^2.$$

Tisserand (1889) suggests the use of a floating coordinate system similar to the $Cuvz$ system employed here. For a purely elastic gyro, e.g., a linearized *Tisserand-type frame* would be at a (small) angle removed from the $Cuvz$ frame. The use of a Tisserand-type frame with the coordinates u', v', and z', leads to

$$J_{v'z'} = J_{z'v'}$$

and does away with the need to use partial inertia products and modified inertia products, since $Y_{v'z'} = I_{v'z'}$, etc. In particular, the equation for the angular momentum becomes a simple

$$\mathbf{H} = [\mathbf{e}_{u'}\ \mathbf{e}_{v'}\ \mathbf{e}_{z'}] \begin{bmatrix} A & 0 & 0 \\ 0 & A & I_{v'z'} \\ 0 & I_{v'z'} & C \end{bmatrix} \begin{bmatrix} 0 \\ \omega'_v \\ \omega'_z \end{bmatrix}.$$

On the other hand, Tisserand coordinates obscure the actual deformation situation, and are not suitable for gyrostats (see chapters 11 and 12) or constrained gyros (section 7.9). For a Tisserand frame when both elasticity and internal damping are present, see Chesser and Rimrott (1988).

7.7 Kinetic Energy of Rotation of a Deforming Body

The kinetic energy T of rotation of a body of mass m is defined by

$$T = \int_m \dot{\mathbf{r}} \cdot \dot{\mathbf{r}}\, dm \tag{7.39}$$

with equation (7.27):

$$\mathbf{r} = \boldsymbol{\rho} + \boldsymbol{\delta}.$$

For an axisymmetric body $I_{uv} = 0$ and $\overset{\circ}{\boldsymbol{\rho}} = \dot{\sigma}\mathbf{e}_z \times \boldsymbol{\rho}$. With $\int \boldsymbol{\rho} \times \overset{\circ}{\boldsymbol{\delta}}\, dm = \int \boldsymbol{\delta} \times \overset{\circ}{\boldsymbol{\rho}}\, dm$, the expression for the kinetic energy of rotation becomes

$$T = \frac{1}{2}((A + \Delta A)\omega_u^2 + (A + \Delta B)\omega_v^2 + (C + \Delta C)\omega_z^2 + 2J_{uz}\omega_u(\omega_z + \dot{\sigma}) + 2J_{vz}\omega_v(\omega_z + \dot{\sigma}) + 2J_{zu}\omega_u(\omega_z - \dot{\sigma}) + 2J_{zv}\omega_v(\omega_z - \dot{\sigma})). \tag{7.40}$$

If ΔA, ΔB, and ΔC are neglected, then

$$T = \frac{1}{2}(A\omega_u^2 + A\omega_v^2 + C\omega_z^2 + 2J_{uz}\omega_u(\omega_z + \dot{\sigma}) + 2J_{vz}\omega_v(\omega_z + \dot{\sigma}) + 2J_{zu}\omega_u(\omega_z - \dot{\sigma}) + 2J_{zv}\omega_v(\omega_z - \dot{\sigma})) \tag{7.41}$$

or

$$T = \frac{1}{2}(A\omega_u^2 + A\omega_v^2 + C\omega_z^2 + 2Y_{uz}\omega_u\omega_z + 2Y_{vz}\omega_v\omega_z) \tag{7.42}$$

with Y_{uz} and Y_{vz} as given by equations (7.24c) and (7.25c).

For the special case of equation (7.35) where $J_{uz} = J_{vz} = 0$,

$$T = \frac{1}{2}(A\omega_u^2 + A\omega_v^2 + C\omega_z^2 + 2I_{uz}\omega_u\Omega_z + 2I_{vz}\omega_v\Omega_z). \tag{7.43}$$

For the special case of equation (7.36) where $J_{zu} = J_{zv} = 0$,

$$T = \frac{1}{2}(A\omega_u^2 + A\omega_v^2 + C\omega_z^2 + 2I_{uz}\omega_u(\omega_z + \dot{\sigma}) + 2I_{vz}\omega_v(\omega_z + \dot{\sigma})). \tag{7.44}$$

In practical applications the terms $A\omega_u^2$ and $Y_{uz}\omega_u\omega_z$ in equation (7.42) are small of second order, and negligible. Furthermore the term $Y_{vz}\omega_v\omega_z$ is small of first order, and introduces only minor adjustments, which can typically also be neglected. This leaves

$$T = \frac{1}{2}(A\omega_v^2 + C\omega_z^2) \tag{7.45}$$

i.e., an expression for the kinetic energy that does not differ from that of the rigid body case. For a situation in which Y_{vz} cannot be neglected, see section 8.5, etc.

As an example, consider an oblate gyro with $C = 300$ m^{2}kg, $A = 200$ m^{2}kg, $J_{uz} = 0.01$ m^{2}kg, $J_{zu} = 0.02$ m^{2}kg, $\omega_u = -0.007$ rad/s, $\omega_v = 30$ rad/s, and $\omega_z = 40$ rad/s. (a) Assume ΔA, ΔB, ΔC, J_{vz}, and J_{vz} are negligible. (b) Repeat the problem with $\Delta A = \Delta B = \Delta C = J_{vz} = J_{zv} =$ 0.01 m^{2}kg. (c) Repeat the problem with everything neglected, except $A\omega_v^2$ and $C\omega_z^2$.

From equation (7.40), for case (a),

$$T = \frac{1}{2}\left(200(0.007^2) + 200(30^2) + 300(40^2) - 2(0.035)0.0007(40)\right)$$

$$= 329\,999.99 \text{ J}$$

for case (b),

$$T = 320\ 999.99 + 33.5 = 330\ 033.49 \text{ J}$$

and for case (c),

$$T = 300\ 000 \text{ J}.$$

7.8 Energy Rates

In general the following relation must hold for a mechanical gyro

$$\dot{U} = \dot{T} + \dot{V} + \dot{D} \tag{7.46}$$

where $\dot{U}$ = rate of work done by externally applied forces (including weight forces); $\dot{T}$ = rate of change of kinetic energy of rotation; $\dot{V}$ = rate of change of elastic energy; and $\dot{D}$ = rate of energy dissipation.

The elastic energy V is typically small compared to the kinetic energy T of rotation. Let us conclude therefore that $\dot{V} << \dot{T}$ as well, and that $\dot{V}$ can be neglected. That leaves

$$\dot{U} = \dot{T} + \dot{D}. \tag{7.47}$$

The rate of external work fed into a dissipative gyro equals the kinetic energy rate plus the energy dissipative rate. In agreement with the collinearity theorems, the cross product $\boldsymbol{\omega} \times \mathbf{H}$ is zero when angular velocity $\boldsymbol{\omega}$ and angular momentum $\mathbf{H}$ are collinear. It is nonzero when they are noncollinear. The cross product $\boldsymbol{\omega} \times \mathbf{H}$ can be looked upon as an internal torque, which rotates at an angular velocity $\boldsymbol{\Omega}$. The power $(\boldsymbol{\omega} \times \mathbf{H}) \cdot \boldsymbol{\Omega}$ then is the energy dissipation rate $\dot{D}$.

According to Euler's law

$$\mathbf{M} = \dot{\mathbf{H}}.$$

If

$$\dot{\mathbf{H}} = \overset{\circ}{\mathbf{H}} + \boldsymbol{\Omega} \times \mathbf{H}$$

then

$$\mathbf{M} = \overset{\circ}{\mathbf{H}} + \boldsymbol{\Omega} \times \mathbf{H}. \tag{7.48}$$

The formation of an inner product with the angular velocity $\boldsymbol{\omega}$ leads to

$$\mathbf{M} \cdot \boldsymbol{\omega} = \overset{\circ}{\mathbf{H}} \cdot \boldsymbol{\omega} + (\boldsymbol{\Omega} \times \mathbf{H}) \cdot \boldsymbol{\omega}. \tag{7.49}$$

If the conditions enumerated here apply, then each term of equation (7.49) has a specific meaning, i.e.,

$$\mathbf{M} \cdot \boldsymbol{\omega} = \dot{U} \tag{7.50}$$

$$\overset{\circ}{\mathbf{H}} \cdot \boldsymbol{\omega} = \dot{T} \tag{7.51}$$

$$(\boldsymbol{\Omega} \times \mathbf{H}) \cdot \boldsymbol{\omega} = \dot{D}. \tag{7.52}$$

7.9 Application to a Constrained System with $\delta_z = 0$

Let a simple example serve to illustrate some of the concepts introduced in this chapter. In Figure 7.4 a mass point m is shown mounted on a flexible shaft. The whole assembly rotates at a (constant) angular velocity $\boldsymbol{\Omega}$. The massless shaft upon which the mass point m is mounted exhibits elastic as well as viscous properties, expressed by k and c.

We neglect any influence of the (small) deflection upon the centrifugal force

$$m\Omega^2 l \sin \nu$$

and consider its cosine component only

$$F = m\Omega^2 l \sin \nu \cos \nu. \tag{7.53}$$

Because of the viscous damping present, the mass m will be taken along through a phase angle ϕ. For the phase angle ϕ,

$$\tan \phi = \frac{c\dot{\sigma}}{k}. \tag{7.54}$$

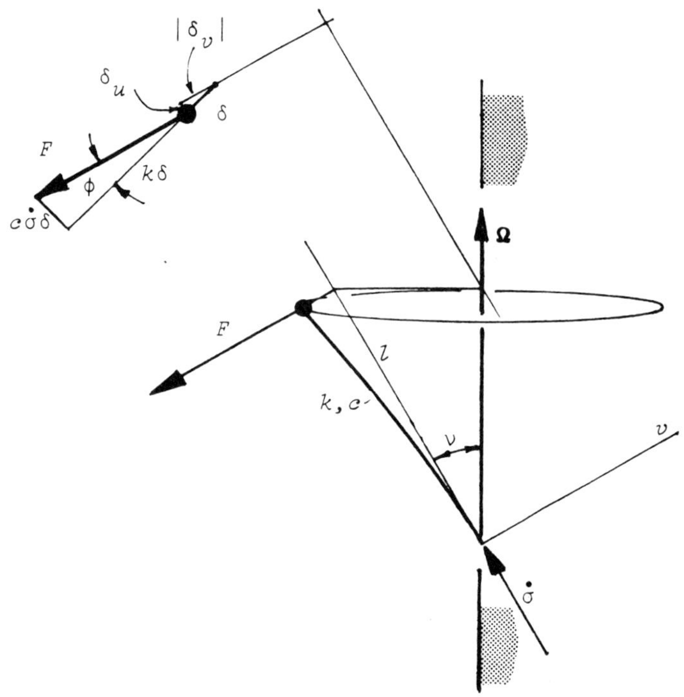

FIGURE 7.4. Small displacements due to centrifugal force.

For the (small) deflection δ_u and the deflection δ_v

$$\frac{\delta_u}{\delta_v} = -\tan\phi. \tag{7.55}$$

With $\delta_v = -\delta\cos\phi$ and $F^2 = (k^2 + c^2\dot{\sigma}^2)\delta^2$

$$\delta = \frac{m\Omega^2 l \sin\nu\cos\nu}{\sqrt{k^2 + c^2\dot{\sigma}^2}} \tag{7.56}$$

one obtains

$$\delta_v = -\frac{m\Omega^2 l \sin\nu\cos\nu}{\sqrt{k^2 + c^2\dot{\sigma}^2}}\cos\phi \tag{7.57}$$

and since

$$\cos\phi = \frac{1}{\sqrt{1+\tan^2\phi}} = \frac{1}{\sqrt{1+(c^2\dot{\sigma}^2)/k^2}} \tag{7.58}$$

$$\delta_v = -\frac{m\Omega^2 l \sin\nu\cos\nu}{k\left(1+\frac{c^2\dot{\sigma}^2}{k^2}\right)}. \tag{7.59}$$

Following a similar approach, with

$$\sin\phi = \frac{\tan\phi}{\sqrt{1+\tan^2\phi}} = \frac{c\dot{\sigma}/k}{\sqrt{1+(c^2\dot{\sigma}^2)/k^2}} \tag{7.60}$$

one obtains

$$\delta_u = \frac{m\Omega^2 l \sin\nu\cos\nu}{k\left(1+\frac{c^2\dot{\sigma}^2}{k^2}\right)}\frac{c\dot{\sigma}}{k}. \tag{7.61}$$

For the partial inertia product

$$J_{zv} = -\int \rho_z\delta_v\,dm = -ml\delta_v \tag{7.62}$$

one obtains, with $\Omega_v = \Omega\sin\nu$ and $\Omega_z = \Omega\cos\nu$,

$$J_{zv} = \alpha\Omega_v\Omega_z \tag{7.63}$$

with an elastic deformation coefficient

$$\alpha = \frac{m^2 l^2}{k\left(1+\frac{c^2\dot{\sigma}^2}{k^2}\right)}. \tag{7.64}$$

For the partial inertia product

$$J_{zu} = -\int \rho_z\delta_u\,dm = -ml\delta_u \tag{7.65}$$

one obtains

$$J_{zu} = -\beta\Omega_v\Omega_z\dot{\sigma} \tag{7.66}$$

with an internal energy dissipation coefficient

$$\beta = \frac{m^2 l^2 c}{k^2 \left(1 + \frac{c^2 \dot{\sigma}^2}{k^2}\right)}. \tag{7.67}$$

Introducing $\dot{\sigma}/\omega_n = r$, $\omega_n = \sqrt{k/m}$, and $\zeta = c\omega_n/2k$, equations (7.64) and (7.67) become

$$\alpha = \frac{m^2 l^2}{k(1 + (2\zeta r)^2)} \tag{7.68}$$

$$\beta = \frac{\alpha c}{k} = \frac{2\zeta\alpha}{\omega_n} = \frac{2\zeta m^2 l^2}{k\omega_n(1 + (2\zeta r)^2)}. \tag{7.69}$$

Note that because of $\delta_z = 0$

$$J_{vz} = J_{uz} = 0$$

for the example of Figure 7.4.

As a numerical example consider a torquefree gyro as shown in Figure 7.1, with $A = 3000$ m^{2}kg, $C = 1000$ m^{2}kg, $k = 1000$ N/m, $c = 20$ Ns/m, $\nu = 30°$, $H = 12000$ Ws^2, $m = 0.5$ kg, and $l = 2$ m.

The velocity components are

$$\Omega_v = \omega_v = \frac{H}{A}\sin\nu = 2 \text{ rad/s}$$

$$\omega_z = \frac{H}{C}\cos\nu = 10.392 \text{ rad/s}$$

$$\Omega_z = \frac{H}{A}\cos\nu = 3.464 \text{ rad/s}$$

$$\dot{\sigma} = \frac{A - C}{A}\Omega_z = 6.928 \text{ rad/s}.$$

The inertial force acting on each mass m is

$$F = ml\,\Omega_v\Omega_z = 6.928\ N.$$

The deflections are

$$\delta_v = -\frac{ml\,\Omega_v\Omega_z}{k\left(1 + \frac{c^2\dot{\sigma}^2}{k^2}\right)} = -0.006797 \text{ m} = -6.797 \text{ mm}$$

$$\delta_u = 0.000942 \text{ m} = 0.942 \text{ mm}.$$

The partial inertia products are

$$J_{vz} = 0$$

$$J_{zv} = -2m\delta_v l = 0.013594 \text{ m}^2\text{kg}$$

$$J_{uz} = 0$$

$$J_{zu} = -2m\delta_u l = -0.001884\ \mathrm{m^2kg}.$$

The modified inertia products are

$$Y_{vz} = 0 + \frac{C}{A}(0.013594) = 0.004531\ \mathrm{m^2kg}$$

$$Y_{uz} = 0 + \frac{C}{A}(-0.001884) = -0.000628\ \mathrm{m^2kg}.$$

7.10 Application to a Constrained System with $\delta_u = \delta_v = 0$

In Figure 7.5 a flexible disk gyro of constant radius such that $\delta_u = \delta_v = 0$ is shown (Magnus, 1971). It is rigidly mounted on a rigid shaft system driven at constant speeds by motors 1 and 2. Motor 1 develops a (constant) angular speed ω_v, motor 2 a (constant) angular speed ω_z. The system is constrained in the sense that the motors serve to keep the two angular velocities constant. The disk is thus subject to an angular velocity

$$\boldsymbol{\omega} = [\mathbf{e}_u\ \mathbf{e}_v\ \mathbf{e}_z] \begin{bmatrix} 0 \\ \omega_v \\ \omega_z \end{bmatrix}. \tag{7.70}$$

The floating $Cuvz$ coordinate system used to describe the motion has an angular velocity

$$\boldsymbol{\Omega} = [\mathbf{e}_u\ \mathbf{e}_v\ \mathbf{e}_z] \begin{bmatrix} 0 \\ \omega_v \\ 0 \end{bmatrix} \tag{7.71}$$

such that $\dot{\sigma} = \omega_z - \Omega_z = \omega_z$ in the present case. The phantom body's inertial force field (6.74b) for a solid gyro such as the one shown in Figure 7.5 can be shown to be

$$d\mathbf{f} = [\mathbf{e}_u\ \mathbf{e}_v\ \mathbf{e}_z] \begin{bmatrix} \rho_u \omega^2 dm \\ \rho_v \omega_z^2 dm \\ \rho_z \omega_v^2 dm - 2\rho_v \omega_v \omega_z dm \end{bmatrix}. \tag{7.72}$$

Any real solid disk is made of an elastic material with internal damping. The elastic properties alone will lead to a small I_{vz}, the damping properties to a small I_{uz}. Yet it is only the I_{uz} that affects the dynamics of the gyro profoundly, as will become apparent in the subsequent equations. With

$$[I] = \begin{bmatrix} A & 0 & I_{uz} \\ 0 & A & I_{vz} \\ I_{uz} & I_{vz} & C \end{bmatrix} \tag{7.73}$$

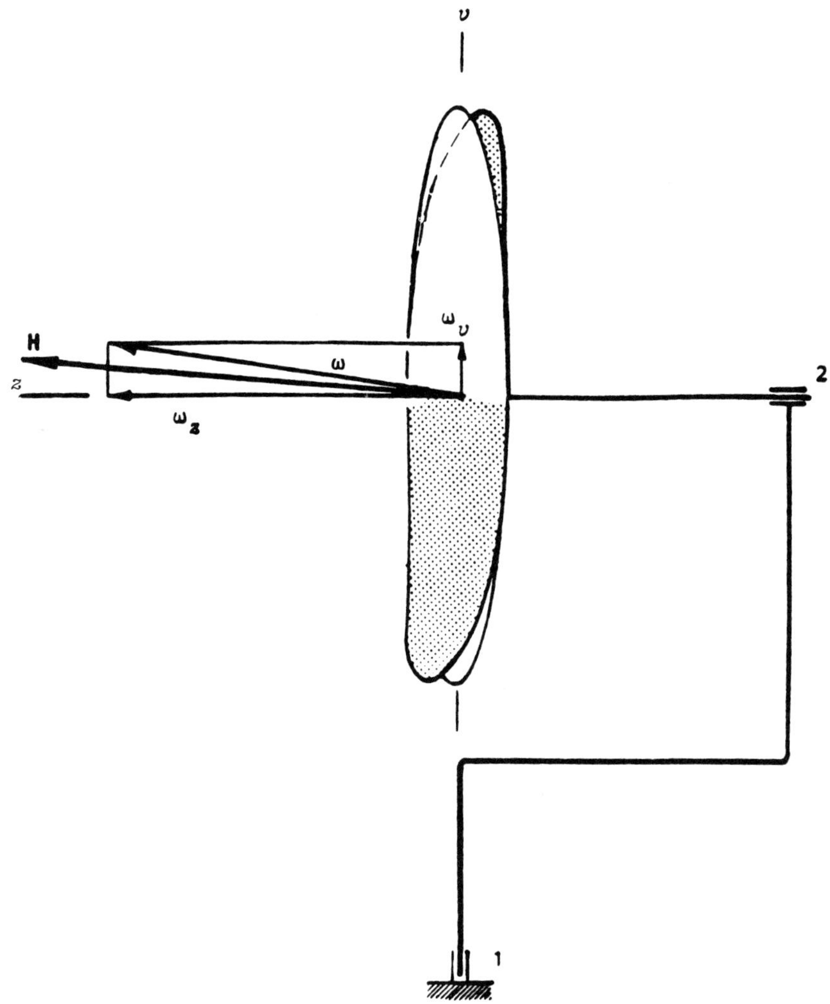

FIGURE 7.5. Deformation of an elastic disk gyro without internal damping.

and the conditions for the applicability of equation (7.36) fulfilled, the gyro's angular momentum becomes

$$\mathbf{H} = [\mathbf{e}_u\ \mathbf{e}_v\ \mathbf{e}_z] \begin{bmatrix} A & 0 & 2I_{uz} \\ 0 & A & 2I_{vz} \\ I_{uz} & I_{vz} & C \end{bmatrix} \begin{bmatrix} 0 \\ \omega_v \\ \omega_z \end{bmatrix} \tag{7.74}$$

$$\mathbf{H} = [\mathbf{e}_u\ \mathbf{e}_v\ \mathbf{e}_z] \begin{bmatrix} 2I_{uz}\omega_z \\ A\omega_v + 2I_{vz}\omega_z \\ C\omega_z + I_{vz}\omega_v \end{bmatrix}. \tag{7.75}$$

The torque transmitted to the gyro is obtained from Euler's law

$$\mathbf{M} = \dot{\mathbf{H}}. \tag{7.76}$$

With

$$\dot{\mathbf{H}} = \overset{\circ}{\mathbf{H}} + \boldsymbol{\Omega} \times \mathbf{H} \tag{7.77}$$

and

$$\overset{\circ}{\mathbf{H}} = 0 \tag{7.78}$$

and

$$\boldsymbol{\Omega} \times \mathbf{H} = \begin{vmatrix} \mathbf{e}_u & \mathbf{e}_v & \mathbf{e}_z \\ 0 & \omega_v & 0 \\ 2I_{uz}\omega_z & A\omega_v + 2I_{vz}\omega_z & C\omega_z + I_{vz}\omega_v \end{vmatrix} \tag{7.79}$$

one obtains

$$\mathbf{M} = [\mathbf{e}_u\ \mathbf{e}_v\ \mathbf{e}_z] \begin{bmatrix} C\omega_v\omega_z + I_{vz}\omega_v^2 \\ 0 \\ -2I_{uz}\omega_v\omega_z \end{bmatrix}. \tag{7.80}$$

The condition $I_{uz} = I_{vz} = 0$ represents the case of a rigid body. The condition $I_{vz} \neq 0$ and $I_{uz} = 0$ would represent the case of a purely elastic body without internal damping. The case $I_{vz} = 0$ and $I_{uz} \neq 0$ would represent a nonelastic body with internal damping. The case $I_{vz} \neq 0$ and $I_{uz} \neq 0$ represents an elastic body with internal damping, i.e., a real solid body.

The power $\dot{U}$ supplied to the gyro is

$$\dot{U} = \mathbf{M} \cdot \boldsymbol{\omega} \tag{7.81}$$

and amounts to

$$\dot{U} = -2I_{uz}\omega_v\omega_z^2. \tag{7.82}$$

Since in the present case all power fed into the system is used to feed the energy dissipation, we may express the energy dissipation rate (which can only be positive) as

$$\dot{D} = (\boldsymbol{\Omega} \times \mathbf{H}) \cdot \boldsymbol{\omega} = -2I_{uz}\omega_v\omega_z^2. \tag{7.83}$$

The kinetic energy stored in the gyro is

$$T = \frac{1}{2}(A\omega_v^2 + C\omega_z^2 + 2I_{vz}\omega_v(\omega_z + \dot{\sigma})). \tag{7.84}$$

The elastic energy stored in the gyro can be shown to be

$$V = \sqrt{I_{uz}^2 + I_{vz}^2}\,\omega_v\omega_z \tag{7.85}$$

such that the total mechanical energy stored in the gyro is

$$T + V = \frac{1}{2}(A\omega_v^2 + C\omega_z^2 + 2I_{vz}\omega_v(\omega_z + \dot{\sigma}) + 2\sqrt{I_{uz}^2 + I_{vz}^2}\,\omega_v\omega_z). \tag{7.86}$$

The kinetic energy rate and the elastic energy rate are both zero, i.e., $\dot{T} = \dot{V} = 0$.

Note that the (elastic) inertia product I_{vz} leads only to minor adjustments compared to a rigid body solution with $I_{vz} = 0$. The (damping) inertia product I_{uz}, on the other hand, leads to novel effects, such as a z-component in the torque (7.80), and the need to have power (7.82) supplied to the system.

In order to obtain the inertia products associated with the deformations we proceed as follows. An inspection of equation (7.72) shows that there are centrifugal forces, e.g., $\rho_u\omega^2 dm$ in the u-direction, $\rho_v\omega_z^2 dm$ in the v-direction, and $\rho_z\omega_v^2 dm$ in the z-direction. They are forces that lead to changes ΔA, ΔB, and ΔC, which are usually of little significance and can be neglected.

That leaves the reversed Coriolis force $-2\rho_v\omega_v\omega_z dm$ in the z-direction, which produces a displacement δ_z directly proportional. Multiplied by the location ρ_v and integrated over the whole body, partial inertia products

$$J_{vz} = -\int_m \rho_v\delta_z dm \tag{7.87}$$

and

$$J_{zv} = -\int \rho_z\delta_v dm = 0 \tag{7.88}$$

are obtained (Figure 7.6). Since $I_{vz} = J_{vz} + J_{zv}$, one may, for this special case, write

$$I_{vz} = 2\alpha\omega_v\omega_z \tag{7.89}$$

where α is an elastic deformation coefficient, measured in Ws^5, which includes all effects other than ω_v and ω_z. It is important to realize that α does not only depend on the gyro body and the material(s) of which it is made, but also the mounting conditions and the proximity of a structural resonance.

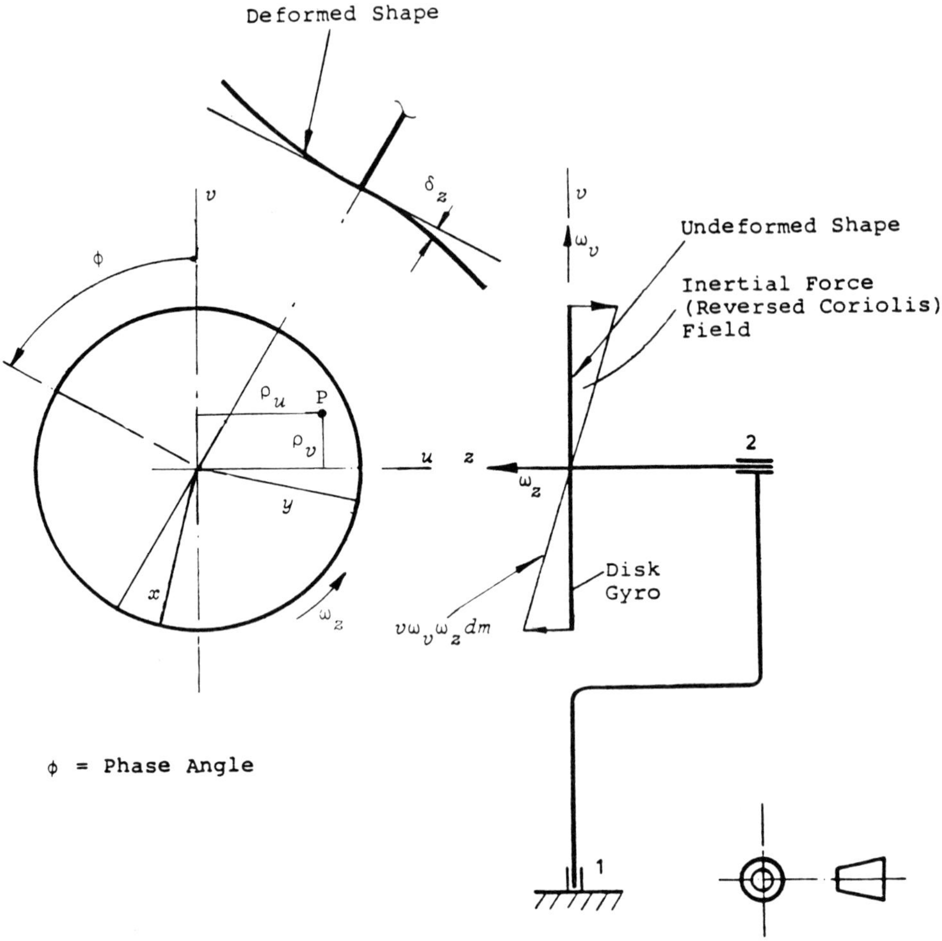

FIGURE 7.6. Deformation of elastic disk gyro with internal damping.

In the vicinity of a structural resonance, with an eigenfrequency of ω_n, and internal damping modeled as viscous, the inertia product I_{vz} becomes (Rimrott and Yu, 1987):

$$I_{vz} = 2\alpha_0 \frac{1 - r^2}{(1 - r^2)^2 + (2\zeta r)^2} \omega_v \omega_z \tag{7.90}$$

such that

$$\alpha = \alpha_0 \frac{1 - r^2}{(1 - r^2)^2 + (2\zeta r)^2} \tag{7.91}$$

with

$$r = \frac{\dot{\sigma}}{\omega_n} = \text{ frequency ratio} \tag{7.92}$$

$$\dot{\sigma} = \omega_z - \Omega_z = \text{ spin speed } (\Omega_z = 0 \text{ in the present example}) \tag{7.93}$$

$$\zeta = \text{ damping ratio.} \tag{7.94}$$

The inertia product I_{vz} changes its sign at resonance, i.e., when $r = 1$. At resonance, the inertia product I_{vz} vanishes. A plot of $I_{vz}/2\alpha_0\omega_v\omega_z$ versus the frequency ratio r is shown in Figure 7.7. Where $\omega_n >> \dot{\sigma}$, then $r^2 <<< 1$, and equation (7.90) assumes the form

$$I_{vz} = 2\alpha_0 \omega_v \omega_z \tag{7.95}$$

i.e., there is then no interaction between structural eigenfrequency ω_n and exciting frequency $\dot{\sigma}$. A similar decoupling takes place when $\omega_n << \dot{\sigma}$, leading to $I_{vz} = 0$ though. Note that at resonance, i.e., when $r = 1$, equation (7.90) becomes

$$I_{vz} = 0. \tag{7.96}$$

Without internal damping, i.e., where $\zeta = 0$, equation (7.90) becomes

$$I_{vz} = 2\alpha_0 \frac{1}{1 - r^2} \omega_v \omega_z . \tag{7.97}$$

Thus I_{vz} is also present in a purely elastic gyro. At resonance, equation (7.97) becomes $I_{vz} \to \pm\infty$.

The presence of internal damping means that the deformation lags the load (Zhang, 1984; Zhang and Ling, 1986) by a phase angle ϕ (Figure 7.6). The displacement δ_z of the mass element dm located at ρ_u, integrated over the whole body mass m, produces partial inertia products

$$J_{uz} = -\int_m \rho_u \delta_z dm \tag{7.98}$$

and, since $\delta_u = 0$,

$$J_{zu} = -\int \rho_z \delta_u dm = 0. \tag{7.99}$$

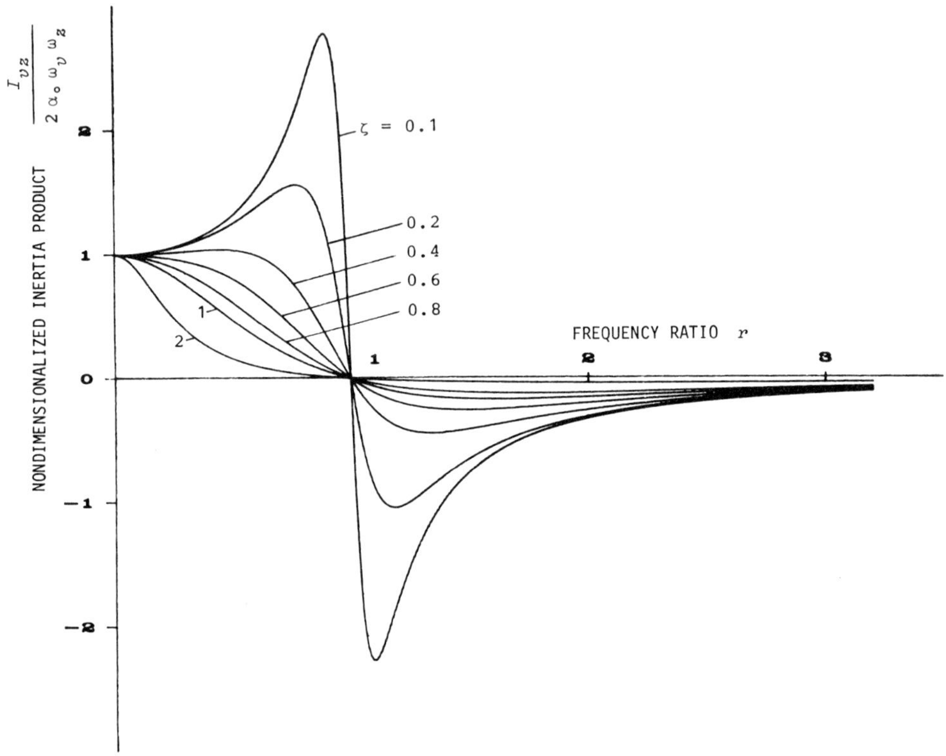

FIGURE 7.7. Elastic inertia product I_{vz} versus frequency ratio.

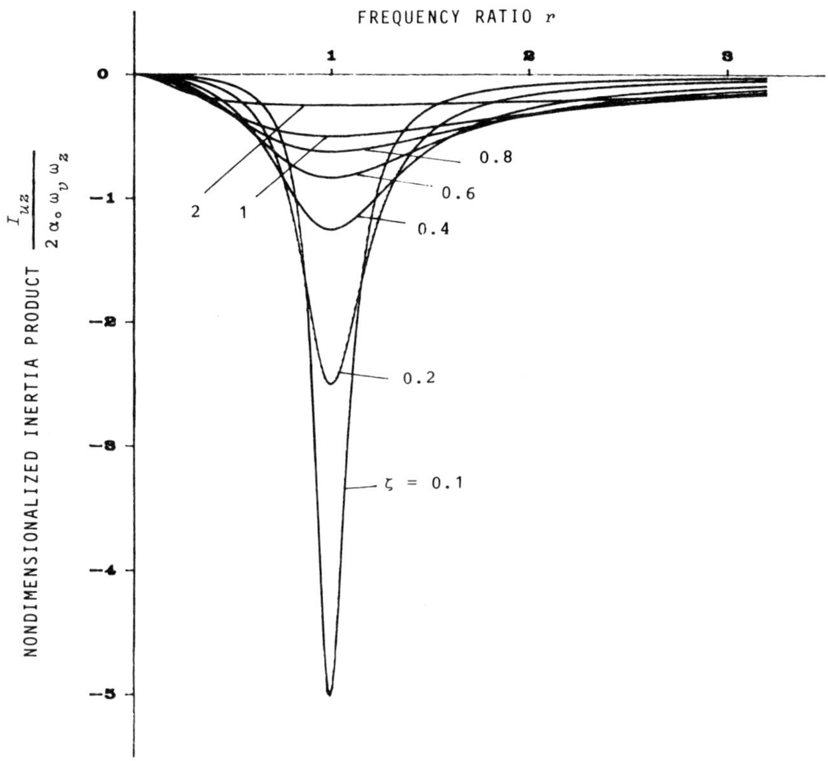

FIGURE 7.8. Damping inertia product I_{uz} versus frequency ratio.

Because

$$I_{uz} = J_{uz} + J_{zu}$$

one can use I_{uz} instead of J_{uz} for the present case.

In the vicinity of a structural resonance with an eigenfrequency of ω_n, and internal damping modeled as viscous, the inertia products can be shown to be related by

$$\frac{I_{uz}}{I_{vz}} = -\tan\phi \tag{7.100}$$

where

$$\tan\phi = \frac{2\zeta r}{1 - r^2} \tag{7.101}$$

is the tangent of the phase lag between the exciting inertial force and the corresponding deformation. Thus

$$I_{uz} = -2\alpha_0 \frac{2\zeta r}{(1 - r^2)^2 + (2\zeta r)^2}\omega_v\omega_z. \tag{7.102}$$

Since $r = \dot{\sigma}/\omega_n$

$$I_{uz} = -\frac{4\alpha_0\zeta}{\omega_n}\frac{1}{(1-r^2)^2 + (2\zeta r)^2}\omega_v\omega_z\dot{\sigma}. \tag{7.103}$$

One can now introduce an internal energy dissipation coefficient β, measured in Ws^6,

$$\beta = \frac{2\alpha_0\zeta}{\omega_n}\frac{1}{(1-r^2)^2 + (2\zeta r)^2} \tag{7.104}$$

such that

$$I_{uz} = -2\beta\omega_v\omega_z\dot{\sigma}. \tag{7.105}$$

When $\omega_n >> \dot{\sigma}$, then $r^2 <<< 1$, and equation (7.103) assumes the form

$$I_{uz} = -2\beta_0\omega_v\omega_z\dot{\sigma} \tag{7.106}$$

with

$$\beta_0 = \frac{2\alpha_0\zeta}{\omega_n} \tag{7.107}$$

and there is no interaction between structural eigenfrequency ω_n and exciting frequency $\dot{\sigma}$. A similar decoupling takes place when $\omega_n << \dot{\sigma}$, leading to $I_{uz} = 0$ though. Note that at resonance, i.e., when $r = 1$, equation (7.103) becomes

$$I_{uz} = -\frac{2\alpha_0\omega_v\omega_z}{\zeta} \tag{7.108}$$

i.e., the inertia product becomes *inversely* proportional to the damping ratio and approaches infinity as the damping ratio approaches zero.

Without internal damping, i.e., when $\zeta = 0$, equation (7.103) becomes

$$I_{uz} = 0. \tag{7.109}$$

Thus I_{uz} is directly associated with internal damping. This circumstance has led to characterizing I_{vz} as the "elastic" inertia product, and I_{uz} as the "(internal) damping" inertia product. It is also interesting to note that the elastic inertial product I_{vz} (equation (7.90)) changes its sign at resonance, while the damping inertia product I_{uz} (equation (7.103)) quite properly does not. A plot of $I_{uz}/2\alpha_0\omega_v\omega_z$ versus the frequency ratio r is shown in Figure 7.8.

An inspection of equations (7.90) and (7.102) indicates that the inertia products I_{vz} and I_{uz} can remain small, if r is small, i.e., exciting frequencies are much smaller than the (lowest) structural eigenfrequency. They can also remain small when damping, i.e., ζ, is present, even at resonance.

For an example, let us pick a two-dimensional dumbbell gyro consisting of four elasto-hysteretic massless arms each of 0.5 m length with four point masses of 0.25 kg each (Figure 7.9). Also let $\omega_v = 2$ rad/s and $\omega_z = 500$

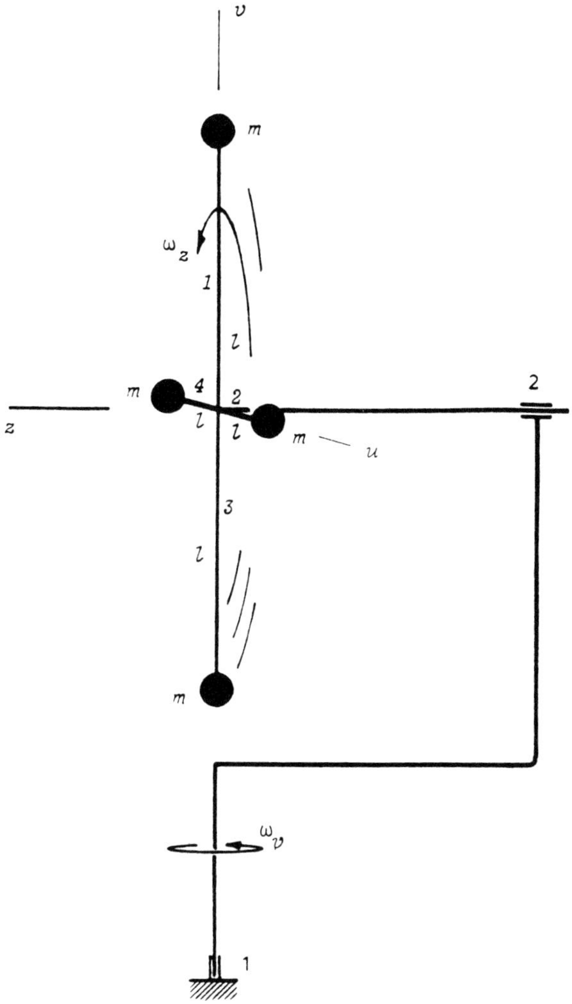

FIGURE 7.9. Model in its undeformed state, used for numerical example.

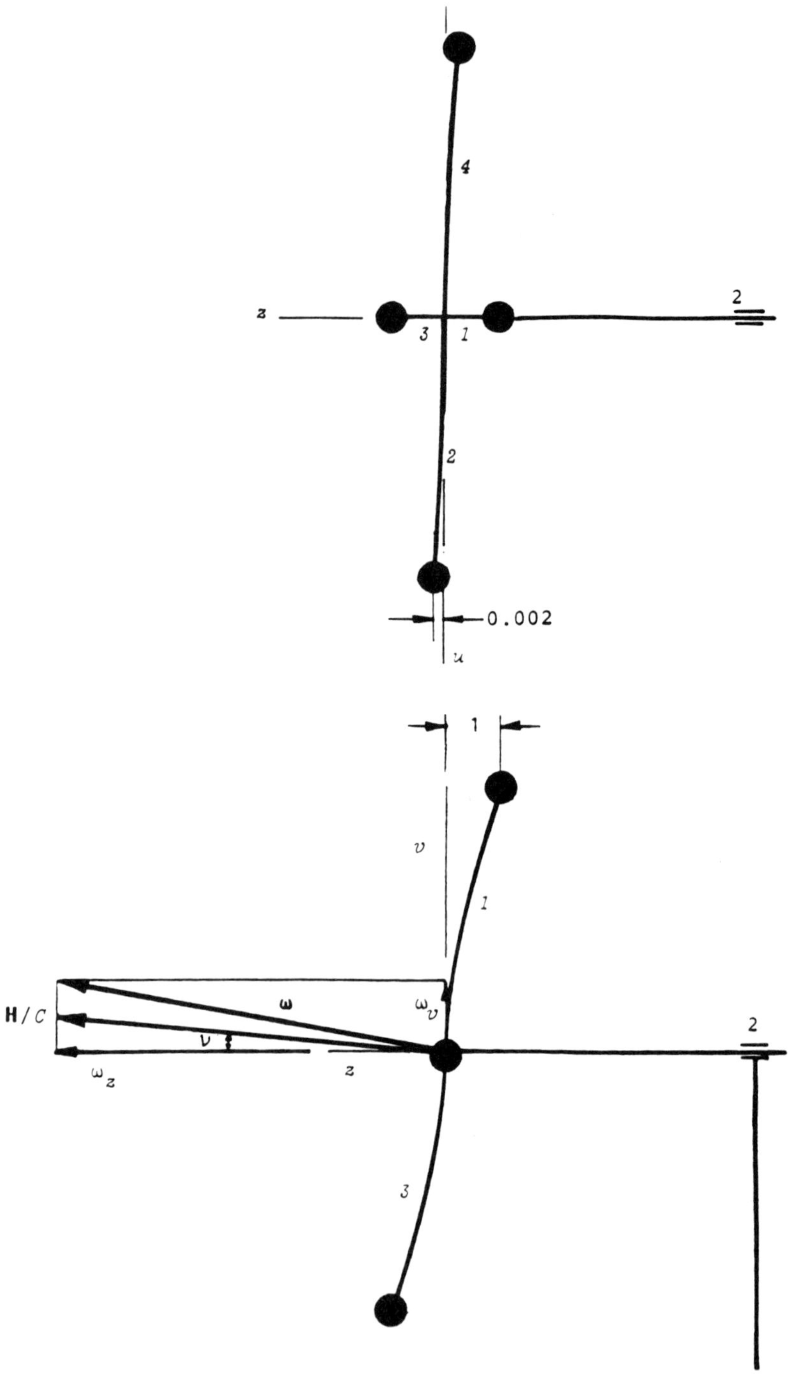

FIGURE 7.10. Displacements leading to I_{vz} and I_{uz}.

rad/s. The (elastic) deformation (of arms 1 and 3) shall be 1 mm and the (damping) deformation (of arms 2 and 4) shall be 0.002 mm (Figure 7.10), both at the moment when $\sigma = 0°$.

We then obtain

$$B = A = 2ml^2 = 2(0.25)0.5^2 = 0.125 \text{ Ws}^3$$

$$C = 4ml^2 = 0.25 \text{ Ws}^3$$

$$J_{vz} = I_{vz} = -2mel = -2(0.25)(-0.001)(0.5) = 0.000\,25 \text{ Ws}^3$$

$$J_{uz} = I_{uz} = -2me_1 l = -2(0.25)0.000\,002(0.5) = -0.000\,0005 \text{ Ws}^3$$

$$J_{zv} = 0$$

$$J_{zu} = 0.$$

The inertia tensor in the deformed state is thus

$$I = \begin{bmatrix} 0.125 & 0 & -0.000\,0005 \\ 0 & 0.125 & 0.000\,25 \\ -0.000\,0005 & 0.000\,25 & 0.25 \end{bmatrix} \text{Ws}^3.$$

The angular velocities are

$$\boldsymbol{\omega} = [\mathbf{e}_u \; \mathbf{e}_v \; \mathbf{e}_z] \begin{bmatrix} 0 \\ 2 \\ 500 \end{bmatrix} \text{rad/s}$$

$$\boldsymbol{\Omega} = [\mathbf{e}_u \; \mathbf{e}_v \; \mathbf{e}_z] \begin{bmatrix} 0 \\ 2 \\ 0 \end{bmatrix} \text{rad/s}$$

$$\dot{\sigma} = \omega_z - \Omega_z = 500 \text{ rad/s}.$$

The angular momentum, from equation (7.31c) is

$$\mathbf{H} = [\mathbf{e}_u \; \mathbf{e}_v \; \mathbf{e}_z] \begin{bmatrix} -0.005 \\ 0.5 \\ 125.005 \end{bmatrix} \text{Ws}^2.$$

The torque is

$$\mathbf{M} = \dot{\mathbf{H}} = \overset{\circ}{\mathbf{H}} + \boldsymbol{\Omega} \times \mathbf{H} = [\mathbf{e}_u \; \mathbf{e}_v \; \mathbf{e}_z] \begin{bmatrix} 250.001 \\ 0 \\ 0.001 \end{bmatrix} \text{Ws}.$$

The power supplied to feed the hysteresis loss is

$$\dot{U} = \mathbf{M} \cdot \boldsymbol{\omega} = 0.001(500) = 0.5 \text{ W}.$$

The Cardan angle

$$\xi = \frac{I_{vz}}{C - A} = \frac{0.00025}{-0.125} = -0.002 \text{ rad} = -0.11459°.$$

The Cardan angle

$$\eta = \frac{I_{uz}}{C - A} = \frac{-0.000\,0005}{0.125} = -0.000\,004 \text{ rad} = -0.000\,229°.$$

The elastic deformation coefficient is

$$\alpha = \frac{I_{vz}}{2\omega_v\omega_z} = \frac{0.00025}{(2)2(500)}\text{Ws}^5 = 0.125\ \mu\text{Ws}^5.$$

The internal damping energy dissipation coefficient, with

$$\dot{\sigma} = \omega_z - \Omega_z = 500 - 0 = 500 \text{ rad/s},$$

is

$$\beta = -\frac{I_{uz}}{2\omega_v\omega_z\dot{\sigma}} = \frac{0.000\,0005}{(2)2(500)500}\text{Ws}^6 = 0.5\,\text{pWs}^6.$$

Using $\dot{D} = 4\beta\omega_v^2\omega_z^2\dot{\sigma}^2$ the energy dissipation rate becomes

$$\dot{D} = 2(10^{-12})2^2(500^2)500^2 = 0.5 \text{ W}$$

and is seen to equal $\dot{U}$.

The gyro's kinetic energy is

$$\begin{aligned} T &= \frac{1}{2}(A\omega_v^2 + C\omega_z^2 + 2I_{vz}\omega_v(\omega_z + \dot{\sigma})) \\ &= \frac{1}{2}0.125(2^2) + 0.25(500^2) + 2(0.00025)2(1000)) \\ &= 31250.75 \text{ Ws} \end{aligned}$$

The gyro's angular momentum has a magnitude of

$$H = \sqrt{0.0005^2 + 0.5^2 + 125.0005^2} = 125.0015 \text{ Ws}^2.$$

The elastic energy V stored in the gyro is

$$V = \sqrt{I_{vz}^2 + I_{uz}^2}\,\omega_v\omega_z = 0.00025(2)500 = 0.25 \text{ Ws}$$

such that the total energy stored in the gyro is

$$T + V = 31\,251 \text{ Ws}.$$

Summing up, the energy rates equation

$$\dot{U} = \dot{T} + \dot{V} + \dot{D}$$

expressed in numbers would read

$$0.5 = 0 + 0 + 0.5.$$

If the elastic deformation would have been ignored, i.e., if $I_{vz} = 0$ had been used throughout, there would have been no effect on the energy rates (Rimrott, 1985).

Note also that the phantom body's angular velocity vector

$$\boldsymbol{\omega} = [\mathbf{e}_u\ \mathbf{e}_v\ \mathbf{e}_z] \begin{bmatrix} 0 \\ 2 \\ 500 \end{bmatrix} \text{rad/s}$$

and the angular momentum vector

$$\mathbf{H} = [\mathbf{e}_u\ \mathbf{e}_v\ \mathbf{e}_z] \begin{bmatrix} -0.0005 \\ 0.5 \\ 125.005 \end{bmatrix} \text{Ws}^2$$

are *not* collinear. Thus, according to the collinearity theorems, there is energy dissipation.

The phase shift is

$$\tan\phi = -\frac{I_{uz}}{I_{vz}} = \frac{0.000\,00005}{0.000\,25} = 0.002$$

$$\phi = 0.115^\circ.$$

The reversed Coriolis force acting on mass 1 is

$$F = 2m\omega_v\omega_z l = 2(0.25)\ 2(500)0.5 = 250\ \text{N}.$$

Since $|e| = 1\,\text{mm} = 0.001$ m, the stiffness of arm 1 is

$$k = \frac{F}{|e|} = \frac{250}{0.001} = 250\,000\ \text{N/m}$$

leading to a structural eigenfrequency of

$$\omega_n = \sqrt{\frac{k}{m}} = \sqrt{\frac{250\,000}{0.25}} = 1000\ \text{rad/s}.$$

The frequency ratio is consequently

$$r = \frac{\dot{\sigma}}{\omega_n} = \frac{500}{1000} = 0.5.$$

The damping ratio ζ is obtained from

$$\tan\phi = \frac{2\zeta r}{1 - r^2} = 0.002$$

$$\zeta = 0.002\frac{1}{2r}(1 - r^2) = 0.0015.$$

7.11 Conclusions

From a study of the constrained system, and in particular from a perusual of the numerical example just presented, the reader will conclude that the elastic deformations (as expressed by J_{vz} and J_{zv}) lead only to very minor adjustments of the results, compared to results obtained by assuming absolute rigidity. This is not the case for the deformations due to internal damping (as expressed by J_{uz} and J_{zu}), which lead to a new component in the angular momentum, a new component in the expression for the applied torque, and eventually to the conclusion that mechanical energy to compensate for the energy dissipation must continually be fed to the gyro.

The long-standing practice of ignoring elastic deformations in many gyrodynamic calculations consequently appears justified. Deformations due to internal damping, however, lead to novel terms and therefore cannot be ignored. The presence of internal damping makes itself felt in the form of partial inertia products J_{zu} and J_{uz}, which both are functions of the spin frequency $\dot{\sigma}$ and the damping ratios. No spin, i.e., $\dot{\sigma} = 0$, or no damping, i.e., $\zeta = 0$, means that $J_{zu} = J_{uz} = I_{uz} = 0$.

In the interest of terminological brevity, the term *dissipative gyro* is often used for solid bodies whose deformations due to internal damping are considered, while their elastic deformations are ignored (Rimrott, 1985).

Suggested Reading

1. Canavin, J.R.; P.W. Likins. "Floating Reference Frames for Flexible Spacecraft," *AIAA Journal of Spacecraft and Rockets,* 14, 12, 1977, 724–732.

2. Chesser, H.C.; F.P.J. Rimrott. "Attitude Drift of Torquefree Flexible Gyros Using the Viscous Damping Model," *Oxford Quarterly Journal of Mechanics and Applied Mathematics,* 41, 4, 1988.

3. Magnus, K. *Kreisel,* Springer-Verlag, 1971, 493 pp.

4. Rimrott, F.P.J. "Dissipative Rigids in Gyrodynamics," *ZAMM,* 65, 7, 1985, 287–300.

5. Rimrott, F.P.J.; Y. Yu. "Attitude Drift of a Torquefree Axisymmetric Gyro through a Structural Resonance," *Proceedings,* COBEM '87, Florianopolis, Vol. 2, 1987, 869–872.

6. Tisserand, F. *Traité de mécanique celéste,* Gauthier-Villars, Paris, 1889, Vol. I, 474 pp.; Vol. II, 552 pp.; Vol. III, 427 pp.; Vol. IV, 548 pp.

7. Zhang, W. "Dynamic Stability of Rotating Viscoelastic Shaft (in Chinese)," *Fudan Journal,* 23, 1984, 425–433.

8. Zhang, W.; F.H. Ling. "Dynamic Stability of the Rotating Shaft Made of Boltzmann Viscoelastic Solid," *ASME Journal of Applied Mechanics*, 53, 1986, 424–429.

Problems

7.1. Given is a gyro, which in its deformed state has an inertia tensor of

$$[I]_{\text{def}} = \begin{bmatrix} 24 & 0 & 0 \\ 0 & 15 & 0 \\ 0 & 0 & 10 \end{bmatrix} \text{m}^2\text{kg}$$

expressed in $Cx'y'z'$ coordinates. A $Cxyz$ coordinate system is located at (small) Cardan angles of $\xi = 0.01$ rad, $\eta = 0.02$ rad, and $\zeta = 0.03$ rad. Express the inertia tensor in $Cxyz$ coordinates.

7.2. Given is a gyro, which in its deformed state has an inertia tensor of

$$[I']_{\text{def}} = \begin{bmatrix} 24 & 0 & 0 \\ 0 & 24 & 0 \\ 0 & 0 & 10 \end{bmatrix} \text{m}^2\text{kg}$$

expressed in $Cx'y'z'$ coordinates. A $Cxyz$ coordinate system is located at (small) Cardan angles of $\xi = 0.01$ rad, $\eta = 0.02$ rad, and at $\zeta = 40°$. Express the inertia tensor in $Cxyz$ coordinates.

7.3. A gyro in its deformed shape, has an inertia tensor that is, e.g., obtained from optical measurements, of

$$[I] = \begin{bmatrix} 40 & 0 & -0.09 \\ 0 & 40 & 0.1 \\ -0.09 & 0.1 & 20 \end{bmatrix} \text{m}^2\text{kg}.$$

(a) Determine the principal inertia tensor. (b) Calculate the (small) Cardan angles ξ and η between the principal axes and the axes used in equation (7.1). (c) Calculate the direction angles α, β, and γ for each of the three principal axes, and make a sketch.

7.4. A gyro, in its deformed shape, has an inertia tensor of

$$[I] = \begin{bmatrix} 40 & 0 & -0.09 \\ 0 & 40 & 0 \\ -0.09 & 0 & 20 \end{bmatrix} \text{m}^2\text{kg}.$$

(a) Determine the principal inertia tensor. (b) Calculate the (small) Cardan angles ξ and η. (c) Calculate the direction angles α, β, and γ for each of the three principal axes, and make a sketch.

7.5. Given is a gyro consisting of a thin rigid disk ($m_1 = 240$ kg, 2 m diameter, and 1 cm thickness (negligible)) and of two massless flexible antennas of 2 m length each, each carrying a tip point mass m_2 of 3 kg. Determine the inertia tensor (a) in the undeformed and (b) in the deformed state if the antenna masts deflect 1 cm each as shown. (c) Find the angle ξ between the principal axes systems in the undeformed and deformed states. (d) Calculate $\Delta A/A$, $\Delta B/A$, and $\Delta C/C$. (e) Determine the partial inertia products J_{vz} and J_{zv}. (f) Determine the modified inertia product Y_{vz} if $\Omega_z/\omega_z = C/A$.

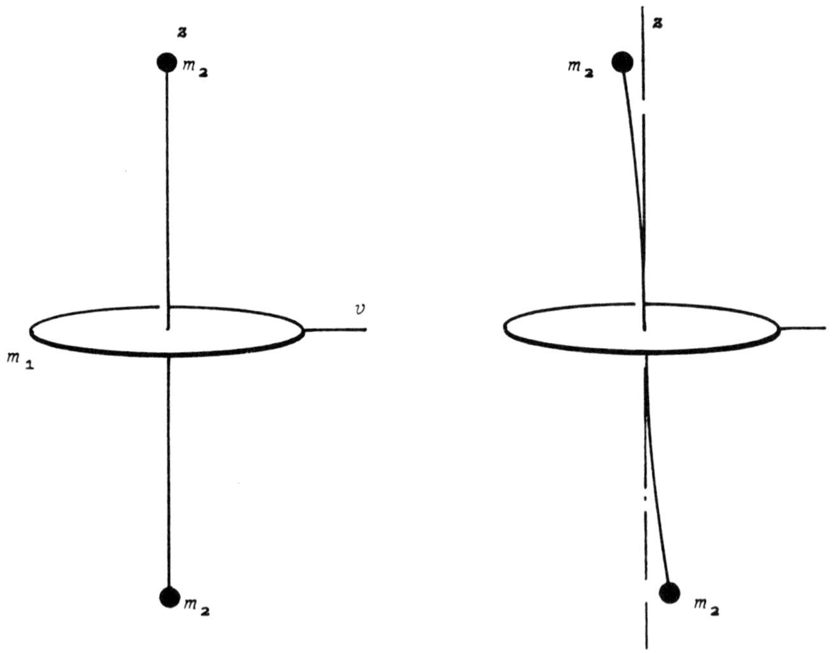

7.6. Given is a gyro consisting of a rigid solid cylinder ($m_1 = 24\,000$ kg, 1.2 m diameter, and 2.6 m height) and of two massless flexible antennas of 0.7 m length each, each carrying a tip point mass m_2 of 3 kg. Determine the inertia tensor (a) in the undeformed and (b) in the deformed state if the antenna masts deflect 1 cm each as shown. (c) Find (and sketch) the angle ξ between the principal axes systems in the undeformed and deformed states. (d) Calculate $\Delta A/A$, $\Delta B/A$, and $\Delta C/C$. (e) Determine the partial inertia products J_{vz} and J_{zv}. (f) Determine the modified inertia product Y_{vz} if $\Omega_z/\omega_z = C/A$.

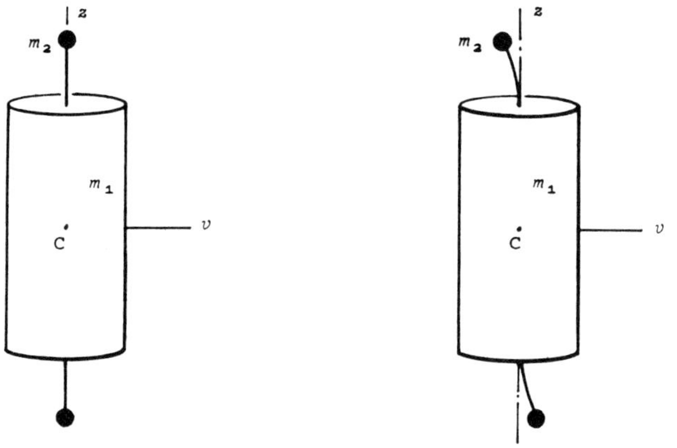

7.7. Given is a gyro consisting of a rigid solid cylinder (m_1 = 24 000 kg, 1.2 m diameter, and 2.6 m height) and of four massless flexible antennas of 0.7 m length each, each carrying a tip point mass m_2 of 3 kg. Determine the inertia tensor (a) in the undeformed and (b) in the deformed state if two antennas masts deflect 1 cm each as shown. (c) Find (and sketch) the angle ξ between the principal axes systems in the undeformed and deformed states. (d) Calculate $\Delta A/A$, $\Delta B/A$, and $\Delta C/C$. (e) Determine the partial inertia products J_{vz} and J_{zv}. (f) Determine the modified inertia product Y_{vz} if $\Omega_z/\omega_z = C/A$.

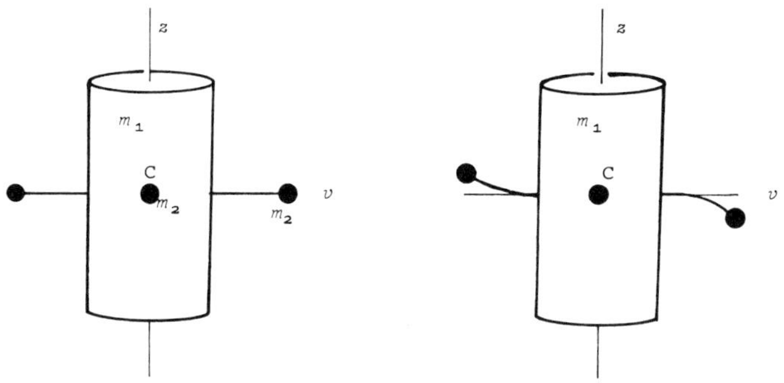

7.8. Given is a gyro consisting of a rigid cylinder ($m_1 = 24\ 000$ kg, 1.2 m diameter, and 2.6 m height) and six massless flexible antenna booms of 0.7 m length each, each carrying a tip point mass $m_2 = m_3 = m_4 = 3$ kg ($m = m_1 + 2m_2 + 2m_3 + 2m_4$). Determine the inertia tensor (a) in the undeformed state and (b) in the deformed state if the two longitudinal antennas deflect 1 cm each as shown and two of the radial antennas deflect 1.6 cm as shown. (c) Find and sketch the angle ξ between the principal axes systems. (d) Calculate $\Delta A/A$, $\Delta B/A$, and $\Delta C/C$. (e) Determine the partial inertia products J_{vz} and J_{zv}. (f) Determine the modified inertia product Y_{vz} if $\Omega_z/\omega_z = C/A$.

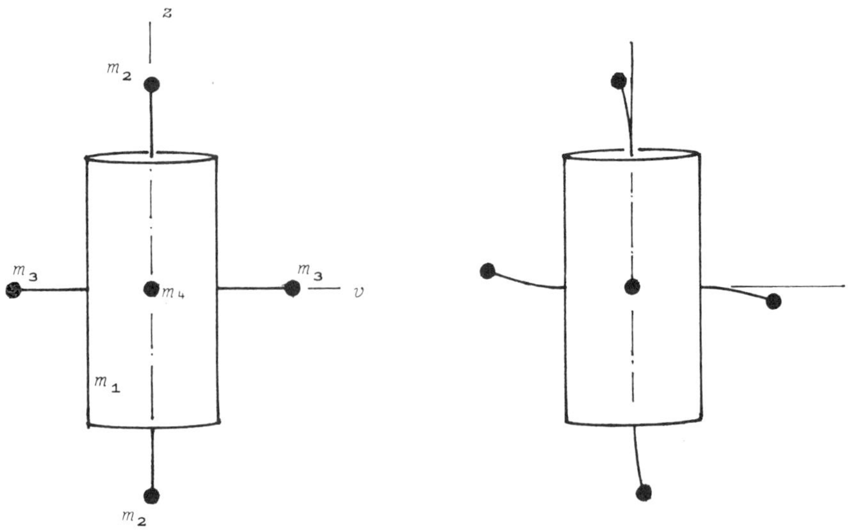

7.9. Given is a gyro consisting of a rigid cylinder of uniform mass distribution ($m_1 = 24\,000$ kg, 1.2 cm diameter, and 2.6 m height) and six massless flexible antenna booms of 0.7 m length each, each carrying a tip point mass of 3 kg. (a) Determine the inertia tensor of the phantom (undeformed) body. (b) Determine the inertia tensor of the deformed body, if $\delta_v = -1$ cm and $\delta_u = 0.1$ cm for the mass m_2 located at $\rho_z = 2$ m, $\delta_z = -1.6$ cm for the mass m_3 located at $\rho_v = 1.3$ cm, and $\delta_z = 0.6$ cm for the mass m_4 located at $\rho_u = 1.3$ m, with the remaining three masses displaced antisymmetrically. (c) Determine and sketch the angles ξ and η between the principal axes systems of the deformed and undeformed states. (d) Calculate $\Delta A/A$, $\Delta B/A$, and $\Delta C/C$. (e) Determine the partial inertia products J_{vz}, J_{zv}, J_{uz}, and J_{zu}. (f) Determine the modified inertia products Y_{vz} and Y_{uz} if $\Omega_z/\omega_z = C/A$.

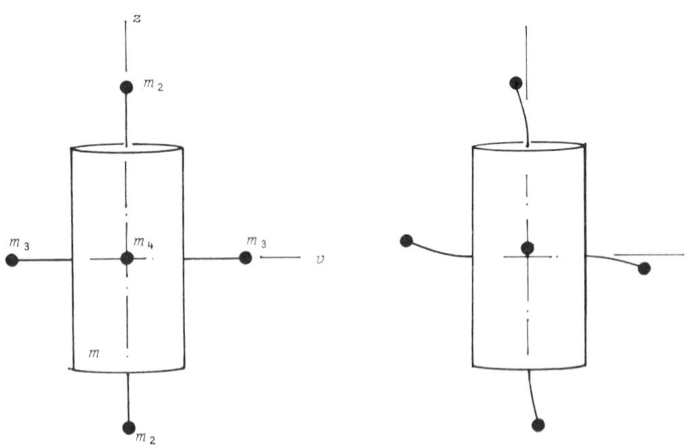

7.10. A thin rigid steel disk of 1.2 m diameter and 3 mm thickness is mounted on a rigid shaft driven by motor 2 at 5 rad/s, which in turn is driven by motor 1 at 2 rad/s. Use the *Cuvz* coordinate system shown and determine and sketch into the drawing: (a) the angular velocity $\boldsymbol{\omega}$ of the disk; (b) the angular momentum **H** of the disk; (c) the angular velocity $\boldsymbol{\Omega}$ of the *Cuvz* coordinate system; (d) the torque **M** required to drive the disk; also determine (e) the power $\dot{U}$ supplied by the motors to the disk; and (f) the power $\dot{U}$ supplied by the motors if the disk was not rigid and $I_{uz} = 0.001$ Ws^3.

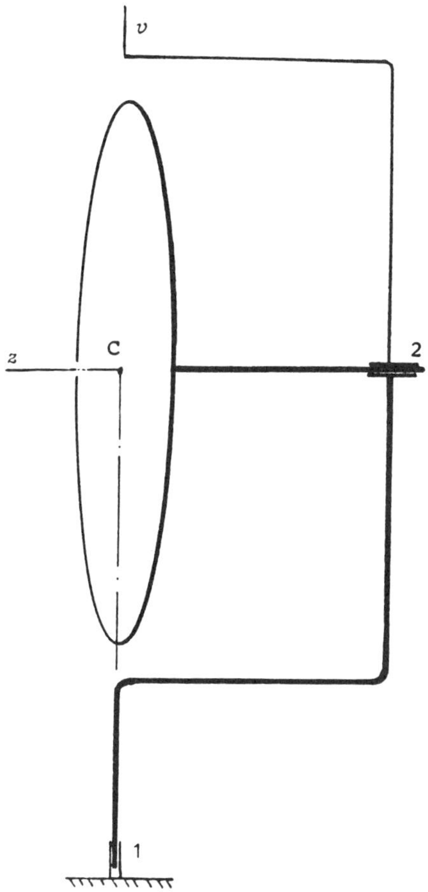

7.11. An axisymmetric gyro, consisting of four point masses mounted at the tip of four massless viscoelastic booms (of stiffness k and damping c), is attached to a rigid shaft driven by motor 2 at a constant angular velocity ω_z, which in turn is attached to a rigid z-shaped frame driven by motor 1 at a constant angular velocity ω_v. (a) Find the inertial Coriolis force field. (b) Establish the differential equation of motion for mass 1 on boom 1. (c) What is the phase lag between force and displacement? (d) Calculate the inertia product I_{vz}. (e) Calculate the inertia product I_{uz}.

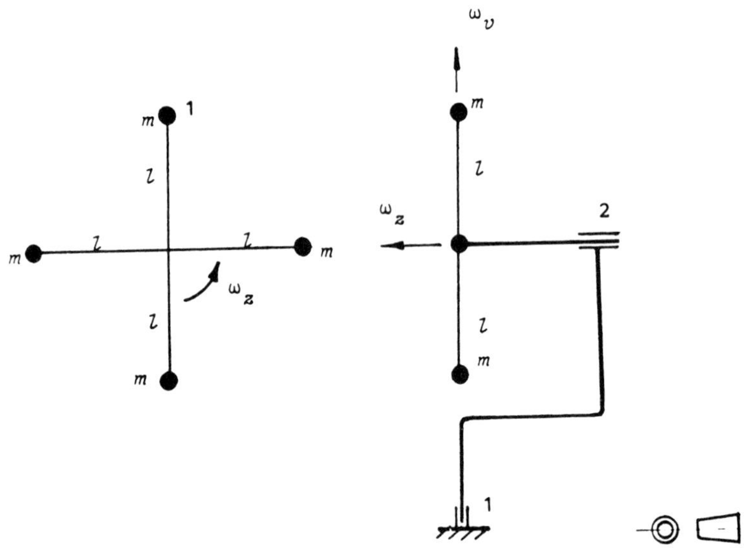

7.12. Show that the $Cuvz$ floating coordinate system used is (a) at an angle of

$$\eta = Y_{uz}/(C - A)$$

from the principal axes system of the deforming body when $J_{uz} = J_{zu}$; (b) at $\eta = (2C - A)Y_{uz}/(C(C - A))$ when $J_{uz} = 0$; (c) at $\eta = AY_{uz}/((2A - C)(C - A))$ when $J_{zu} = 0$; and (d) at $\eta = 0$ when $J_{uz} = -J_{zu}$.

8

Secular Attitude Drift of a Torquefree Dissipative Axisymmetric Gyro

8.1 Attitude Drift Rate and Attitude Stability

In order to investigate attitude drift and attitude stability of a torquefree solid gyro, we begin with the general case (equation (7.38)). The angular velocity of the phantom body is

$$\boldsymbol{\omega} = [\mathbf{e}_u\ \mathbf{e}_v\ \mathbf{e}_z] \begin{bmatrix} \omega_u \\ \omega_v \\ \omega_z \end{bmatrix} \tag{8.1}$$

and the angular momentum of the deforming body is, according to equation (7.38),

$$\mathbf{H} = [\mathbf{e}_u\ \mathbf{e}_v\ \mathbf{e}_z] \begin{bmatrix} A\omega_u + Y_{uz}\omega_z \\ A\omega_v \\ C\omega_z \end{bmatrix}. \tag{8.2}$$

The floating $Cuvz$ coordinate system is chosen such that $H_u = 0$. From equation (8.2)

$$H_u = A\omega_u + Y_{uz}\omega_z = 0. \tag{8.3}$$

Since $\omega_u = \dot{\nu}$, one obtains a relationship between drift rate $\dot{\nu}$ and modified inertia product Y_{uz}, amounting to

$$\dot{\nu} = -\frac{Y_{uz}}{A}\omega_z. \tag{8.4}$$

Equation (8.4) establishes the fundamental relationship between attitude drift rate $\dot{\nu}$ and gyro deformation.

The energy dissipation rate can be obtained from equation (7.52)

$$\dot{D} = (\boldsymbol{\Omega} \times \mathbf{H}) \cdot \boldsymbol{\omega}. \tag{8.5}$$

With

$$\boldsymbol{\Omega} = [\mathbf{e}_u\ \mathbf{e}_v\ \mathbf{e}_z] \begin{bmatrix} \dot{\nu} \\ \omega_v \\ \frac{C}{A}\omega_z \end{bmatrix} \tag{8.6}$$

$$\dot{D} = (A - C)\omega_v\omega_z\dot{\nu} \tag{8.7}$$

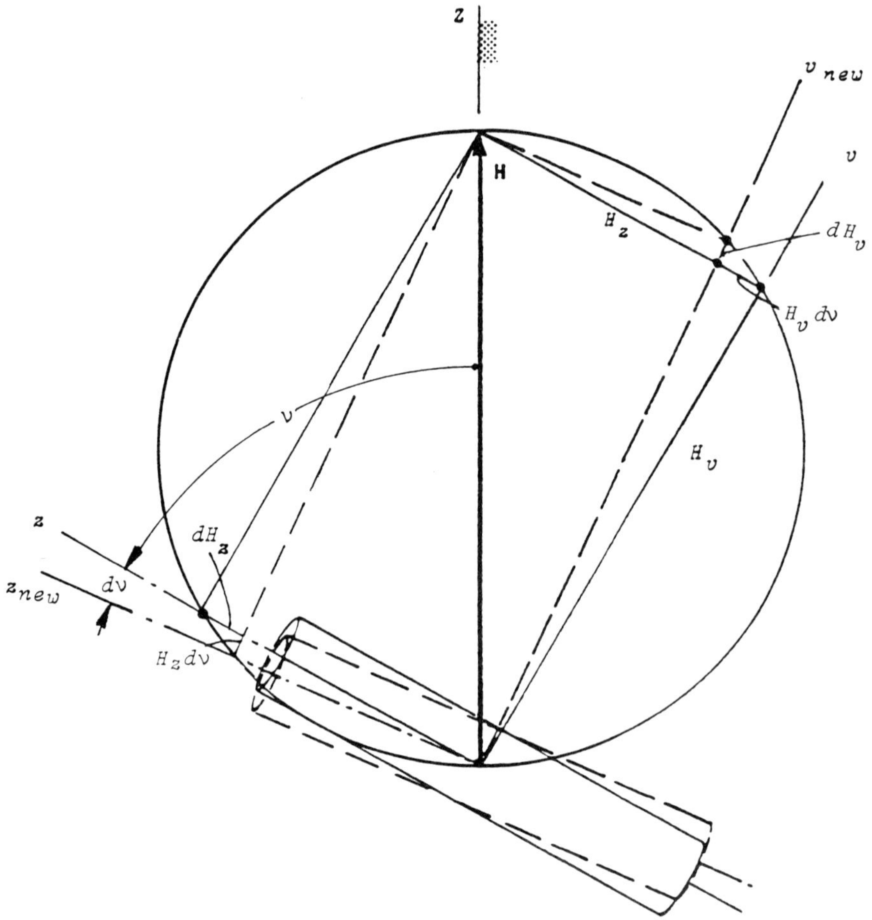

FIGURE 8.1. The vz-plane during an increase of nutation angle while the angular momentum remains constant.

or (Figure 8.1)

$$\dot{\nu} = \frac{\dot{D}}{(A - C)\omega_v \omega_z}. \tag{8.8}$$

Combining equations (8.4) and (8.8), a general relationship between energy dissipation rate $\dot{D}$ and modified inertia product Y_{uz} can be obtained.

$$Y_{uz} = -\frac{A}{(A - C)\omega_v \omega_z^2}\dot{D} \tag{8.9}$$

with $\dot{D}$ positive definite, if the case $A = C$ is excluded. Since, from equations (6.27),

$$\omega_v = \frac{H \sin \nu}{A} \tag{8.10}$$

and

$$\omega_z = \frac{H \cos \nu}{C} \tag{8.11}$$

equation (8.9) becomes

$$Y_{uz} = -\frac{2A^2C^2}{(A - C)H^2 \sin 2\nu \cos \nu}\dot{D}. \tag{8.12}$$

In the first quadrant, i.e., when $0 < \nu < 90°$, $Y_{uz} < 0$ for prolate ($A > C$) gyros and $Y_{uz} > 0$ for oblate ($C > A$) gyros. A negative Y_{uz} in equation (8.4) means a negative $\dot{\nu}$, i.e., an oblate gyro drifts towards $\nu = 0°$. If $\nu = 0°$ is the reference attitude, then oblate gyros are stable. Note that $\dot{D}$ cannot be negative.

From equation (8.12) it can also be concluded that the occurrence of internal damping implies the presence of a Y_{uz}, and that vice versa, the presence of a Y_{uz} implies that internal damping occurs.

With the spin $\dot{\sigma}$ from equation (6.29) and precession $\dot{\psi}$ from equation (6.28), the modified inertia product (8.12) becomes

$$Y_{uz} = -\frac{2C}{H\dot{\sigma}\dot{\psi}\sin 2\nu}\dot{D}. \tag{8.13}$$

Since all terms on the right hand side of equation (8.13) are positive with the exception of the spin $\dot{\sigma}$, one can conclude that Y_{uz} is positive when $\dot{\sigma}$ is negative. Thus, attitude stability, i.e., $\dot{\nu} < 0$, is assured when the spin

$$\dot{\sigma} < 0. \tag{8.14}$$

Some authors use the ratio $\dot{\sigma}/\dot{\psi}$ as attitude stability indicator. For a reference attitude of $\nu = 0°$, $\dot{\sigma}/\dot{\psi} < 0$ means a stable gyro (Landon's rule, 1968), $\dot{\sigma}/\dot{\psi} > 0$ an unstable gyro, and $\dot{\sigma}/\dot{\psi} = 0$ represents the boundary between stable and unstable gyro configurations.

The kinetic energy is given by equation (7.45). Differentiated with respect to time using equations (8.10) and (8.11),

$$\dot{T} = -(A - C)\omega_v \omega_z \dot{\nu}. \tag{8.15}$$

The elastic energy rate is typically so small that $\dot{V} = 0$ can be assumed. Since there is no torque, there is no external energy supply, thus $\dot{U} = 0$. From equation (7.47) one can then conclude that

$$\dot{T} = -\dot{D} \tag{8.16}$$

as an inspection of equations (8.15) and (8.7) also shows.

The preceding equations become simpler and more instructive when the ansatz

$$Y_{uz} = -\beta \omega_v \omega_z \dot{\sigma} \tag{8.17}$$

is made, where β is an internal energy dissipation coefficient with the dimension Ws^6.

Expressed in terms of the attitude angle ν, by means of equations (8.10), (8.11), and (8.17), the modified inertia product becomes

$$Y_{uz} = -\beta \frac{(A - C)H^3}{2A^2C^2} \cos\nu \sin 2\nu. \tag{8.18}$$

It vanishes when $\nu = 0°,\ 90°,\ 180°, \cdots$, and for a spherical gyro, i.e., when $C = A$.

The drift rate (8.4) becomes

$$\dot{\nu} = \beta \frac{\omega_v \omega_z^2 \dot{\sigma}}{A} = \beta \frac{(A - C)H^4}{2A^3C^3} \cos^2\nu \sin 2\nu \tag{8.19}$$

and moves the attitude towards $\nu = 0°,\ 180°,\ 360°, \cdots$, when the gyro is oblate ($C > A$), and towards $\nu = 90°,\ 270°, \cdots$, when the gyro is prolate ($A > C$).

Using spin $\dot{\sigma}$ and precession $\dot{\psi}$, from equations (6.29) and (6.28), the drift rate can be expressed as

$$\dot{\nu} = \beta \frac{H^2}{2AC^2} \dot{\sigma}\dot{\psi} \cos\nu \sin 2\nu. \tag{8.20}$$

The energy dissipation rate (8.7) becomes

$$\begin{aligned} \dot{D} &= -Y_{uz}\omega_v\omega_z\dot{\sigma} = \beta \omega_v^2 \omega_z^2 \dot{\sigma}^2 \\ &= \beta \frac{(A - C)^2 H^6}{4A^4C^4} \cos^2\nu \sin^2 2\nu. \end{aligned} \tag{8.21}$$

It is seen to be positive and semi-definite as required. It vanishes at attitudes of $\nu = 0°,\ 90°,\ 180°,\ 270°, \cdots$, and for spherical gyros, i.e., when $C = A$.

When both $J_{uz} = -\beta_1(2 - C/A)\omega_v\omega_z\dot{\sigma}$ and $J_{zu} = -\beta_2(C/A)\omega_v\omega_z\dot{\sigma}$, are present, then with $\dot{\sigma} = ((A-C)/A)\omega_z$ and $\Omega_z = (C/A)\omega_z$, and equation (7.24c),

$$\beta = \left(2 - \frac{C}{A}\right)^2 \beta_1 + \left(\frac{C}{A}\right)^2 \beta_2. \tag{8.22}$$

As an example, consider a torquefree oblate gyro with $A = 200\,\text{m}^2\text{kg}$, $C = 300\,\text{m}^2\text{kg}$, and $\beta = 5(10^{-6})\text{Ws}^6$. The angular momentum is $H = 600\,\text{Ws}^2$ and the attitude $\nu = 3°$.

The angular velocity components are

$$\omega_v = \frac{H}{A}\sin\nu = \frac{600}{200}\sin 3° = 0.157 \text{ rad/s}$$

$$\omega_z = \frac{H}{C}\cos\nu = \frac{600}{300}\cos 3° = 1.997 \text{ rad/s}$$

$$\dot{\sigma} = \frac{A-C}{A}\omega_z = \frac{200-300}{200}1.997 = -0.999 \text{ rad/s}.$$

The drift rate is

$$\dot{\nu} = \beta\frac{\omega_v\omega_z^2\dot{\sigma}}{A} = -0.1564\,(10^{-9})\,\text{rad/s}$$

$$\dot{\nu} = -0.000774\,\text{deg/day} = 0.283\,\text{deg/a}.$$

The gyro drifts towards $\nu = 0°$. The gyro's distortion is measured by its modified inertia product

$$Y_{uz} = -\beta\omega_v\omega_z\dot{\sigma} = -0.940 \text{ m}^2\text{kg}$$

which is much smaller than either of the two inertia moments, A or C. The energy dissipation rate is

$$\dot{D} = \beta\omega_v^2\omega_z^2\dot{\sigma}^2 = 0.49\ \mu\text{W}.$$

8.2 The Attitude Drift Process

When the attitude angle ν of a torquefree gyro changes, there are many side effects. No torque means that the angular momentum remains constant. Thus

$$\dot{\mathbf{H}} = \overset{\circ}{\mathbf{H}} + \boldsymbol{\Omega} \times \mathbf{H} = 0. \tag{8.23}$$

With

$$\boldsymbol{\Omega} = [\mathbf{e}_u\ \mathbf{e}_v\ \mathbf{e}_z]\begin{bmatrix} \dot{\nu} \\ \omega_v \\ \frac{C}{A}\omega_z \end{bmatrix} \tag{8.24}$$

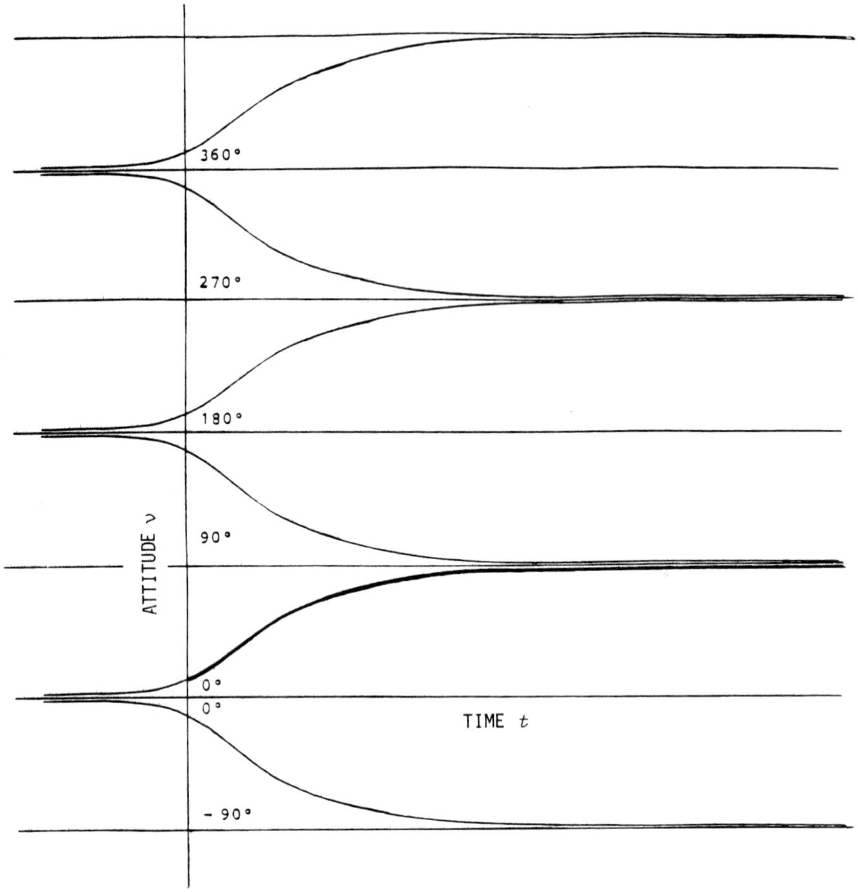

FIGURE 8.2. Ogives of attitude drift of prolate gyros $(A > C)$.

as angular velocity of the floating $Cuvz$ coordinate system, and $\mathbf{H}$ as given by equation (8.2), the components of $\dot{\mathbf{H}}$ become

$$\begin{bmatrix} 0 \\ A\dot{\omega}_v - C\omega_z\dot{\nu} \\ C\dot{\omega}_z + A\omega_v\dot{\nu} \end{bmatrix} = \begin{bmatrix} 0 \\ 0 \\ 0 \end{bmatrix} \tag{8.25}$$

and consequently one obtains for the velocity component rates

$$\dot{\omega}_v = \frac{C}{A}\omega_z\dot{\nu} \tag{8.26a}$$

$$\dot{\omega}_z = -\frac{A}{C}\omega_v\dot{\nu}. \tag{8.26b}$$

Equations (8.26a) and (8.26b) can be rearranged to express the angular momentum component rates

$$\dot{H}_v = H_z\dot{\nu} \tag{8.27a}$$

$$\dot{H}_z = -H_v\dot{\nu}. \tag{8.27b}$$

Note that

$$H_v\dot{H}_v + H_z\dot{H}_z = 0 \tag{8.28}$$

or

$$H_v^2 + H_z^2 = H^2 = \text{ constant} \tag{8.29}$$

as required.

If the internal energy dissipation coefficient β is constant, i.e., $\beta = \beta_0$, then equation (8.19) can be rearranged to

$$\int_0^t dt = \frac{2A^3C^3}{(A-C)H^4\beta_0}\int_{\nu_0}^{\nu}\frac{d\nu}{\cos^2\nu\sin 2\nu} \tag{8.30}$$

and integrated in closed form to give

$$t = \frac{A^3C^3}{4(A-C)H^2\beta_0}\left(\tan^2\nu - \tan^2\nu_0 + \ln\frac{\tan^2\nu}{\tan^2\nu_0}\right). \tag{8.31}$$

Equation (8.31) is plotted in Figure 8.2, for a prolate gyro. For an oblate gyro

$$t = \frac{A^3C^3}{4(C-A)H^4\beta_0}\left(\tan^2\nu_0 - \tan^2\nu + \ln\frac{\tan^2\nu_0}{\tan^2\nu}\right). \tag{8.32}$$

Equation (8.32) is plotted in Figure 8.3.

When the internal energy dissipation coefficient β is a function of the attitude ν, i.e., when e.g.,

$$\beta = \alpha_0\frac{2\zeta/\omega_n}{(1-r^2)^2 + (2\zeta r)^2} \tag{8.33}$$

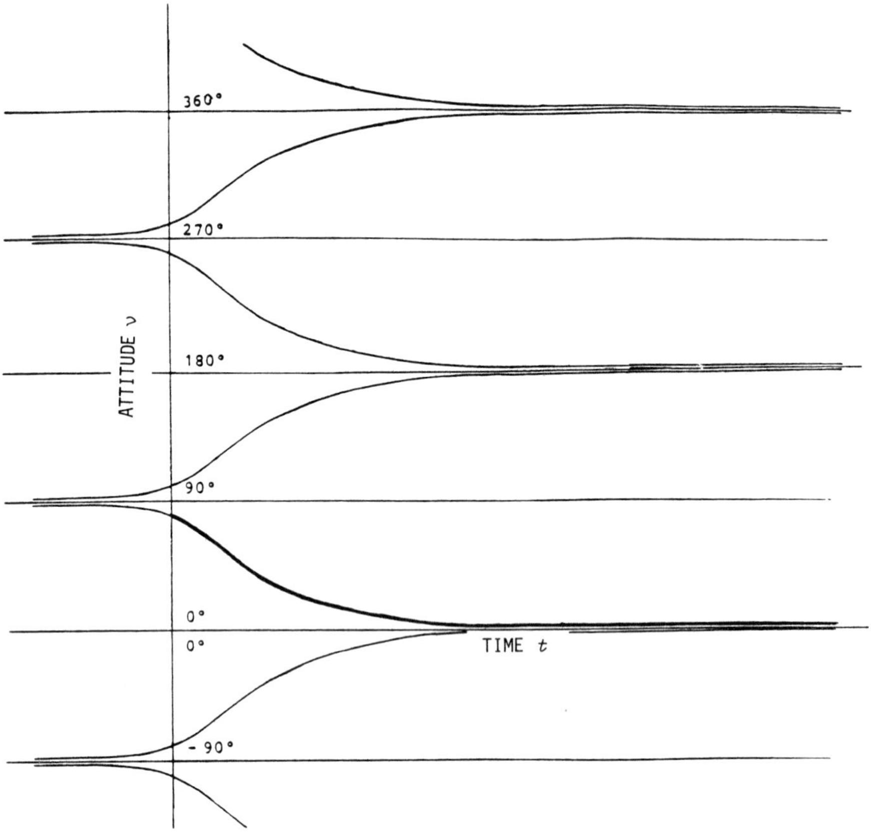

FIGURE 8.3. Ogives of attitude drift of oblate gyros ($A < C$).

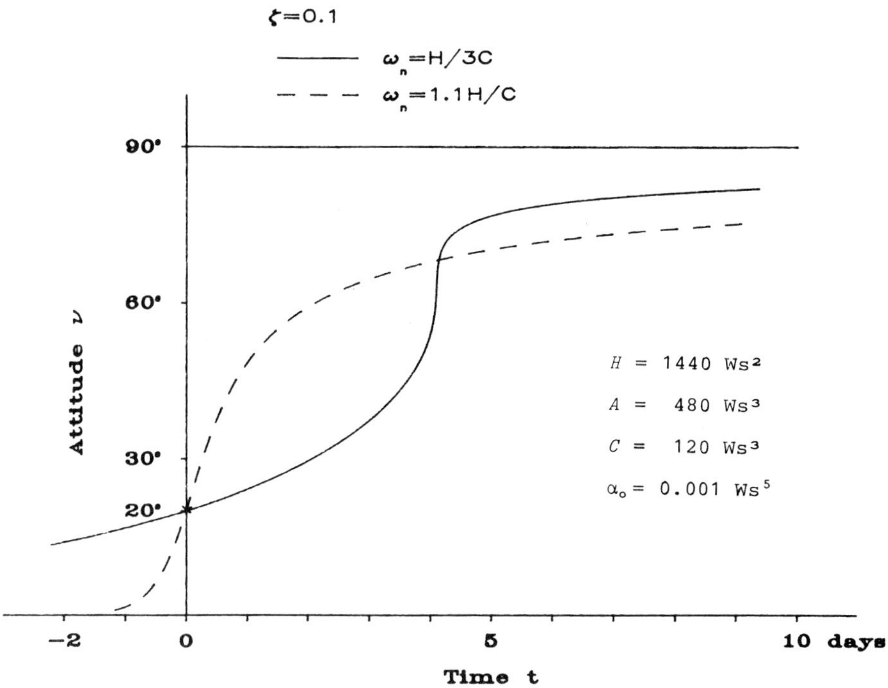

FIGURE 8.4. Attitude as function of time for $\zeta = 0.1$.

with

$$r = \frac{\dot{\sigma}}{\omega_n} = \frac{(A-C)H}{AC\omega_n} \cos \nu \tag{8.34}$$

and equation (8.19) is rearranged to

$$\int_0^t dt = \int_{\nu_0}^{\nu} \frac{d\nu}{f(\nu)} \tag{8.35}$$

then $\nu = \nu(t)$ can be obtained numerically. The result for a specific case is plotted in Figure 8.4.

During the drifting of the attitude, i.e., when $\dot{\nu} \neq 0$, the locations of the various vectors, i.e., $\mathbf{H}$, $\boldsymbol{\Omega}$, and $\boldsymbol{\omega}$ as given by equations (8.2), (8.6), and (8.1), are depicted in Figure 8.5.

The magnitude and direction of the angular velocity also changes during attitude drifting. From

$$\omega^2 = \omega_u^2 + \omega_v^2 + \omega_z^2 \tag{8.36}$$

one obtains for the magnitude rate of the gyro's angular velocity:

$$\dot{\omega} = \frac{\omega_u \dot{\omega}_u}{\omega} + \frac{\omega_v \dot{\omega}_v}{\omega} + \frac{\omega_z \dot{\omega}_z}{\omega}. \tag{8.37}$$

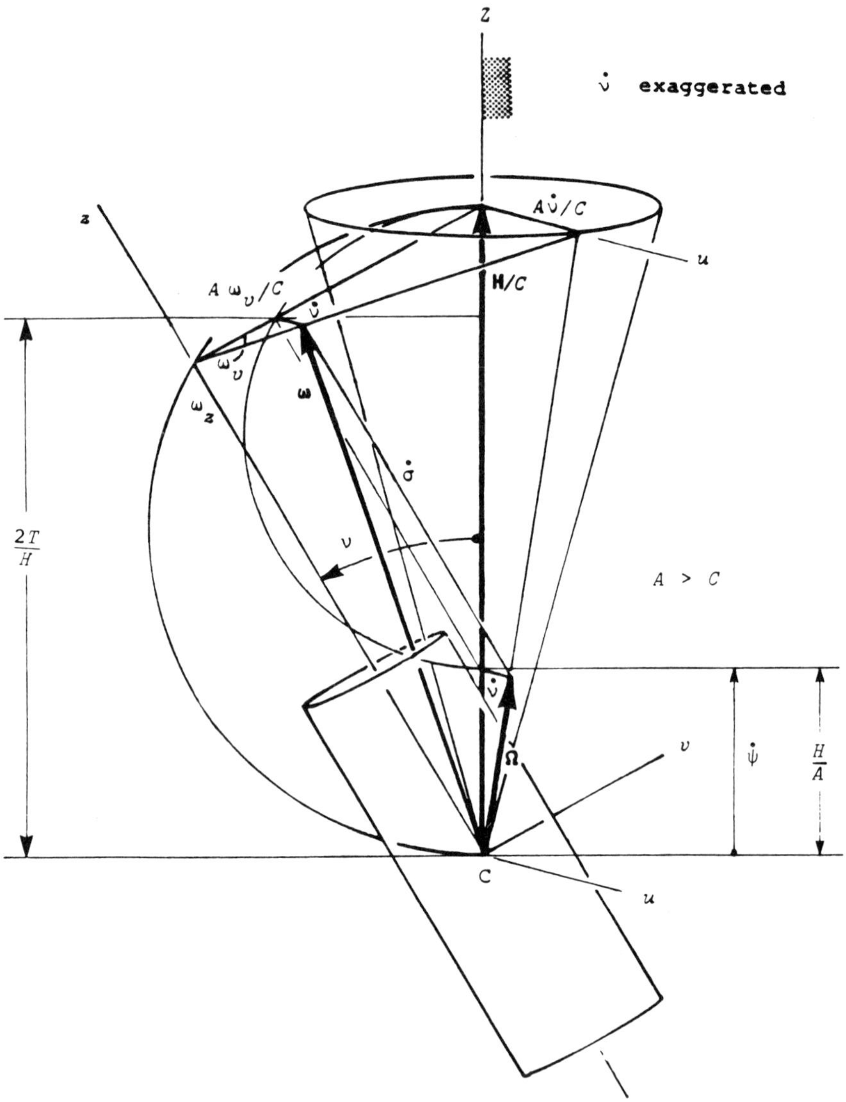

FIGURE 8.5. The torquefree dissipative gyro ($\mathbf{M} = 0$, $\dot{\nu} \neq 0$).

Since $\omega_u = \dot{\nu}$ and $|\dot{\nu}| << \omega$ the first term on the right side of equation (8.37) may be neglected. With the help of equations (8.26a) and (8.26b),

$$\dot{\omega} = \frac{C}{A\omega}\left(1 - \frac{A^2}{C^2}\right)\omega_v\omega_z\dot{\nu}. \tag{8.38}$$

Introducing equation (8.8),

$$\dot{\omega} = -\frac{1}{\omega}\frac{A+C}{AC}\dot{D}. \tag{8.39}$$

Since the energy dissipation rate $\dot{D}$ is positive semi-definite, one can conclude that the magnitude of the gyro's angular velocity is steadily decreasing as long as there is energy dissipation (Figure 8.6).

Returning to the previous example, the magnitude of the angular velocity changes at a rate of

$$\dot{\omega} = -\frac{1}{\sqrt{0.157^2 + 1.997^2}}\frac{200+300}{200(300)}0.49(10^{-6}) = -2.038(10^{-9})\ \text{rad/s}^2.$$

The angle μ (Figure 8.6) between angular momentum vector $\mathbf{H}$ and the projection into the vz-plane of the angular velocity vector $\boldsymbol{\omega}$ can be obtained from

$$\tan\mu = \frac{\left(\frac{A}{C} - 1\right)\tan\nu}{\frac{A}{C} + \tan^2\nu}. \tag{8.40}$$

The angle μ is positive, i.e., in the same direction as ν, for prolate gyros and negative for oblate gyros.

With the help of equations (8.10) and (8.11), the kinetic energy (7.45) may be expressed as a function of the attitude angle ν:

$$T = \frac{H_2}{2C}\left(1 + \left(\frac{C}{A} - 1\right)\sin^2\nu\right) = \frac{H^2}{2A}\left(1 + \left(\frac{A}{C} - 1\right)\cos^2\nu\right). \tag{8.41}$$

See Figures 8.7 and 8.8.

8.3 Circles in the vz-Plane

Since according to equation (8.29)

$$H_v^2 + H_z^2 = H^2 = \text{constant}$$

we find that H_v and H_z *describe a circle in the vz-plane during kinetic energy decline* (Figure 8.1).

From equation (6.28) we conclude that if I_{vz} is disregarded,

$$\dot{\psi} = \frac{H}{A} \tag{8.42}$$

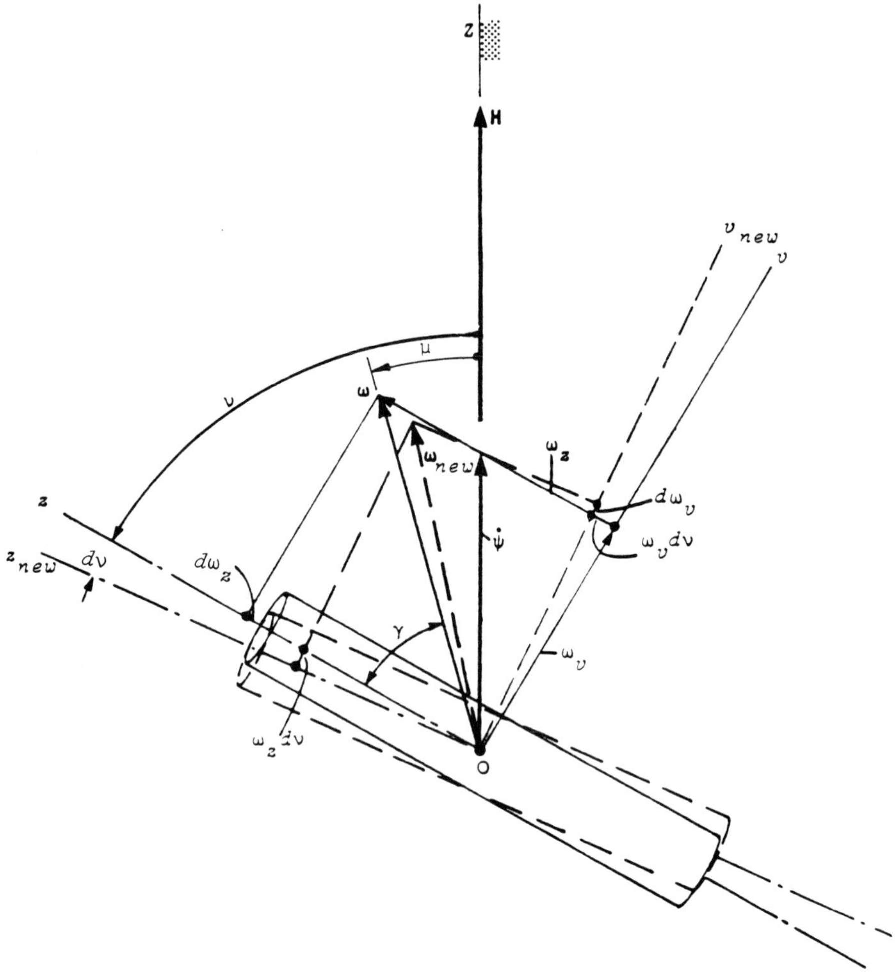

FIGURE 8.6. The vz-plane during an increase of nutation angle while the angular momentum remains constant, there is a change of direction and a decrease in magnitude of the angular velocity.

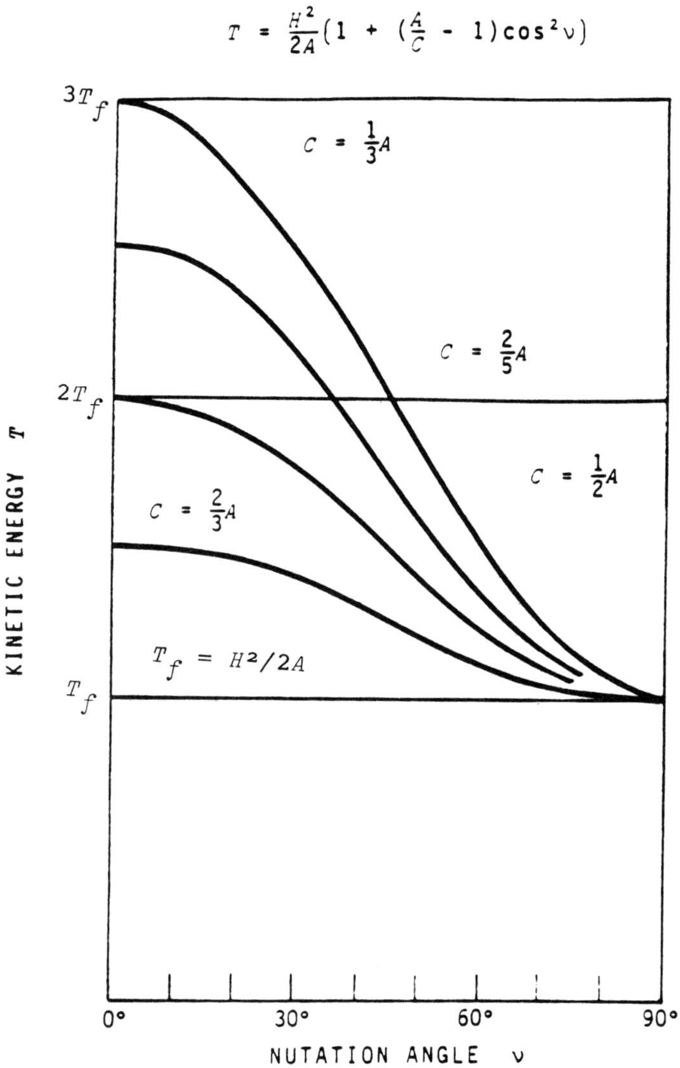

FIGURE 8.7. Kinetic energy decline as function of nutation angle for elongated gyro ($C < A$).

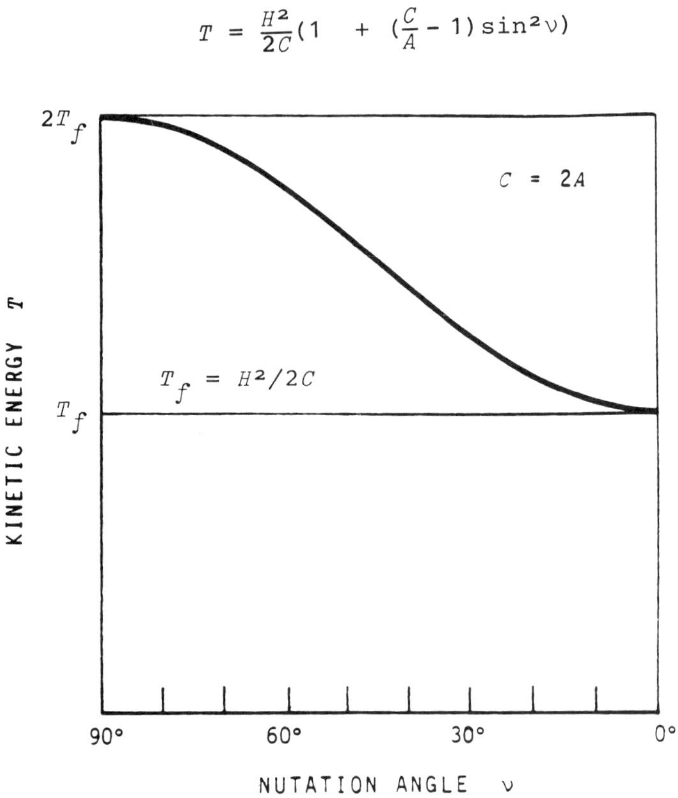

FIGURE 8.8. Kinetic energy decline as function of nutation angle for flattened gyro ($C > A$).

i.e., the *the precession remains constant throughout the kinetic energy decline.* With the help of equation (8.10), we conclude that

$$\omega_v^2 + \left(\frac{\omega_v}{\tan\nu}\right)^2 = \dot{\psi}^2 = \text{ constant} \tag{8.43}$$

and that consequently *the angular velocity component ω_v describes a circle in the vz-plane during energy decline* (ω_v-circle in Figures 8.9 and 8.10).

From equation (6.29), we find that

$$\dot{\sigma}^2 = (\dot{\sigma}\tan\nu)^2 = \frac{H^2(A-C)}{A^2C^2} = \text{ constant} \tag{8.44}$$

and conclude that *the spin $\dot{\sigma}$ describes a circle in the vz-plane during kinetic energy decline* ($\dot{\sigma}$-circle in Figures 8.9 and 8.10).

After squaring equation (8.11) we find that

$$\omega_z^2 + (\omega_z \tan\nu)^2 = \frac{H^2}{C^2} \tag{8.45}$$

from which we conclude that *the angular velocity component ω_z describes a circle in the vz-plane during kinetic energy decline* (ω_z-circle in Figures 8.9 and 8.10).

8.4 Arbitrary Initial and Stable Final State

We shall study a flattened gyro ($C > A$) first. We begin by assuming that the gyro was free of internal damping in the past. The *initial* state 0 thus represents that attitude, which the gyro has *before* nutation damping commences and before the attitude begins to drift.

Initially the gyro satellite thus carries out a steady precession at ν_0 (Figures 8.11 and 8.12). The angular velocity is then of a magnitude $\omega_0 = \sqrt{\omega_{0v}^2 + \omega_{0z}^2}$ or

$$\omega_0 = \omega_{0z}\sqrt{1 + \frac{C^2}{A^2}\tan^2\nu_0} \tag{8.46}$$

from equations (8.10) and (8.11). Further

$$\omega_v = \frac{C\tan\nu}{A}\omega_z = \frac{H}{A}\sin\nu. \tag{8.47}$$

The angular momentum is of magnitude

$$H_0 = \frac{C\omega_{0z}}{\cos\nu_0}. \tag{8.48}$$

The kinetic energy is

$$T_0 = \frac{1}{2}C\omega_{0z}^2\left(1 + \frac{C}{A}\tan^2\nu_0\right). \tag{8.49}$$

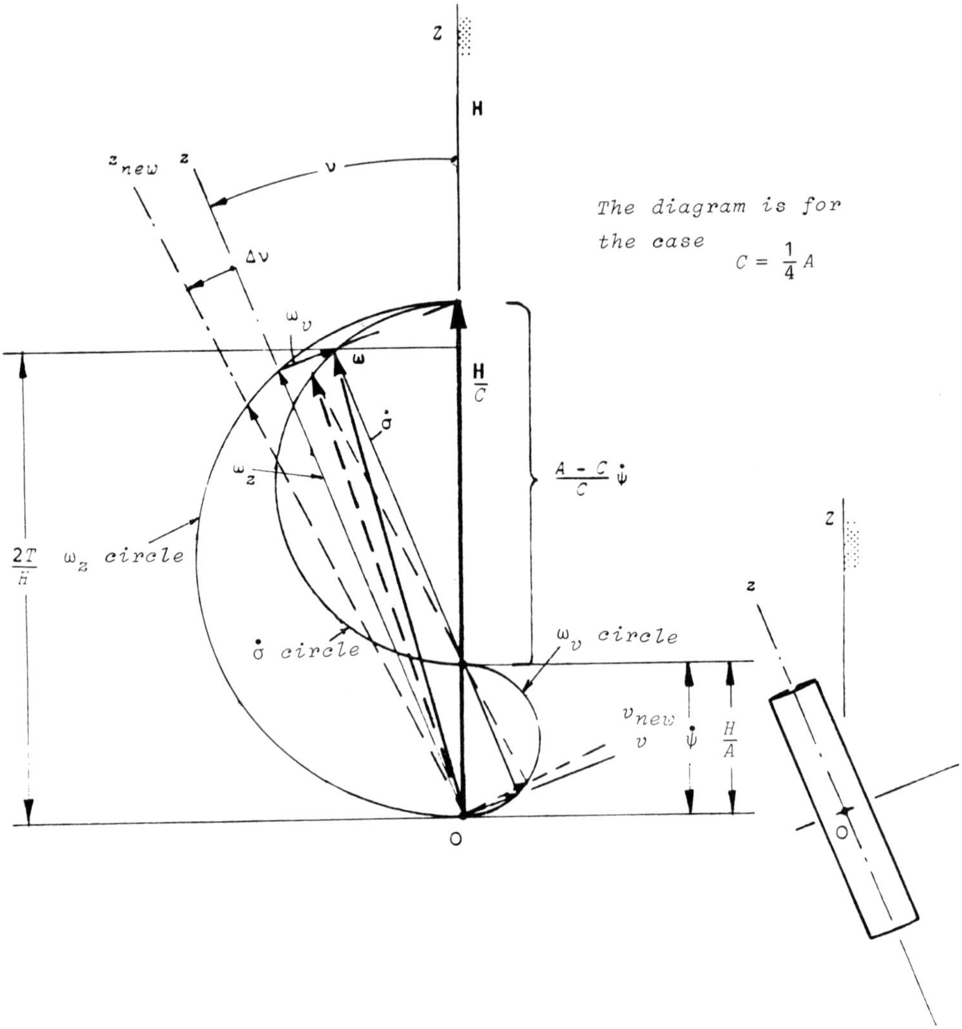

FIGURE 8.9. The three angular velocity component circles during nutation angle increase towards 90° (for $C < A$).

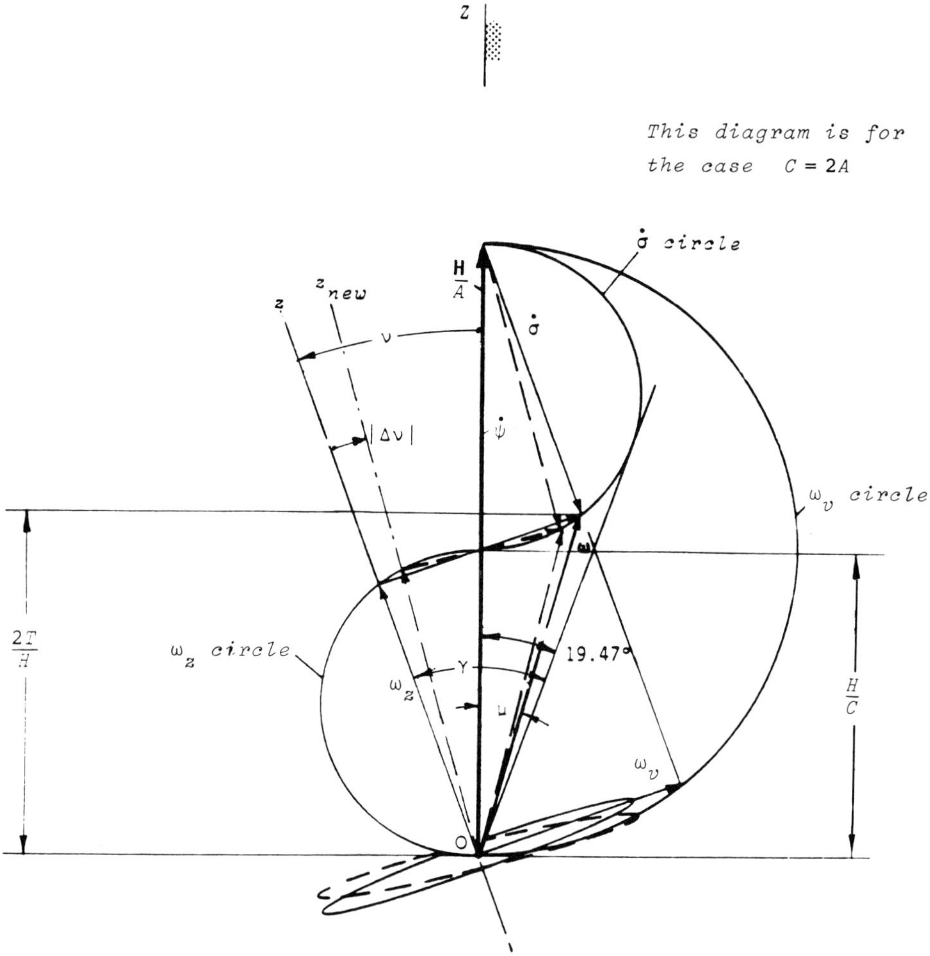

FIGURE 8.10. The three angular velocity component circles during nutation angle decrease (for $C > A$).

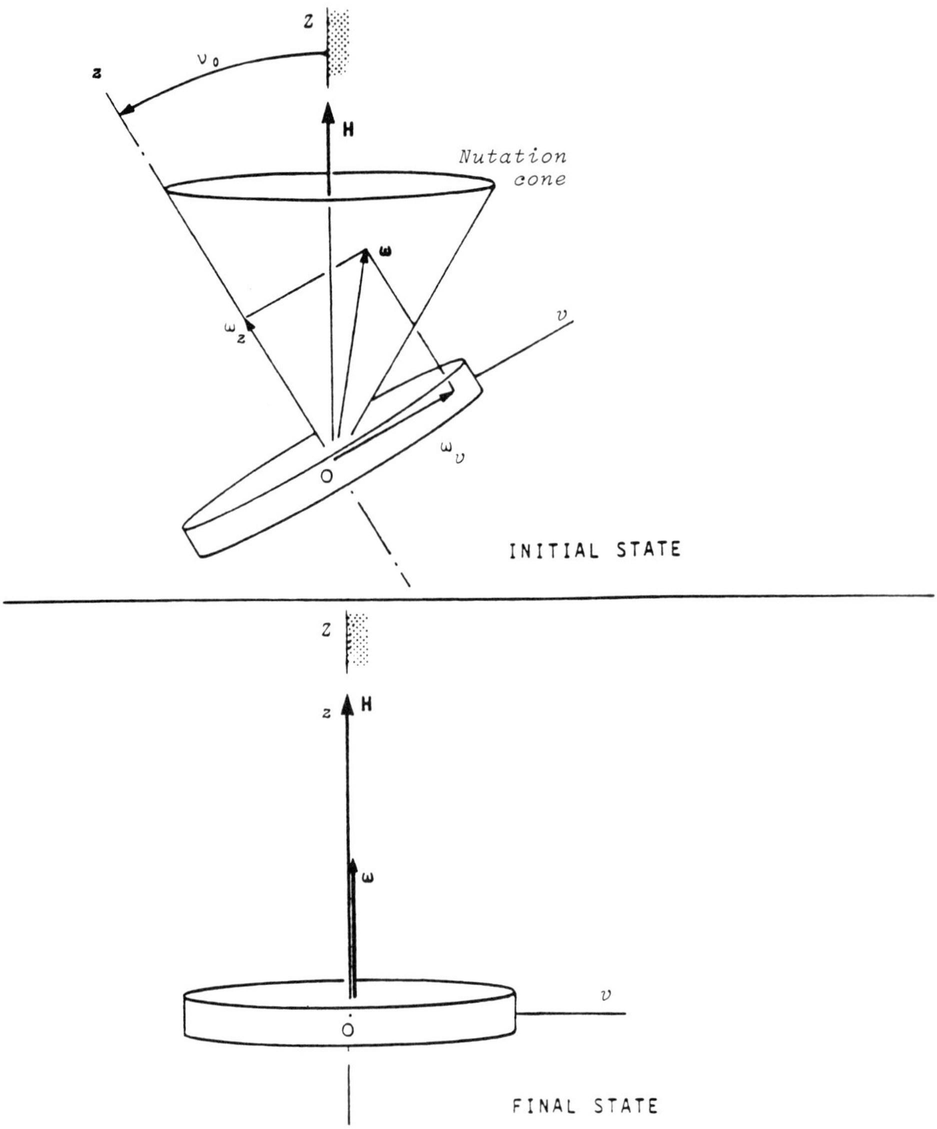

FIGURE 8.11. Decrease of nutation angle for flattened gyro ($C > A$).

The spin is

$$\dot{\sigma}_0 = -\frac{C-A}{A}\omega_{0z}. \tag{8.50}$$

The precession is

$$\dot{\psi}_0 = \frac{C}{A\cos\nu_0}\omega_{0z} = \frac{H}{A}. \tag{8.51}$$

As we have initially assumed a gyro free of internal energy dissipation, we have for the initial state

$$\dot{\nu}_0 = 0. \tag{8.52}$$

After internal damping has commenced, the stable mode, i.e., the final state f, is reached eventually, when the nutation angle has become

$$\nu_f = 0. \tag{8.53}$$

The angular momentum has remained the same

$$\mathbf{H}_f = \mathbf{H}_0 \tag{8.54}$$

which means that

$$C\omega_{fz} = \frac{1}{\cos\nu_0}C\omega_{0z} \tag{8.55}$$

for which

$$\omega_{fz} = \frac{\omega_{0z}}{\cos\nu_0} = \omega_f \tag{8.56}$$

and

$$\omega_{fv} = \frac{C\tan 0}{A}\omega_{fz} = 0. \tag{8.57}$$

The magnitude of the new angular velocity in terms of the original angular velocity and the original nutation angle is

$$\omega_f = \omega_0\sqrt{1 - \frac{(C^2 - A^2)\sin^2\nu_0}{A^2 + (C^2 - A^2)\sin^2\nu_0}}. \tag{8.58}$$

The new angular velocity can be thought of as consisting of the two Euler frequencies (Figure 8.10)

$$\dot{\sigma}_f = \frac{C-A}{A}\omega_f \tag{8.59}$$

and

$$\dot{\psi}_f = \frac{C}{A}\omega_f. \tag{8.60}$$

The final kinetic energy is (Figure 8.8)

$$T_f = T_0\left(1 - \frac{(C-A)\sin^2\nu_0}{A + (C-A)\sin^2\nu_0}\right). \tag{8.61}$$

It cannot decline any further.

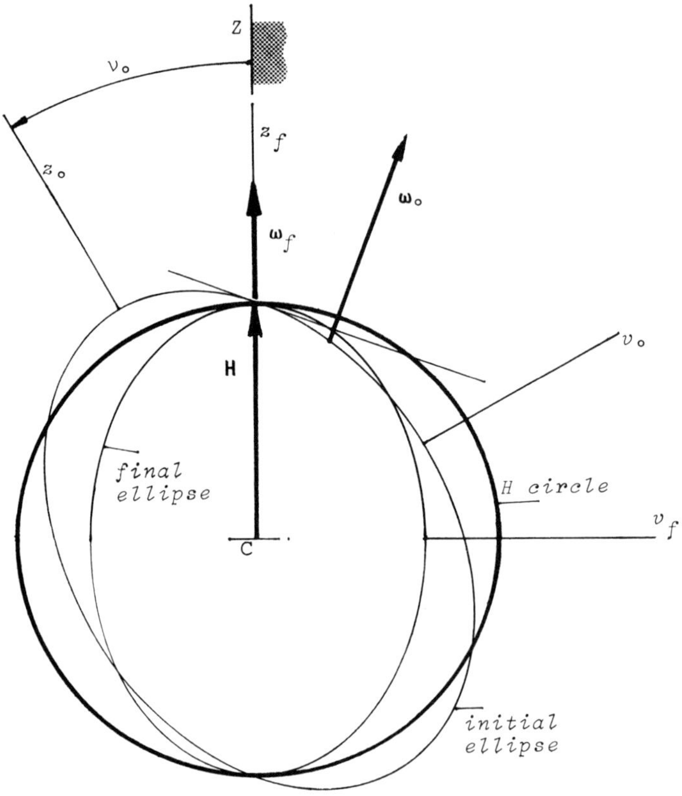

FIGURE 8.12. MacCullagh ellipses for an arbitrary initial and for the final stable attitude for a disk gyro.

For the flattened gyro there is a limit to the angle between the (constant) angular momentum **H** and the angular velocity $\boldsymbol{\omega}$.

An inspection of Figure 8.10 indicates that

$$\tan\mu = \tan(\gamma - \nu) = \frac{\left(\frac{C}{A} - 1\right)\tan\nu}{1 + \frac{C}{A}\tan^2\nu}. \tag{8.62}$$

It has a maximum for

$$\tan\nu = \sqrt{\frac{A}{C}} \tag{8.63}$$

which for the extreme case of $C = 2A$ results in

$$\tan\mu = \frac{1}{2\sqrt{2}} \tag{8.64a}$$

or

$$\mu = 19.47°. \tag{8.64b}$$

For the case of the elongated gyro ($C < A$) we consider the same initial state ν_0, and note that equations (8.46)–(8.52) apply. For the elongated gyro, the final state is characterized by (Figure 8.13)

$$\nu_f = 90° \tag{8.65}$$

and, of course,

$$\mathbf{H}_f = \mathbf{H}_0. \tag{8.66}$$

Thus

$$\omega_f = \dot{\psi}_f = \frac{C}{A}\frac{\omega_{0z}}{\cos\nu_0} \tag{8.67}$$

or

$$\omega_f = \dot{\psi}_f = \omega_0\sqrt{1 - \frac{(A^2 - C^2)\cos^2\nu_0}{C^2 + (A^2 - C^2)\cos^2\nu_0}}. \tag{8.68}$$

The final kinetic energy is (Figure 8.7)

$$T_f = T_0\left(1 - \frac{(A - C)\cos^2\nu_0}{C + (A - C)\cos^2\nu_0}\right). \tag{8.69}$$

It cannot decline any further.

8.5 Elastic Deformation and Attitude Stability

In the attitude drift process studied in sections 8.1–8.4, the elastic inertia products were ignored. This practice becomes less and less acceptable the closer a gyro approaches a terminal attitude, i.e., $\nu = 0°$ or $\nu = 90°$.

In order to demonstrate the influence of elastic deformation, take a look at an oblate dissipative gyro ($C > A$), such as the one shown in Figure

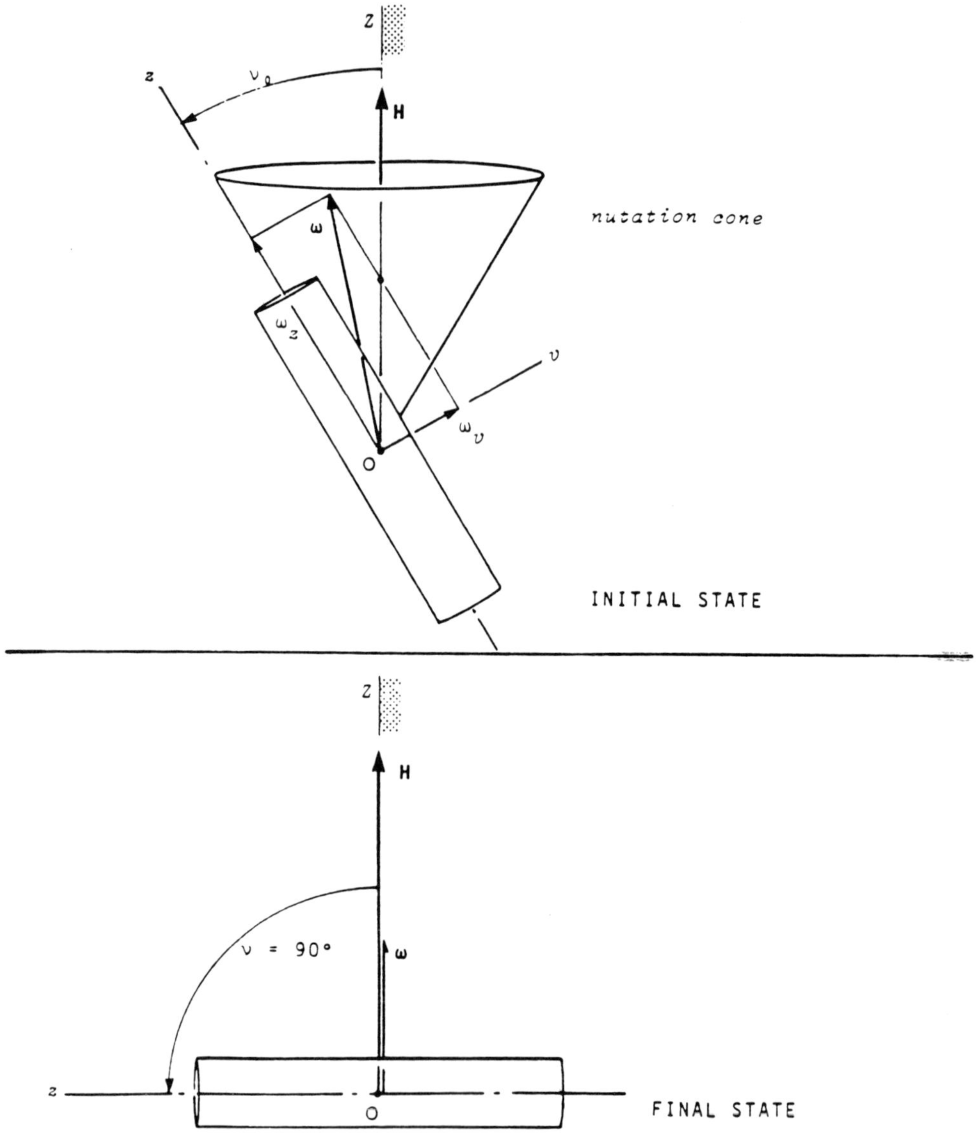

FIGURE 8.13. Increase of nutation angle for elongated gyro ($C < A$).

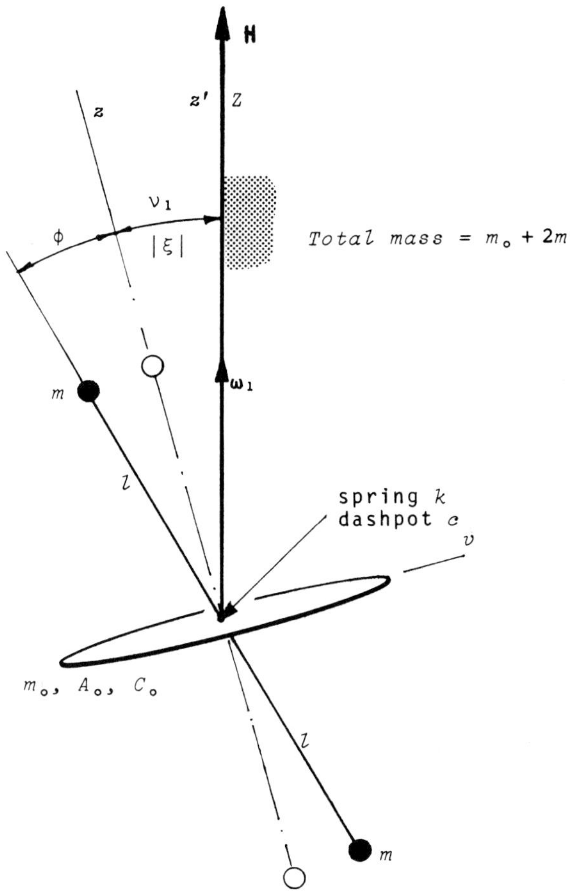

FIGURE 8.14. Deformed equilibrium configuration with collinear angular momentum and angular velocity.

8.14, which is drifting and approaching $\nu = 0°$. Because of its elasticity, it will have an I_{vz}, which in turn means that one of its principal axes is, according to equation (7.18b), located at

$$\xi = -\frac{I_{vz}}{C - A}. \tag{8.70}$$

If this principal axis becomes collinear with the angular momentum $\mathbf{H}$, then there is no longer any spin ($\dot{\sigma} = 0$), nor any energy dissipation ($\dot{D} = 0$). The gyro body has become kinetically rigid and all parts are moving at one and the same angular velocity $\boldsymbol{\omega}_1$, collinear with the angular momentum

H. The kinetic energy of the gyro is now

$$T_1 = \frac{1}{2}\frac{H^2}{C_1} = \frac{1}{2}\frac{H^2}{A_0 \sin^2 \nu + C_0 \cos^2 \nu + 2ml^2 \sin^2(\nu + \phi)} \tag{8.71}$$

where A_0 and C_0 are the principal inertia moments of the rigid disk, and m is the mass at the tip of each of the two massless rigid rods, which are joined to the disk and have flexed through an angle ϕ. The elastic energy stored in the springs is

$$V_1 = 2\frac{1}{2}kl^2\phi^2 = ml^2\omega_n^2\phi^2. \tag{8.72}$$

A relationship between ϕ and ν can be established, either by relating the centrifugal force to the spring force, or by differentiating the Lagrangian $L = T_1 - V_1$ partially with respect to ϕ and setting the result equal to zero:

$$\frac{\partial L}{\partial \phi} = \frac{H^2}{C_1^2}(2ml^2 \sin(\nu + \phi)\cos(\nu + \phi)) - 2ml^2\omega_n^2\phi = 0. \tag{8.73}$$

With $\omega_1 = H/C_1$, $r_1 = \omega_1/\omega_n$, and assuming that the angles ν and ϕ are small, the result is

$$\phi = \frac{r_1^2}{1 - r_1^2}\nu. \tag{8.74}$$

The inertia product I_{vz}, by the way, is now

$$I_{vz} = 2ml^2\phi = 2ml^2\frac{r_1^2}{1 - r_1^2}\nu. \tag{8.75}$$

Partial differentiation of the Lagrangian with respect to ν must also equal zero, and leads to

$$\frac{\partial L}{\partial \nu} = \frac{H^2}{C_1^2}(A_0 \sin\nu\cos\nu - C_0 \sin\nu\cos\nu$$

$$+ 2ml^2 \sin(\nu + \phi)\cos(\nu + \phi)) = 0. \tag{8.76}$$

With ϕ from equation (8.74) and for small angles

$$r_1^2 = \frac{C_0 - A_0 - 2ml^2}{C_0 - A_0} \tag{8.77}$$

from which it can be concluded that

$$r_1^2 < 1 \tag{8.78}$$

or

$$\omega_1 < \omega_n \tag{8.79}$$

is a requirement of the motion depicted in Figure 8.14. Attitude equilibrium (i.e., $\nu = \nu_1 =$ constant) is only possible when the structural eigenfrequency ω_n is greater than the rotational speed ω_1 of the gyro.

The gyro will be stable at this attitude if its total energy level at this attitude is lower than that at any other attitude, where there is also collinearity of $\mathbf{H}$ and $\boldsymbol{\omega}$. If, on the other hand, $\nu = 0°$ is the reference attitude and stability about this attitude is desired, then the total mechanical energy E_0 at this attitude must be less than the total mechanical energy E_1 at $\nu = \nu_1$, for the gyro to drift on towards $\nu = 0°$.

The total energy at ν_1 consists of kinetic and elastic energy and can be obtained as

$$E_1 = T_1 + V_1 = \frac{H^2}{2C_0} + \frac{H^2}{2C_0}\left(1 - \frac{A_0}{C_0} - \frac{2ml^2}{C_0}\frac{1}{1-r_1^2}\right)\nu_1^2. \tag{8.80}$$

The total mechanical energy at $\nu = 0°$ is simply

$$E_0 = T_0 = \frac{H^2}{2C_0}. \tag{8.81}$$

For E_1 to be greater than E_0, the following condition must hold:

$$1 - \frac{A_0}{C_0} - \frac{2ml^2}{C_0}\frac{1}{1-r_1^2} > 0 \tag{8.82}$$

or

$$r_1^2 < \frac{C_0 - A_0 - 2ml^2}{C_0 - A_0}. \tag{8.83}$$

Now, unfortunately the $r_1 = \omega_1/\omega_n$ of inequality (8.83) is not available. But the angular velocity $\omega_0 = H/C_0$ at $\nu = 0°$ is readily available, and since the gyro is close to $\nu = 0°$, ω_0 is close to ω_1. For small angles, the two are related by

$$\begin{aligned}\omega_1 = \frac{H}{C_1} &= \frac{H}{A_0\sin^2\nu + C_0\cos^2\nu + 2ml^2\sin^2(\nu+\phi)}\\ &= \omega_0\left(1 + \left(1 - \frac{A_0}{C_0} - \frac{2ml^2}{C_0}\frac{1}{(1-r_1^2)^2}\right)\nu_1^2\right).\end{aligned} \tag{8.84}$$

Since because of inequality (8.82), the expression in the inner bracket on the left-hand side of the equation is positive, one may conclude that

$$r_1 = \frac{\omega_1}{\omega_n} > \frac{\omega_0}{\omega_n} = r_0 \tag{8.85}$$

such that inequality (8.83) can be expanded to

$$r_0^2 < r_1^2 < \frac{C_0 - A_0 - 2ml^2}{C_0 - A_0}. \tag{8.86}$$

Inequality (8.86) can be rearranged to

$$C_0 > A_0 + 2ml^2 \frac{1}{1 - r_0^2} \tag{8.87}$$

with the introduction of $C = C_0$ and $A = A_0 + 2ml^2$ for the model of Figure 8.14, there is finally obtained

$$C > A + 2ml^2 \frac{r_0^2}{1 - r_0^2} \tag{8.88}$$

or, in other words, consideration of the gyro's elastic deformation requires a more stringent rule (8.88) for attitude stability than merely $C > A$. Note again that the magnitude of the damping present in the system plays no role. For very stiff appendages $r_0 \to 0$, and the stability criterion approaches $C > A$.

Stability criteria, such as (8.88), make no allowance for *almost* stable situations, in the literature referred to as *benign* instabilities, which for the present example would occur if inequality (8.88) was indeed not satisfied, but the stable equilibrium attitude ν_1 was so close to 0° as to be virtually indistinguishable from it. For further discussion of attitude stability criteria see, e.g., Dong and Schlack (1974).

Because the system is scleronomic, the total energy E_1, equation (8.80), equals the *Hamiltonian* (Leipholz, 1987). The Hamiltonian may, in the present circumstances, also serve as *Liapunov function* (Meirovitch, 1970), which must be positive definite for asymptotic stability, and must have a negative total time derivative. Considering that A_0, C_0, and m are all positive quantities, equation (8.80) is indeed positive definite and has a negative time derivative if

$$\frac{A_0}{C_0} + \frac{2ml^2}{C_0} \frac{1}{1 - r_1^2} < 1$$

which represents the same condition as inequality (8.82).

8.6 Effect of Deformation on Angular Velocities

It has been stated previously that satisfactory results can often be obtained in attitude drift studies, when the inertia product I_{vz} is ignored. If, on the other hand, it is to be included, several adjustments must be made. With I_{vz}, J_{vz}, and J_{zv}, the components of the angular momentum are

$$H \sin \nu = A\omega_v + J_{vz}(\omega_z + \dot{\sigma}) + J_{zv}(\omega_z - \dot{\sigma}) \tag{8.89}$$

$$H \cos \nu = C\omega_z + I_{vz}\omega_v . \tag{8.90}$$

Neglecting small terms leads to

$$\omega_v = \frac{H}{A} \sin \nu \left(1 - \frac{I_{vz}}{C \tan \nu}\right) + \frac{J_{zv} - J_{vz}}{A} \dot{\sigma}. \tag{8.91}$$

In a similar fashion, one obtains

$$\omega_z = \frac{H}{C} \cos \nu \left(1 - \frac{I_{vz}}{A} \tan \nu\right). \tag{8.92}$$

Since the angular velocity $\boldsymbol{\Omega}$ of the floating $Cuvz$ coordinate system is collinear with the angular momentum $\mathbf{H}$

$$\frac{\Omega_z}{\Omega_v} = \frac{\Omega_z}{\omega_v} = \frac{H_z}{H_v} = \frac{1}{\tan \nu} \tag{8.93}$$

and consequently

$$\Omega_z = \frac{\omega_z}{\tan \nu} = \frac{H}{A} \cos \nu \left(1 - \frac{I_{vz}}{C \tan \nu}\right) + \frac{J_{zv} - J_{zv}}{A} \frac{\dot{\sigma}}{\tan \nu}. \tag{8.94}$$

Of special interest is the spin frequency

$$\dot{\sigma} = \omega_z - \Omega_z = \frac{H}{1 + \frac{J_{zv} - J_{vz}}{A \tan \nu}} \left(\frac{A - C}{AC} \cos \nu + \frac{I_{vz}}{AC} \frac{\cos 2\nu}{\sin \nu}\right). \tag{8.95}$$

An inspection of equation (8.95) indicates that the rigid body spin $\dot{\sigma} = ((A - C)/AC) H \cos \nu$ is a good approximation for near-rigid bodies as long as the attitude angle ν is of reasonable size. When ν becomes small, however, the rigid-body approximation is no longer acceptable.

For small angles ν

$$\dot{\sigma} = -\frac{H}{1 + \frac{J_{zv} - J_{vz}}{A \nu}} \left(\frac{C - A}{AC} - \frac{I_{vz}}{AC} \frac{1}{\nu}\right). \tag{8.96}$$

The spin $\dot{\sigma}$ vanishes when

$$\nu = \nu_1 = \frac{I_{vz}}{C - A}. \tag{8.97}$$

The attitude angle ν_1 is the closer to 0°, the smaller the inertia product I_{vz}. With I_{vz} from equation (8.75), $J_{vz} = 0$, and $J_{zv} = I_{vz}$, the spin for the model of Figure 8.14 in the vicinity of $\nu = 0°$ can be written

$$\dot{\sigma} = \frac{H}{1 + \frac{2ml^2}{A} \frac{r^2}{1 - r^2}} \left(\frac{C - A}{AC} \cos \nu - \frac{2ml^2}{AC} \frac{r^2}{1 - r^2} \cos 2\nu\right). \tag{8.98}$$

At $\nu = 0°$ the spin is negative, i.e., it has not changed its sign throughout its approach to 0°, if

$$C - A > 2ml^2 \frac{r_0^2}{1 - r_0^2}. \tag{8.99}$$

A negative spin means a negative $\dot{\nu}$ (equation (8.20)), i.e., the gyro drifts towards $\nu = 0°$, or in other words the attitude $\nu = 0°$ is stable.

On the other hand, a sign change has occurred (at $\nu = \nu_1$), when

$$0 < C - A < 2ml^2 \frac{r_0^2}{1 - r_0^2} \tag{8.100}$$

which means that the gyro will drift with $\dot{\nu} < 0$ towards ν_1 when $90° > \nu > \nu_1$, and it will drift with $\dot{\nu} > 0$ towards ν_1 with $0 < \nu < \nu_1$. In other words, when inequality (8.100) applies, then the attitude ν_1 is stable, while the attitude $\nu = 0°$ has become unstable.

The precession at $\nu = 0°$, by the way, for the model of Figure 8.14 is

$$\dot{\psi} = \frac{H}{A} \frac{1 - \frac{4ml^2}{C} \frac{r_0^2}{1-r_0^2}}{1 + \frac{2ml^2}{A} \frac{r_0^2}{1-r_0^2}} \tag{8.101}$$

while its spin at $\nu = 0°$ is

$$\dot{\sigma} = -\frac{H}{A} \frac{\frac{C-A}{C} - \frac{2ml^2}{C} \frac{r_0^2}{1-r_0^2}}{1 + \frac{2ml^2}{A} \frac{r_0^2}{1-r_0^2}}. \tag{8.102}$$

As long as inequality (8.99) applies, spin and precession have opposite signs.

8.7 Effect of Deformation on Kinetic Energy and Strain Energy

When there is deformation, then, from equation (7.42) with the terms $A\omega_u^2$ and $2Y_{uz}\omega_u\omega_z$ neglected because they are small and of second order, and from equations (8.91) and (8.92),

$$T = \frac{1}{2}\left(\frac{H^2}{A}\sin^2\nu + \frac{H^2}{C}\cos^2\nu - 2I_{vz}\frac{H^2}{AC}\sin\nu\cos\nu\right) \tag{8.103}$$

at an attitude $\nu = \nu' + \Delta\nu$. Expressed in terms of ν' and a small $\Delta\nu$,

$$T = \frac{1}{2}\left(\frac{H^2}{A}\sin^2\nu' + \frac{H^2}{C}\cos^2\nu' - 2\left[(A-C)\Delta\nu + I_{vz}\right]\frac{H^2}{AC}\sin\nu'\cos\nu'\right). \tag{8.104}$$

A rigid gyro at an attitude ν' has a kinetic energy of

$$T_{rigid} = \frac{1}{2}\left(\frac{H^2}{A}\sin^2\nu' + \frac{H^2}{C}\cos^2\nu'\right). \tag{8.105}$$

Compared to the rigid gyro, the deforming gyro has traded some kinetic energy by deforming as expressed by I_{vz} and by readjusting its attitude by a (small) amount $\Delta\nu$. Both are absorbed by the elastic strain energy V stored in the gyro, such that

$$T_{rigid} = T + V \tag{8.106}$$

or

$$V = ((A - C)\Delta\nu + I_{vz})\frac{H^2}{AC}\sin\nu'\cos\nu'. \tag{8.107}$$

The attitude angle adjustment is

$$\Delta\nu = \frac{1}{A - C}\left[\frac{ACV}{H^2\sin\nu'\cos\nu'} - I_{vz}\right] \tag{8.108}$$

and remains small as long as both I_{vz} and $ACV/H^2\sin\nu'\cos\nu'$ remain small in comparison to $A - C$. When $\Delta\nu << \nu$, then $\nu' \simeq \nu$ and ν may be used instead of ν' throughout. Equation (8.106) is an obvious invitation to use $\dot{T}_{rigid}$ instead of $\dot{T} + \dot{V}$ in the energy rate balance (5.60)

$$\dot{U} = \dot{T} + \dot{V} + \dot{D} \tag{8.109}$$

which becomes, thus, a much simpler

$$\dot{U} = \dot{T}_{rigid} + \dot{D}. \tag{8.110}$$

In the interest of notational brevity, the subscript *rigid* is frequently dropped. The use of equation (8.105) in attitude drift studies instead of equation (8.103) requires that the attitude adjustment $\Delta\nu$ be small.

As long as $\Delta\nu$ remains small, the present theory can then also be applied during passage through a resonance of a substructure. Since I_{vz} is small by definition, it remains to ensure that the strain energy V is small as well. This is the case, e.g., when a small flexible antenna boom is mounted on a large "rigid" gyro bus, i.e., when the sole elastic energy storage is in small substructures.

Suggested Reading

1. Dong, W.N.; A.L. Schlack, Jr. "Stability of a Spinning Satellite with Flexible Antennas," *AIAA Journal,* 12, 2, 1974, 1737–1739.

2. Landon, V.D. "Early Evidence of Stabilization of a Vehicle Spinning about its Axis of Least Inertia," *Proceedings,* Symposium on Attitude Stabilization and Control of Dual-Spin Spacecraft, USAF Report No. SAMSO-TR-68-191, 1968, 9–10.

3. Leipholz, H.H.E. *Stability Theory,* John Wiley, 1987, 359 pp.

4. Meirovitch, L. *Methods of Analytical Dynamics,* McGraw-Hill, 1970, 524 pp.

5. Rimrott, F.P.J. "Nutation Accretion or Decay due to Internal Energy Dissipation," *Transactions of the CSME,* 10, 4, 1986, 219–232.

6. Rimrott, F.P.J.; Y. Yu. "Deviation from Rigidity in Gyrodynamics," *Rakenteiden Mechaniikka,* Helsinki, 20, 2, 1987, 47–68.

Problems

8.1. A satellite ($m = 1200$ kg) has the shape of a rectangular parallelepiped and its mass is uniformly distributed ($a = 3$ m, $b = 4$ m, $c = 12$ m). It is located in space such that no external moment acts on the satellite. It happens to spin at 3 rpm about its body-fixed z-axis. (a) Compute the kinetic energy of the spin motion of the satellite. After one year, the satellite is found to spin about its body-fixed x-axis. (b) Determine the new angular velocity, (c) compute the kinetic energy, and (d) calculate the energy loss. (e) Show the angular momentum vector and sketch the satellite's attitude before and after.

8.2. A torquefree symmetrical gyro ($m = 1200$ kg) has the shape of a square parallelepiped and its mass is uniformly distributed. The dimensions are $a = b = 2$ m, $c = 6$ m. The gyro was absolutely rigid and had a velocity vector with the components $\omega_x = 3$, $\omega_y = 4$, $\omega_z = 12$ rad/s at some time $t = 0$ in the past. (a) Determine the inertia tensor of the gyro if x, y, and z are the principal axes. (b) Determine the velocity component ω_ν, at $t = 0$. (c) Determine the nutation angle $\nu = \nu_0$, at $t = 0$. (d) Determine the kinetic energy T_0, at $t = 0$. (e) What angle ν_f characterizes the stable mode in the long run for this gyro? What is the kinetic energy T_f? (f) At time $t = 0^+$, an internal damper of a power of 0.001 W became active, which remained constant until the stable mode had been reached, at what time $t = t_f$ would the gyro have reached its stable mode?

8.3. Knowing that $\mathbf{H} =$ constant, show that $\dot{T} = -(A-C)\omega_v\omega_z\dot{\nu}$ by using

$$\mathbf{H} = [\mathbf{e}_u\ \mathbf{e}_v\ \mathbf{e}_z]\begin{bmatrix} 0 \\ A\omega_v \\ C\omega_z \end{bmatrix}, \qquad \boldsymbol{\omega} = [\mathbf{e}_u\ \mathbf{e}_v\ \mathbf{e}_z]\begin{bmatrix} \dot{\nu} \\ \omega_v \\ \omega_z \end{bmatrix},$$

$$\boldsymbol{\Omega} = [\mathbf{e}_u\ \mathbf{e}_v\ \mathbf{e}_z]\begin{bmatrix} \dot{\nu} \\ \omega_v \\ \frac{C}{A}\omega_z \end{bmatrix}.$$

8.4. Given is a flattened symmetrical gyro, precessing steadily at $\nu = \nu_0$ (no energy sink). Then an energy sink becomes effective, until the stable reference attitude ($\nu = 0°$) is reached. Determine the final spin $\dot{\sigma}_f$ in terms of the initial spin and the initial nutation angle, and the final precession $\dot{\psi}_f$ in terms of the initial precession and the initial nutation angle.

8.5. A symmetrical gyro of 400 kg mass, and gyration radii $i_x = i_y = 0.5$ m and $i_z = 0.7$ m, precesses steadily at $\nu_0 = 50°$. Its angular velocity has a magnitude of $\omega_0 = 0.9$ rad/s. Determine (a) angular momentum, (b) kinetic energy, (c) spin, and (d) precession. (e) Make a sketch. Then an internal energy sink causes the torquefree gyro to reach its reference attitude ($\nu = 0°$) eventually, for which you are asked to find its (f) angular velocity, (g) kinetic energy, (h) spin, and (i) precession. (j) Make a sketch.

8.6. A symmetrical gyro of 200 kg mass, and gyration radii $i_x = i_y = 0.4$ m and $i_z = 0.5$ m precesses steadily at a nutation angle of $\nu_0 = 40°$. Its angular velocity has a magnitude of $\omega_0 = 1.2$ rad/s. Determine (a) angular momentum, (b) kinetic energy, (c) spin, and (d) precession. (e) Make a sketch.

Then an internal energy sink saps away energy until a stable attitude is reached, for which you are asked to find (f) the angular velocity, (g) kinetic energy, (h) how much energy was lost to the internal energy sink, (i) the spin, and (j) the precession. (k) Make a sketch.

8.7. Given is

$$\dot{\psi} = \frac{C}{A \cos \nu} \omega_z .$$

If ν and ω_z are considered as varying with time, while C and A are both constant, show that $\ddot{\psi} = 0$ during nutation decay.

8.8. A torquefree gyro ($m = 600$ kg) has the shape of a square parallelepiped and its mass is considered uniformly distributed. The external dimensions are $a = b = 5$ m, $c = 1$ m. The gyro can be assumed to be absolutely rigid for $t \leq 0$. At $t = 0$ it has an angular velocity of

$$\boldsymbol{\omega} = [\mathbf{e}_x \ \mathbf{e}_y \ \mathbf{e}_z] \begin{bmatrix} 3 \\ 4 \\ 12 \end{bmatrix} \text{rad/s}.$$

(a) Determine the inertia tensor of the gyro for $Oxyz$ as principal axes system, and assuming the gyro as rigid. (b) Determine the magnitude of the angular velocity, and of the component ω_v. (c) Determine the angular momentum. (d) Determine the nutation angle. (e) Determine the gyro's kinetic energy.

For the time interval $0 < t < t_f$ a massless damper is acting, whose power is $\dot{T} = 0.001$ W $=$ constant, until the stable attitude is reached at $t = t_f$. (f) What is the gyro's final kinetic energy? (g) How long does the gyro take to reach the stable attitude?

8.9. A torquefree disk gyro with

$$[I] = \begin{bmatrix} 105 & 0 & 0 \\ 0 & 105 & 0 \\ 0 & 0 & 200 \end{bmatrix} \text{m}^2\text{kg}$$

has originally an $\omega_z = 30$ rad/s and a nutation angle of $\nu = 30°$. (a) Determine ω'_z, ω_v, ω, $\dot{\sigma}$, $\dot{\psi}$, H_v, H_z, H, and T. Make a sketch showing the gyro's attitude. (b) After a long period of time, the gyro has reached its stable attitude with $\nu = 0°$. Determine ω_z, ω'_z, ω_v, ω, $\dot{\sigma}$, $\dot{\psi}$, H_v, H_z, H and T. Make a sketch.

8.10. Show that the magnitude change $\dot{\omega}$ of the angular velocity and the damping modified inertia product Y_{uz} are related by

$$\dot{\omega} = \frac{A^2 - C^2}{A^2 C^2 \omega} \omega_v \omega_z^2 Y_{uz}.$$

8.11. Calculate the maximum value that the angle μ between $\mathbf{H}$ = constant and $\boldsymbol{\omega}$ can assume if

$$[I] = \begin{bmatrix} A & 0 & 0 \\ 0 & A & 0 \\ 0 & 0 & 2A \end{bmatrix}.$$

8.12. Calculate the maximum value that the angle μ between $\mathbf{H}$ = constant and $\boldsymbol{\omega}$ can assume if

$$[I] = \begin{bmatrix} 5C & 0 & 0 \\ 0 & 5C & 0 \\ 0 & 0 & C \end{bmatrix}.$$

8.13. Given is a rigid gyro bus of 1.2 m diameter, 6 m length, and 600 kg mass uniformly distributed. Four massless booms of 1 m length each and $k = 10\,000$ N/m and $c = 100$ Ns/m each carry a tip mass of 1 kg, as shown. The gyro is torquefree and has an angular momentum of $H = 12\,000\,\text{Ws}^2$. Determine (a) the eigenfrequency ω_n of each boom; (b) the damping ratio ζ; (c) the principal inertia tensor of the undeformed gyro; (d) the velocity components ω_v, ω_z, $\dot{\psi}$, and $\dot{\sigma}$, for $\nu = 30°$; (e) the inertial forces acting on each tip mass; (f) the partial inertia products J_{vz}, J_{zv}, J_{uz}, J_{zu}, and the phase angle ϕ; (g) the elastic strain energy V stored in the four booms; (h) the nutation angle adjustment $\Delta\nu$; (i) the attitude drift rate $\dot{\nu}$; (j) the energy dissipation rate $\dot{D}$; and (k) the kinetic energy of the rigid gyro and of the deforming gyro.

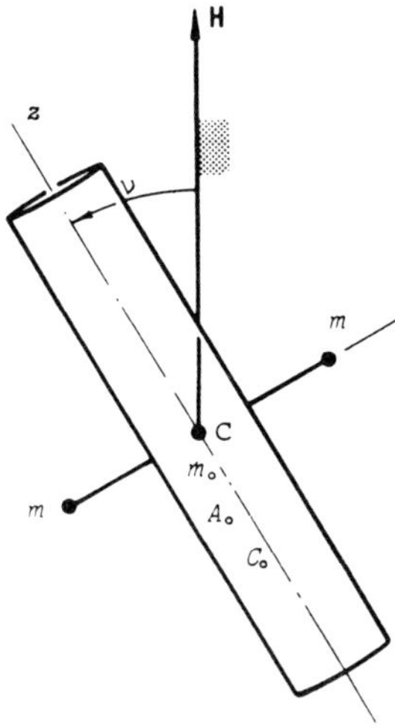

8.14. Given is a rigid gyro bus of 1.2 m diameter, 6 m length, and 600 kg mass. Four massless booms of 1 m length each and $k = 6773$ N/m and $c = 82.3$ Ns/m each carry a tip mass of 1 kg. The gyro is torquefree and has an angular momentum of $H = 12\,000$ Ws^2. Determine (a) the eigenfrequency ω_n of each boom; (b) the damping ratio ζ; (c) the principal inertia tensor of the undeformed gyro; (d) the velocity components ω_v, ω_z, $\dot{\psi}$, and $\dot{\sigma}$, for $\nu = 30°$; (e) the inertial forces acting on each tip mass; (f) the partial inertia products J_{uz}, J_{zv}, J_{uz}, J_{zu}, and the phase angle ϕ; (g) the elastic strain energy V stored in the four booms; (h) the nutation angle adjustment $\Delta\nu$; (i) the attitude drift rate $\dot{\nu}$; (j) the dissipation rate $\dot{D}$; and (k) the kinetic energy of the rigid gyro and of the deforming gyro.

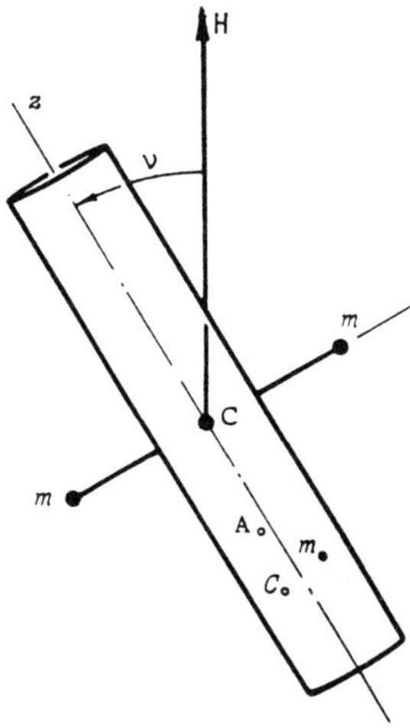

8.15. Repeat problem 8.14, this time with $k = 3600$ N/m and $c = 60$ Ns/m.

8.16. Show that for a torquefree gyro with $J_{vz} = 0$,

$$\omega_v = \frac{H}{A}\sin\nu\left(1 - \frac{J_{zv}}{A\tan\nu}\right)$$

$$\omega_z = \frac{H}{C}\cos\nu\left(1 - \frac{J_{zv}}{A}\tan\nu\right)$$

and that

$$T = \frac{1}{2}\left(\frac{H^2}{A}\sin^2\nu + \frac{H^2}{C}\cos^2\nu - 2\frac{H^2 J_{zv}}{AC}\sin\nu\cos\nu\right).$$

8.17. Given is a rigid gyro bus of 1.2 m diameter, 6 m length, and 600 kg mass uniformly distributed. Two massless booms of 1 m length each and $k = 10^6$ N/m and $c = 1$ Ns/m each carry a tip point mass of 1 kg, as shown. The gyro is torquefree and has an angular momentum of 12 000Ws^2. Determine (a) the eigenfrequency ω_n of each boom; (b) the damping ratio ζ; (c) the principal inertia tensor of the undeformed gyro; (d) the velocity components ω_v, ω_z, $\dot{\psi}$, and $\dot{\sigma}$, for $\nu = 30°$; (e) the inertial force acting on each tip mass; (f) the partial inertia products J_{vz}, J_{zv}, J_{uz}, J_{zu}, and the phase angle ϕ; (g) the elastic strain energy V stored in the two booms; (h) the nutation angle adjustment $\Delta\nu$; (i) the attitude drift rate $\dot{\nu}$; (j) the dissipation rate $\dot{D}$; and (k) the kinetic energy of the rigid gyro and of the deforming gyro.

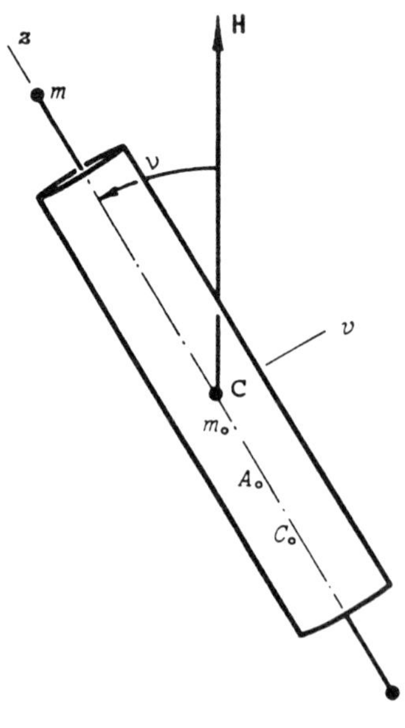

8.18. A torquefree spacecraft consists of an axisymmetric rigid bus with two longitudinal antennas, spring- and dashpot-mounted very close to the gyro's mass center. When undeformed, the gyro's inertia moments are $B = A = 1750\,\mathrm{m^2kg}$ and $C = 1800\,\mathrm{m^2kg}$. Each longitudinal antenna consists of a rigid massless boom of 10 m length and a tip mass of 1.2 kg, and has a structural eigenfrequency of $\omega_n = 1.2$ rad/s. The gyro has an angular momentum of $H = 1080\,\mathrm{Ws^2}$, and rotates about the axis of symmetry. Is this rotation stable?

9

Despin

Satellites are often carried to the insertion point of their orbit on rockets, which spin at a certain rate for reasons of stability.

After the satellite is separated from the last rocket stage, its spin must be adjusted to the rate prescribed by the mission profile. This spin is usually less than the rocket spin. A reduction of the spin frequency is called *despin.* Despin can be achieved by several means, such as by the firing of small rockets, by an extension of booms, or by unwinding and subsequent jettisoning of yo-yo masses. The latter two are discussed in the present chapter. The effect of the firing of rockets will be dealt with in Chapter 10.

9.1 Despin by Yo-Yo Masses

Figure 9.1 shows two yo-yo masses, by means of which the spin of a gyrosatellite can be reduced, made to vanish, or even reversed, while the total angular momentum is conserved.

The two yo-yo masses, each of mass m, are released simultaneously. The cables of length l unwind from the gyro body. During the unwinding, the total inertia moment increases. The principle of a conserved angular momentum then demands that the spin decreases. At the conclusion of the despin process, the yo-yo masses are let go by disengaging the cables. The gyrosatellite has assumed the spin frequency desired.

We work with floating unit vectors $\mathbf{e}_x$ and $\mathbf{e}_y$. The z-axis of the gyro is the spin axis. There are no external forces or moments acting on the system. The angle between the original location of each yo-yo mass and the momentary location is θ (see Figure 9.1). The radius of the gyro from which the yo-yo masses unwind is b. The mass of the gyro without yo-yos is M, its inertia moment about the z-axis is C.

The position of each yo-yo mass is given by the position vector

$$\mathbf{r} = [\mathbf{e}_x \ \mathbf{e}_y \ \mathbf{e}_z] \begin{bmatrix} b \\ -b\theta \\ 0 \end{bmatrix} . \tag{9.1}$$

The unit vectors rotate about the z-axis at a frequency $\dot{\theta}$, while the gyro itself spins at ω. We thus obtain for the absolute angular velocity of each yo-yo mass, as well as for the absolute angular velocity of the coordinate

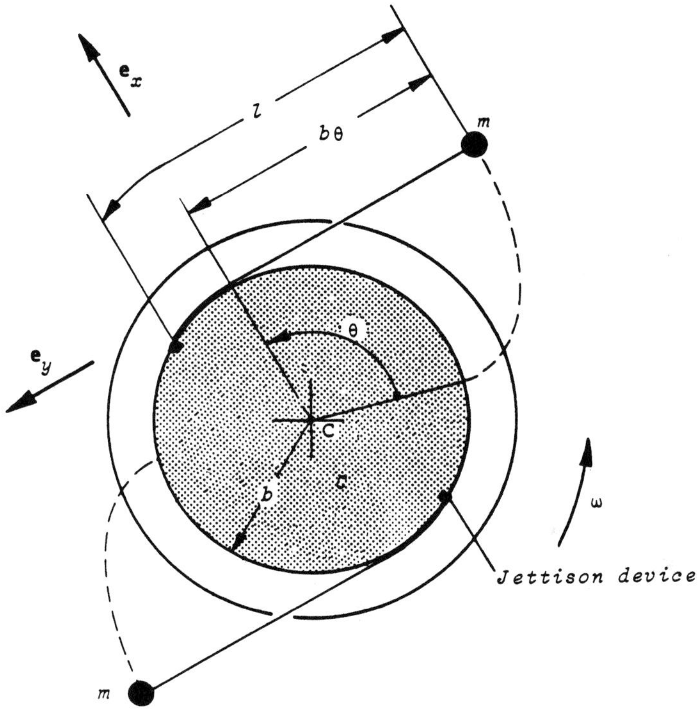

FIGURE 9.1. Axisymmetric gyro bus with yo-yo masses.

system used,

$$\boldsymbol{\Omega} = [\mathbf{e}_x \ \mathbf{e}_y \ \mathbf{e}_z] \begin{bmatrix} 0 \\ 0 \\ \omega + \dot{\theta} \end{bmatrix}. \tag{9.2}$$

The velocity of one yo-yo mass is

$$\dot{\mathbf{r}} = \overset{\circ}{\mathbf{r}} + \boldsymbol{\Omega} \times \mathbf{r} = [\mathbf{e}_x \ \mathbf{e}_y \ \mathbf{e}_z] \begin{bmatrix} b\theta(\omega + \dot{\theta}) \\ b\omega \\ 0 \end{bmatrix}. \tag{9.3}$$

The angular momentum of one yo-yo mass is

$$\mathbf{r} \times m\dot{\mathbf{r}} = mb^2(\omega + \theta^2(\omega + \dot{\theta}))\mathbf{e}_z \tag{9.4}$$

and the angular momentum of the gyro (without yo-yos) is

$$C\omega\mathbf{e}_z \tag{9.5}$$

giving a total angular momentum of

$$\mathbf{H} = [\mathbf{e}_x \ \mathbf{e}_y \ \mathbf{e}_z] \begin{bmatrix} 0 \\ 0 \\ C\omega + 2mb^2(\omega + \theta^2(\omega + \dot{\theta})) \end{bmatrix}. \tag{9.6}$$

If we use for the initial spin ω_0 at $\theta = 0$ and observe that the angular momentum is conserved, then

$$C\omega_0 + 2mb^2\omega_0 = C\omega + 2mb^2\omega + 2mb^2\theta^2(\omega + \dot{\theta})$$

or

$$\omega_0 - \omega = \frac{2mb^2}{C_0 + 2mb^2}\theta^2(\omega + \dot{\theta}). \tag{9.7}$$

The kinetic energy is

$$T = \frac{1}{2}C\omega^2 + \frac{1}{2}(2m)b^2(\omega^2 + \theta^2(\omega + \dot{\theta})^2). \tag{9.8}$$

It is also conserved and we can equate the kinetic energy at ω_0 and $\theta = 0$ with that at any other instant

$$(C + 2mb^2)\omega_0^2 = (C + 2mb^2)\omega^2 + 2mb^2\theta^2(\omega + \dot{\theta})^2$$

or

$$\omega_0^2 - \omega^2 = \frac{2mb^2}{C + 2mb^2}\theta^2(\omega + \dot{\theta})^2. \tag{9.9}$$

We now divide equation (9.9) by equation (9.7) to get

$$\omega_0 + \omega = \omega + \dot{\theta}$$

or

$$\dot{\theta} = \omega_0. \tag{9.10}$$

Equation (9.10) shows that the relative velocity of cable unwinding is constant and equal to the original spin frequency of the gyro.

We now use equations (9.7) and (9.10) to calculate the spin frequency of the gyro:

$$\omega = \omega_0 \frac{C + 2mb^2 - 2mb^2\theta^2}{C + 2mb^2 + 2mb^2\theta^2}. \tag{9.11}$$

For a cable of length l which disengages itself at the end of the unwinding process, the angle at the moment of disengagement is

$$\theta_f = \frac{l}{b}. \tag{9.12}$$

The time between yo-yo mass release ($\theta = 0°$) and cable disengagement ($\theta = \theta_f$) is

$$t = \frac{l}{b\omega_0} \tag{9.13}$$

and the final spin, at that moment and thereafter, is

$$\omega_f = \omega_0 \frac{C + 2mb^2 - 2ml^2}{C + 2mb^2 + 2ml^2}. \tag{9.14}$$

If initial and final spin frequency are prescribed, the cable length required is

$$l = \sqrt{\frac{C + 2mb^2}{2m} \frac{\omega_0 - \omega_f}{\omega_0 + \omega_f}}. \tag{9.15}$$

If the final spin is to vanish, i.e., if $\omega_f = 0$, that can be achieved by means of a cable of length

$$l = \sqrt{\frac{C + 2mb^2}{2m}}. \tag{9.16}$$

An inspection of equation (9.14) also indicates that the spin direction can be reversed. The (theoretical) limit is a cable of infinite length, $l = \infty$, for which equation (9.14) gives indeed

$$\omega_f = -\omega_0$$

indicating that the spin would be fully reversed.

Differentiation of equation (9.11) gives the angular acceleration of the satellite during yo-yo mass unwinding and is

$$\dot{\omega} = -8mb^2\theta\omega_0^2 \frac{C + 2mb^2}{(C + 2mb^2 + 2mb^2\theta^2)^2}. \tag{9.17}$$

It reaches its largest magnitude just prior to cable disengagement, i.e., when $b\theta_f = l$:

$$\dot{\omega}_f = -8mbl\omega_0^2 \frac{C + 2mb^2}{(C + 2mb^2 + 2ml^2)^2}. \tag{9.18}$$

The force F in each cable is

$$F = -\frac{C\dot{\omega}}{2b} = 4mb\theta\omega_0^2 \frac{C(C + 2mb^2)}{(C + 2mb^2 + 2mb^2\theta^2)^2}. \tag{9.19}$$

It rises from $F = 0$ when $\theta = 0$ to

$$F_f = 4ml\omega_0^2 \frac{C(C + 2mb^2)}{(C + 2mb^2 + 2ml^2)^2}$$

where $\theta = \theta_f$.

A perfect symmetry of the yo-yo masses, the cables, and the attachments, and perfect simultaneity of yo-yo mass releases and cable disengagements are required to maintain the attitude of the gyro satellite. This is a tall order and thus we conclude that yo-yo masses are only marginally suitable for satellite despin. A single yo-yo mass is, however, a very common solution in cases where, e.g., a spent rocket casing is to be thrown off its attitude, to prevent interference with the recently separated spacecraft.

Because of general weight restrictions, yo-yo masses must be kept small.

9.2 Despin by Boom Extension

In order to develop some idea as to how despin can be achieved by extending booms, we shall study the case of a symmetrical extension of two masses m located at the tip of massless booms. The gyrosatellite has a mass M and an inertia moment C, both without the tip masses m. The gyrosatellite initially spins at a frequency ω_0. The tip masses are initially located at the radius b.

In order to despin the satellite, i.e., to reduce its spin frequency, the two booms, each with a mass m at the tip, are slowly extended (Figure 9.2). The total angular momentum of satellite plus tip masses is conserved since no external torque is acting.

We thus have

$$H = (C + 2mb^2)\omega_0 = (C + 2mx^2)\omega \tag{9.20}$$

giving

$$\omega = \frac{C + 2mb^2}{C + 2mx^2}\omega_0 \tag{9.21}$$

for the spin frequency of the satellite.

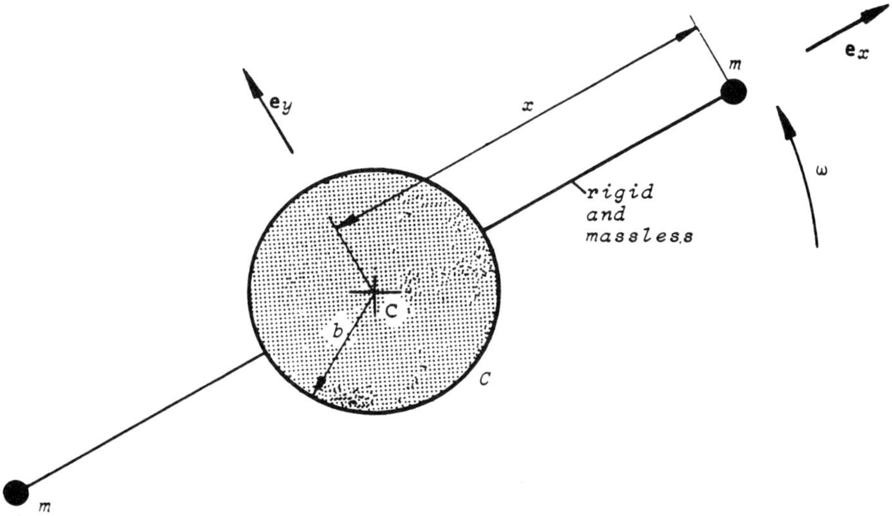

FIGURE 9.2. Axisymmetric gyro bus with extending boom.

As an example we consider a satellite of 200 kg mass, plus 2 × 1 kg for the two masses to be extended. The satellite has a gyration radius of 0.2 m and an outer radius of 0.5 m. Initially it has a spin of 100 rev/min. What is the spin after the booms have been extended 9.5 m each? Equation (9.21) gives

$$\omega = \frac{200(0.2^2) + 2(1)0.5^2}{200(0.2^2) + 2(1)10^2} 100 = 4.086 \text{ rev/min}.$$

The method of extending booms (in contradistinction to yo-yo masses) is well suited to achieve an accurate adjustment of the spin frequency, which can be reduced or increased, depending upon whether the booms are extended or retracted. On the other hand, it is not possible to despin satellites completely by extending booms.

The booms are subjected to bending during the extension process. Since they have to be light, they are usually thin-walled tubes (MacNaughton, 1963; Rimrott, 1965), which can only withstand a limited bending moment. To calculate the bending moment, assume absolutely rigid and massless booms. The acceleration of one tip mass is

$$\ddot{\mathbf{r}} = [\mathbf{e}_x \ \mathbf{e}_y \ \mathbf{e}_z] \begin{bmatrix} \ddot{x} - x\omega^2 \\ x\dot{\omega} + 2\dot{x}\omega \\ 0 \end{bmatrix}. \tag{9.22}$$

Only the y-component leads to a bending moment, which at the built-in end has the magnitude

$$M = \frac{2mC(C + 2mb^2)}{(C + 2mx^2)^2}(x - b)\omega_0 \dot{x}. \tag{9.23}$$

This bending moment is, of course, very small when the extension progress begins, i.e., when $x \simeq b$. Later on, it reaches a maximum. Setting $dM/dx = 0$, we find that the maximum occurs at an extension of

$$x = \frac{2}{3} b \left[1 + \sqrt{1 + \frac{3}{4} \frac{C}{2mb^2}} \right] . \tag{9.24}$$

Applying equation (9.24) for the case of the previous example, we find that the maximum bending moment occurs when

$$x = 1.535\,\mathrm{m}.$$

9.3 Eccentric Installation

By installing the extension booms eccentrically, i.e., such that they are displaced some distance from the gyro centerline, as, e.g., indicated on the square, symmetrical gyro of Figure 9.3 by the distance e, it is possible to reduce (but also to increase!) the bending moment.

We assume massless booms and a concentrated mass m at the tip of two symmetrically extending booms. The angular momentum of the gyro, including tip masses, is conserved, since no external torque is assumed to be acting. Initially, while one tip mass is located at $x = b$, $y = e$, and $z = 0$, and the other at $x = -b$, $y = -e$, and $z = 0$, the angular momentum is

$$H = [C + 2m(e^2 + b^2)]\omega_0. \tag{9.25}$$

When the tip masses are extending at the relative velocity $\dot{x}$, the angular momentum is

$$H = [C + 2m(e^2 + x^2)]\omega - 2me\dot{x}. \tag{9.26}$$

Equating equations (9.25) and (9.26) one obtains for the gyro's angular velocity

$$\omega = \frac{[C + 2m(e^2 + b^2)]\omega_0 + 2me\dot{x}}{C + 2m(e^2 + x^2)}. \tag{9.27}$$

The acceleration of one tip mass is

$$\mathbf{a} = \overset{\circ}{\mathbf{v}} + \boldsymbol{\omega} \times \mathbf{v}.$$

With

$$\overset{\circ}{\mathbf{v}} = [\mathbf{e}_x \; \mathbf{e}_y \; \mathbf{e}_z] \begin{bmatrix} \ddot{x} - e\dot{\omega} \\ \dot{x}\omega + x\dot{\omega} \\ 0 \end{bmatrix}$$

and

$$\boldsymbol{\omega} \times \mathbf{v} = \begin{vmatrix} \mathbf{e}_x & \mathbf{e}_y & \mathbf{e}_z \\ 0 & 0 & \omega \\ \dot{x} - e\omega & x\omega & 0 \end{vmatrix} = [\mathbf{e}_x \; \mathbf{e}_y \; \mathbf{e}_z] \begin{bmatrix} -x\omega^2 \\ \dot{x}\omega - e\omega^2 \\ 0 \end{bmatrix} \tag{9.28}$$

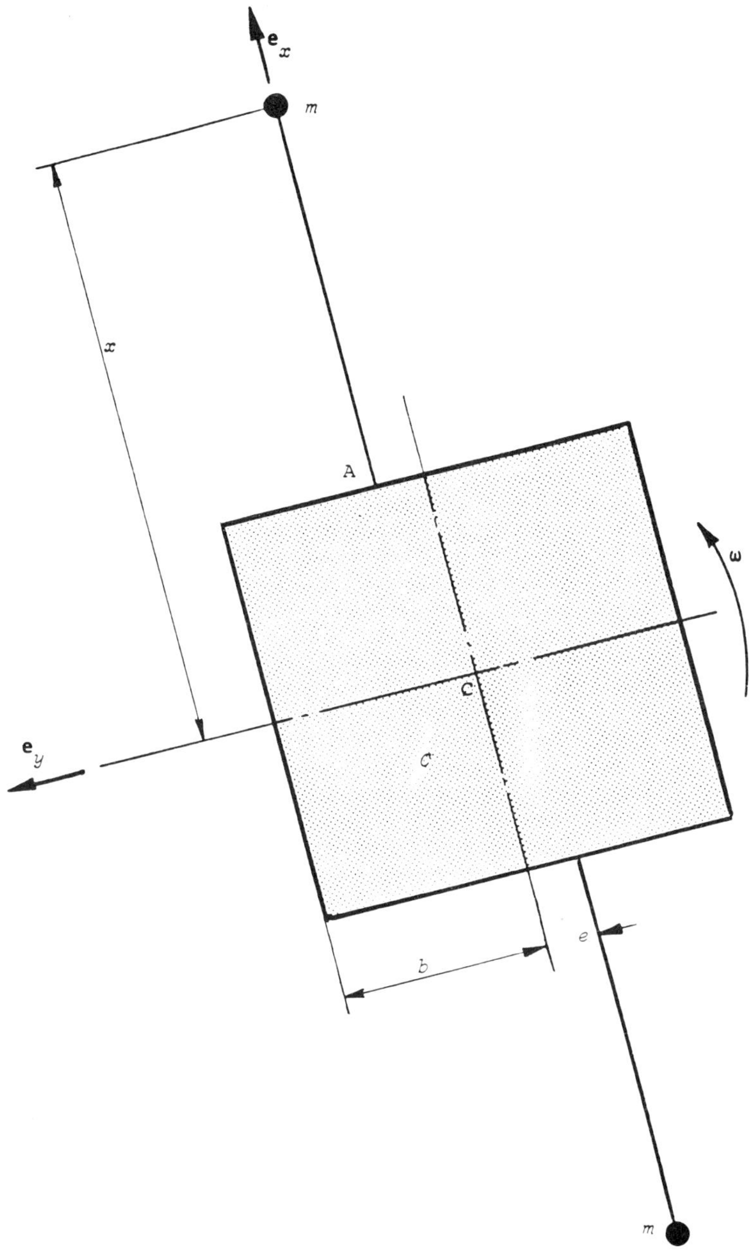

FIGURE 9.3. Eccentrically mounted extension booms.

the acceleration of one tip mass becomes

$$\mathbf{a} = [\mathbf{e}_x \ \mathbf{e}_y \ \mathbf{e}_z] \begin{bmatrix} \ddot{x} - x\omega^2 - e\dot{\omega} \\ x\dot{\omega} + 2\dot{x}\omega - e\omega^2 \\ 0 \end{bmatrix}. \tag{9.29}$$

Only the y-component contributes to the bending moment (as long as the booms are assumed rigid). Then the bending moment at the built-in end, i.e., point A in Figure 9.3, amounts to

$$M = m(x\dot{\omega} + 2\dot{x}\omega - e\omega^2)(x - b). \tag{9.30}$$

We now restrict the problem to a constant extension speed, $\ddot{x} = 0$. From equation (9.27), the angular acceleration can then be shown to be

$$\dot{\omega} = -\frac{[C + 2m(e^2 + b^2)]\omega_0 + 2me\dot{x}}{[C + 2m(e^2 + x^2)]^2} 4mx\dot{x}. \tag{9.31}$$

With the help of equation (9.31), the bending moment of equation (9.30) can now be expressed by

$$M = \frac{2m(C + me^2)\{[C + 2m(e^2 + b^2)]\omega_0 + 2me\dot{x}\}}{[C + 2m(e^2 + x^2)]^2}(x - b)\cdot$$
$$\left[\dot{x} - \frac{C + 2m(e^2 + b^2)}{2(C + me^2)} e\omega_0\right]. \tag{9.32}$$

Forming the derivative dM/dx (with $\dot{x}$ = constant) and equating it to zero provides us with the amount of extension, at which the bending moment at the built-in end becomes a maximum:

$$x = \frac{2}{3} b \left[1 + \sqrt{1 + \frac{3}{4} \frac{C + 2me^2}{2mb^2}}\right]. \tag{9.33}$$

An inspection of equation (9.32) indicates that by choosing an excentricity e that satisfies

$$\dot{x} = \frac{C + 2m(e^2 + b^2)}{2(C + me^2)} e\omega_0 \tag{9.34}$$

the bending moment at the built-in end vanishes. Equation (9.33) also shows that a sign reversal for e, or for ω_0, leads to a sign reversal for $\dot{x}$. Or, in other words, if the same boom on the same satellite is retracted, rather than extended, the bending moment will double, rather than vanish.

It will also be appreciated that a more complete analysis of the boom bending ought to take into consideration the distributed mass of the boom, the flexibility of the boom, changes in boom extension velocity, any rotation of the gyro's axis, and possibly other effects, such that the problem may become very complicated indeed.

9.4 Configuration Change Terms in Euler Equations

The angular momentum expression (9.26) in the preceding article points to a shortcoming in the derivation of the angular momentum as given by equation (1.5), for cases where the size and/or shape of the gyro changes, such as, e.g., shown in Figure 9.4 where two masses A and B are extending at a relative velocity $\dot{z}$ with respect to the rigid main body. Note that the common mass center does not shift within the main body. Let us return to the determination of the angular momentum, as shown in Chapter 1, but this time we use as velocity vector for the mass element dm:

$$\dot{\mathbf{r}} = \overset{\circ}{\mathbf{r}} + \boldsymbol{\omega} \times \mathbf{r} \tag{9.35}$$

in order to include changes in size and shape of the gyro. We used a body-fixed $Cxyz$ coordinate system (assuming for convenience that part of the gyro is and remains rigid) originating at the mass center C.

The angular momentum $\mathbf{H} = \int_m \mathbf{r} \times \dot{\mathbf{r}}\, dm$ now becomes

$$\mathbf{H} = \int_m \{(\mathbf{r} \times \overset{\circ}{\mathbf{r}})dm + \mathbf{r} \times (\boldsymbol{\omega} \times \mathbf{r})dm\}. \tag{9.36}$$

Since

$$\overset{\circ}{\mathbf{r}} = [\mathbf{e}_x\ \mathbf{e}_y\ \mathbf{e}_z] \begin{bmatrix} \dot{x} \\ \dot{y} \\ \dot{z} \end{bmatrix} \tag{9.37}$$

the first cross product in equation (9.36) becomes

$$\mathbf{r} \times \overset{\circ}{\mathbf{r}} = \begin{vmatrix} \mathbf{e}_x & \mathbf{e}_y & \mathbf{e}_z \\ x & y & z \\ \dot{x} & \dot{y} & \dot{z} \end{vmatrix} = [\mathbf{e}_x\ \mathbf{e}_y\ \mathbf{e}_z] \begin{bmatrix} y\dot{z} - \dot{y}z \\ z\dot{x} - \dot{z}x \\ x\dot{y} - \dot{x}y \end{bmatrix} \tag{9.38}$$

and upon integration over the whole mass an *internal angular momentum* $\mathbf{K}$ is obtained

$$\int_m (\mathbf{r} \times \overset{\circ}{\mathbf{r}})dm = [\mathbf{e}_x\ \mathbf{e}_y\ \mathbf{e}_z] \begin{bmatrix} K_x \\ K_y \\ K_z \end{bmatrix} = \mathbf{K} \tag{9.39}$$

with

$$K_x = \int_m (y\dot{z} - \dot{y}z)dm \tag{9.40}$$

$$K_y = \int_m (z\dot{x} - \dot{z}x)dm \tag{9.41}$$

$$K_z = \int_m (x\dot{y} - \dot{x}y)dm. \tag{9.42}$$

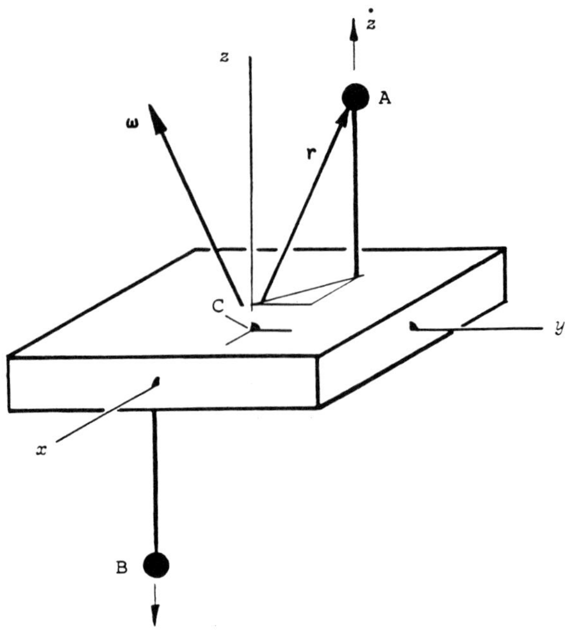

FIGURE 9.4. Gyro of changing size/shape.

The triple cross product in equation (9.36), upon integration over the whole mass, leads to the same terms as those appearing in equation (1.5). Combining the two, the angular momentum now becomes

$$\mathbf{H} = [\mathbf{e}_x \ \mathbf{e}_y \ \mathbf{e}_z] \left(\begin{bmatrix} K_x \\ K_y \\ K_z \end{bmatrix} + \begin{bmatrix} I_x & I_{xy} & I_{xz} \\ I_{xy} & I_y & I_{yz} \\ I_{xz} & I_{yz} & I_z \end{bmatrix} \begin{bmatrix} \omega_x \\ \omega_y \\ \omega_z \end{bmatrix} \right). \tag{9.43}$$

Writing

$$\mathbf{H} = \{\mathbf{e}_i\}^T \left(\{K_i\} + [I_{ij}]\{\omega_j\} \right) \tag{9.44}$$

the time derivative of the angular momentum then becomes

$$\dot{\mathbf{H}} = \{\mathbf{e}_i\}^T \left(\{\dot{K}_i\} + [\dot{I}_{ij}]\{\omega_j\} + [I_{ij}]\{\dot{\omega}_j\} \right) + \{\dot{\mathbf{e}}_i\}^T \left(\{K_j\} + [I_{ij}]\{\omega_j\} \right) \tag{9.45}$$

or

$$\dot{\mathbf{H}} = \overset{\circ}{\mathbf{H}} + \boldsymbol{\Omega} \times \mathbf{H} \tag{9.46}$$

where $\overset{\circ}{\mathbf{H}}$ represents the first term on the right side of equation (9.45), and $\boldsymbol{\Omega} \times \mathbf{H}$ the second term.

As an example, consider the satellite shown in Figure 9.5. Its total mass is $M + 2m$, consisting of the rigid main body mass M, and two masses m located at the tip of two massless booms.

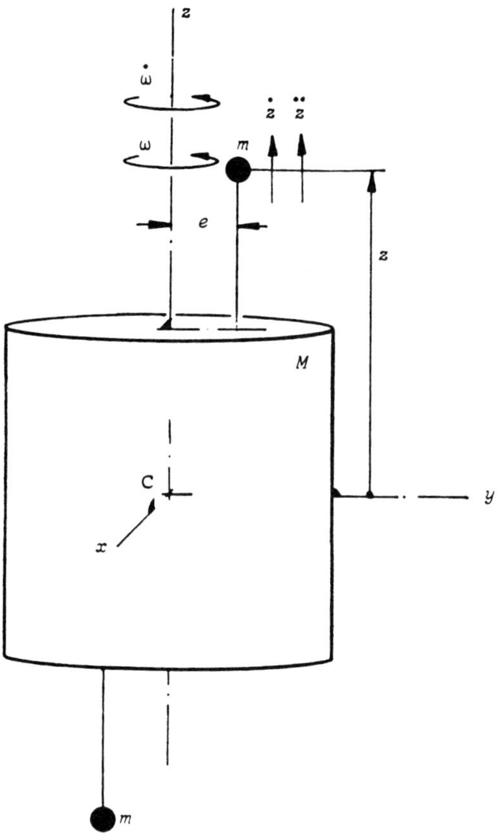

FIGURE 9.5. Axisymmetric gyro bus with moving extension masses.

While the booms extend at equal rates $\dot{z}$ and equal accelerations $\ddot{z}$, the lengths z of extension are also equal, such that the satellite's mass center remains at C. The main body is axisymmetric, i.e., the inertia moments are A, A, and C. The inertia tensor of the whole satellite, thus, is

$$[I] = \begin{bmatrix} A' & 0 & 0 \\ 0 & B' & -2mez \\ 0 & -2mez & C' \end{bmatrix}$$

with

$$A' = A + 2m(e^2 + z^2)$$
$$B' = A + 2mz^2$$
$$C' = C + 2me^2.$$

Let the satellite carry out a pure rotation about its z-axis, such that the angular velocity

$$\boldsymbol{\omega} = [\mathbf{e}_x \; \mathbf{e}_y \; \mathbf{e}_z] \begin{bmatrix} 0 \\ 0 \\ \omega \end{bmatrix}$$

and let it further be specified that the satellite is to continue to rotate purely about its z-axis with an angular speed increasing at a rate $\dot{\omega}$, while the two masses m are extending at $\dot{z}$ and $\ddot{z}$. Which external torque has to be applied to bring this about?

We commence the solution by establishing the angular momentum $\mathbf{H}$. For this, we need the terms

$$K_x = \int (y\dot{z} - \dot{y}z)dm = 2me\dot{z}$$

$$K_y = \int (z\dot{x} - \dot{z}x)dm = 0$$

$$K_z = \int (x\dot{y} - \dot{x}y)dm = 0$$

such that, according to equation (9.42),

$$\mathbf{H} = [\mathbf{e}_x \; \mathbf{e}_y \; \mathbf{e}_z] \left(\begin{bmatrix} 2me\dot{z} \\ 0 \\ 0 \end{bmatrix} + \begin{bmatrix} A' & 0 & 0 \\ 0 & B' & -2mez \\ 0 & -2mez & C' \end{bmatrix} \begin{bmatrix} 0 \\ 0 \\ \omega \end{bmatrix} \right)$$

or

$$\mathbf{H} = [\mathbf{e}_x \; \mathbf{e}_y \; \mathbf{e}_z] \begin{bmatrix} 2me\dot{z} \\ -2mez\,\omega \\ C'\,\omega \end{bmatrix}.$$

The time derivative is

$$\dot{\mathbf{H}} = \overset{\circ}{\mathbf{H}} + \boldsymbol{\omega} \times \mathbf{H}$$

which, with

$$\overset{\circ}{\mathbf{H}} = [\mathbf{e}_x \ \mathbf{e}_y \ \mathbf{e}_z] \begin{bmatrix} 2me\ddot{z} \\ -2me\dot{z}\omega - 2mez\dot{\omega} \\ C'\dot{\omega} \end{bmatrix}$$

and

$$\boldsymbol{\omega} \times \mathbf{H} = \begin{vmatrix} \mathbf{e}_x & \mathbf{e}_y & \mathbf{e}_z \\ 0 & 0 & \omega \\ 2me\dot{z} & -2mez\omega & C'\omega \end{vmatrix} = [\mathbf{e}_x \ \mathbf{e}_y \ \mathbf{e}_z] \begin{bmatrix} 2mez\omega^2 \\ 2me\dot{z}\omega \\ 0 \end{bmatrix}$$

yields

$$\dot{\mathbf{H}} = [\mathbf{e}_x \ \mathbf{e}_y \ \mathbf{e}_z] \begin{bmatrix} 2me(\ddot{z} + z\omega^2) \\ -2mez\dot{\omega} \\ (C + 2me^2)\dot{\omega} \end{bmatrix}.$$

Since according to Euler's law

$$\mathbf{M} = \dot{\mathbf{H}}$$

we conclude that an external moment

$$\mathbf{M} = [\mathbf{e}_x \ \mathbf{e}_y \ \mathbf{e}_z] \begin{bmatrix} 2me(\ddot{z} + 2\omega^2) \\ -2mez\dot{\omega} \\ (C + 2me^2)\dot{\omega} \end{bmatrix}$$

has to act on the satellite to keep it rotating about the z-axis.

Suggested Reading

1. MacNaughton, J.D. "Unfurlable Metal Structures for Spacecraft," *Canadian Aeronautical and Space Journal,* 9, 4, 1963, 103–106.

2. Rimrott, F.P.J. "Storable Tubular Extendible Member—A Unique Machine Element," *Machine Design,* 37, 28, 1965, 156–165.

Problems

9.1. A symmetrical satellite of 500 kg mass (without yo-yo weights) and gyration radius $i_z = 2$ m has an outside radius of 3 m. Two yo-yo masses of 1 kg each are installed around the outside and are to be released to reduce the satellite spin from 300 rev/min to 1 rev/min. (a) What is the length of each (massless) yo-yo cable? (b) What is the force transmitted in each yo-yo cable, just prior to release?

9.2. A perfectly symmetric yo-yo mass arrangement is proposed for a cylindrical gyro with uniform mass distribution. The gyro has a mass of 2400 kg (without yo-yo masses) and an outside radius of 2 m. Initially the gyro rotates at 10 rev/min. If it is to rotate at 3 rev/min in the opposite direction after the two yo-yo masses of 10 kg each have been released, (a) how long would each of the (massless) connecting cables have to be? (b) How much time does the process take? Name several reasons (c) in favor of and (d) against the proposed method of achieving rotation reversal.

9.3. If two concentrated masses of 1 kg each, located at the tip of massless, inflexible, extendible booms are employed to achieve a reduction of spin from 300 rev/min to 1 rev/min of a 500-kg gyro with an outside radius of 3 m and a gyration radius $i_z = 2$ m, how far must the booms be extended?

9.4. A gyroscope body changes in size and shape, such that for an arbitrary mass point dm located at $\mathbf{r}$ the velocity must be expressed by

$$\dot{\mathbf{r}} = \overset{\circ}{\mathbf{r}} + \boldsymbol{\omega} \times \mathbf{r}.$$

The gyroscope's mass center remains at $\mathbf{r} = 0$. (a) Determine the angular momentum of the body in terms of I_x, I_{xy}, etc.; ω_x, etc.; and $K_x = \int_m (y\dot{z} - \dot{y}z)dm$, etc. (b) Derive the Euler equations in terms of the same quantities and their time derivatives.

9.5. A gyro is freely spinning in space. The $Cxyz$ coordinate system is body-fixed. The original angular velocity is

$$\boldsymbol{\omega}_0 = [\mathbf{e}_x \; \mathbf{e}_y \; \mathbf{e}_z] \begin{bmatrix} 0 \\ 0 \\ \omega_0 \end{bmatrix}.$$

There is no external force and no external moment acting. The total mass of the satellite is $(M + 2m)$. It consists of an inner portion M of inertia moment C (about the z-axis) and a pair of (rigid) rods. Each rod has a mass m and is of length l when fully extended. It is possible to withdraw the rods and store them at $x = 0$ (i.e., the rod portion stored at $x = 0$ does not contribute to the inertia moment). Originally, the rods are fully withdrawn. (a) Determine the angular velocity when each rod is extended to a length x. (b) Determine the angular velocity when each rod is fully extended.

9.6. Plot the function

$$M = \frac{2mC(C + 2mb^2)}{(C + 2mx^2)^2}(x - b)\omega_0 v$$

for $m = 3$ kg, $C = 1000\,\mathrm{m^2kg}$, $b = 2$ m, $\omega_0 = 30\mathrm{rad/s^2}$, $v = 4$ m/s; for $b \leq x \leq 20$ m.

9.7. From a satellite originally spinning at ω_0, two massless, rigid booms are extended at a constant velocity v, each carrying a point mass m. The booms are mounted at an eccentricity e from the centerline as shown. (a) Calculate the bending moment M at the built-in end. (b) Plot the bending moment at the built-in end as a function of the extension x, for $m = 3$ kg, $C = 1000\,\mathrm{m^2kg}$, $b = 2$ m, $e = 1$ m, $\omega_0 = 30$ rad/s, $v = 4$ m/s; for $b \leq x \leq 20$ m. (c) Repeat the plot for $e = 0.25$ m. (d) For which value of e would $M = 0$ throughout the extension process?

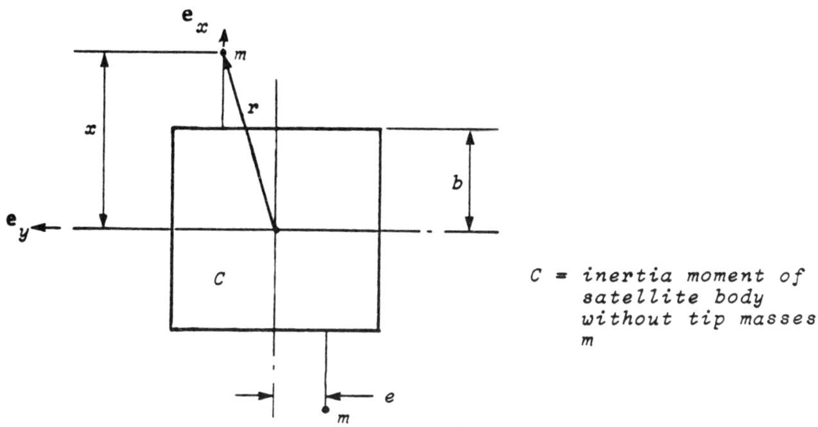

10

Torque about Body-Fixed Axes of an Axisymmetric Gyro

Satellites can be, and frequently are, equipped with minirockets for attitude and/or orbit adjustments. By their very nature, these minirockets are body-fixed.

Grammel (1950) refers to gyros under the action of a torque, whose components are given in the body-fixed coordinate system and are in no way dependent on the gyro's momentary attitude in space, as *self-excited.* A gyro's motion under the influence of a torque produced by body-fixed minirockets can be determined solely by means of the Euler equations. The mass loss of the satellite caused by the minirocket operation is normally neglected. Furthermore, minirockets typically produce a more or less constant thrust of, say, anywhere between 100 N and 4000 N, and that for a very short duration, say 80 ms.

In this chapter, only the attitude dynamics of axisymmetric gyros of constant inertia tensor will be discussed. For the case of a varying inertia tensor see, e.g., Jarmolow (1957).

10.1 Torque Perpendicular to Axis of Symmetry

Let us begin our investigation by looking at an axisymmetric gyro, with an angular momentum of

$$\mathbf{H} = [\mathbf{e}_x \ \mathbf{e}_y \ \mathbf{e}_z] \begin{bmatrix} A\omega_x \\ A\omega_y \\ C\omega_z \end{bmatrix} . \tag{10.1}$$

The gyro is subjected to a constant torque of magnitude M about, say, the x-axis, which is perpendicular to the axis of symmetry.

$$\mathbf{M} = [\mathbf{e}_x \ \mathbf{e}_y \ \mathbf{e}_z] \begin{bmatrix} M \\ 0 \\ 0 \end{bmatrix} . \tag{10.2}$$

We assume that the mass loss from the minirockets which produce the torque is so minute that the gyro's mass and consequently its inertia tensor remain constant. Euler's equations then assume the form

$$A\dot{\omega}_x + (C - A)\omega_z\omega_y = M \tag{10.3a}$$

$$A\dot{\omega}_y - (C - A)\omega_z\omega_x = 0 \tag{10.3b}$$

$$C\dot{\omega}_z = 0 \tag{10.3c}$$

and immediately lead to the conclusion that

$$\omega_z = \omega_{z0} = \text{constant}. \tag{10.4}$$

Differentiation of equation (10.3a) with respect to time and substitution of $\dot{\omega}_y$ from equation (10.3b) lead to

$$\ddot{\omega}_x + \left(\frac{C - A}{A}\omega_{z0}\right)^2 \omega_x = 0. \tag{10.5}$$

We write

$$\frac{C - A}{A}\omega_{z0} = \omega'_{z0} \tag{10.6}$$

and solve equation (10.5)

$$\omega_x = C_1 \cos\omega'_{z0}t + C_2 \sin\omega'_{z0}t. \tag{10.7}$$

Entered into equation (10.3b) and integrating, one obtains

$$\omega_y = C_1 \sin\omega'_{z0}t - C_2 \cos\omega'_{z0}t + C_3. \tag{10.8}$$

The three integration constants C_1, C_2, and C_3 are now chosen such that

$$\omega_y = \omega_{y0} \quad \text{when} \quad t = 0$$

$$\omega_x = \omega_{x0} \quad \text{when} \quad t = 0$$

and furthermore such that equation (10.3a) is satisfied.

One obtains

$$C_1 = \omega_{x0}$$

$$C_2 = -\omega_{y0} + \frac{M}{A\omega'_{z0}}$$

$$C_3 = \frac{M}{A\omega'_{z0}}$$

and consequently

$$\omega_x = \omega_{x0}\cos\omega'_z t - \left(\omega_{y0} - \frac{M}{A\omega'_{z0}}\right)\sin\omega'_{z0}t \tag{10.9}$$

$$\omega_y = \omega_{x0}\sin\omega'_{z0}t + \left(\omega_{y0} - \frac{M}{A\omega'_{z0}}\right)\cos\omega'_{z0}t + \frac{M}{A\omega'_{z0}}. \tag{10.10}$$

Because of the symmetry of the gyro about the z-axis, the x-axis can always be located such that a torque perpendicular to the axis of symmetry can be represented with an $M_y = 0$ without loss of generality.

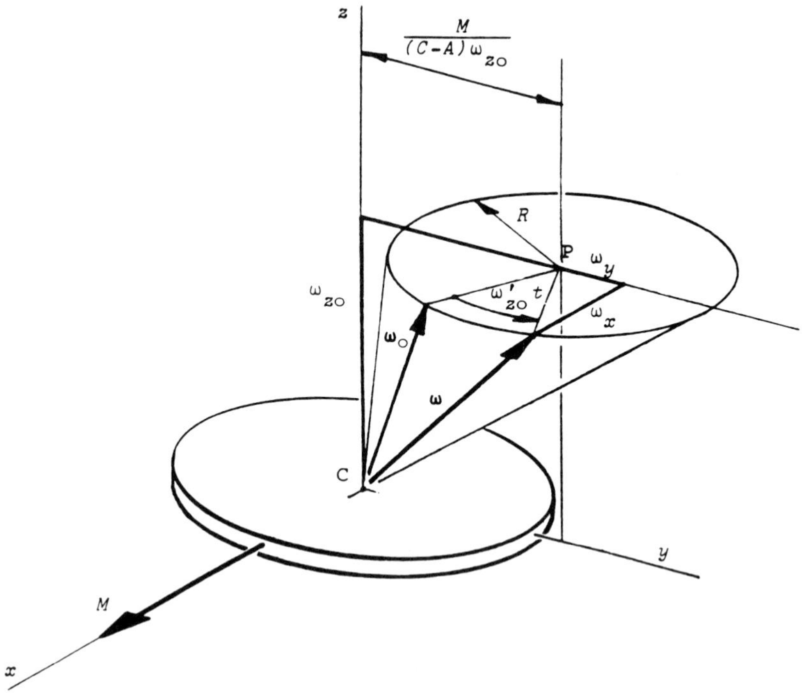

FIGURE 10.1. Flattened axisymmetric gyro with constant torque about x-axis.

An inspection of equations (10.9) and (10.10) indicates that the angular velocity vector $\boldsymbol{\omega}$ describes an oblique circular cone in the body-fixed $Cxyz$ coordinate system, with a base circle parallel to the gyro's xy-plane, and located at a distance ω_{z0} from that plane. The center P of the base circle is located in the yz-plane at a distance of

$$\frac{M}{(C-A)\omega_{z0}} \tag{10.11}$$

from the z-axis, as shown in Figure 10.1.

It is interesting to note that the angular velocity vector returns to its original location (in the $Cxyz$ coordinate system) in regular cycles, after a period of

$$\tau = \frac{2\pi}{\omega'_{z0}} = \frac{2\pi A}{(C-A)\omega_{z0}}. \tag{10.12}$$

The radius of the base circle of the oblique cone is

$$R = \sqrt{\omega_{x0}^2 + \left(\omega_{y0} - \frac{M}{(C-A)\omega_{z0}}\right)^2}. \tag{10.13}$$

It depends not only on the magnitude of the torque, but also on the components of the initial angular velocity. The radius vanishes, i.e., the cone collapses into its centerline CP, when

$$\omega_{x0} = 0$$

and

$$\omega_{y0} = \frac{M}{(C-A)\omega_{z0}}. \tag{10.14}$$

10.2 Torque about Axis of Symmetry

We begin again with a gyro whose angular momentum is as given by equation (10.1). The gyro is subjected to a constant torque M about the z-axis

$$\mathbf{M} = [\mathbf{e}_x \ \mathbf{e}_y \ \mathbf{e}_z] \begin{bmatrix} 0 \\ 0 \\ M \end{bmatrix}. \tag{10.15}$$

We again assume that the mass loss caused by the minirockets used to supply the torque is so minute that the gyro's mass and its inertia tensor can be assumed to remain constant. Euler's equations then assume the form

$$A\dot{\omega}_x + (C-A)\omega_z\omega_y = 0 \tag{10.16a}$$

$$A\dot{\omega}_y - (C-A)\omega_z\omega_x = 0 \tag{10.16b}$$

$$C\dot{\omega}_z = M \tag{10.16c}$$

and immediately lead to

$$\omega_z = \omega_{z0} + \frac{M}{C}t. \tag{10.17}$$

Multiplication of equation (10.16a) with ω_x, multiplication of equation (10.16b) with ω_y, and subsequent addition of the two, leads to

$$\omega_x\dot{\omega}_x + \omega_y\dot{\omega}_y = 0. \tag{10.18}$$

Integration gives

$$\omega_x^2 + \omega_y^2 = \omega^2 = \text{constant}$$

or

$$\omega_v = \sqrt{\omega_x^2 + \omega_y^2} = \text{constant} \tag{10.19}$$

just the same result as obtained in equation (6.13) for the torque-free gyro of rotational symmetry.

A relative angular velocity ω_z' of motion of the ω_v-component in the xy-plane can also be determined. For the present case, however, it will not be constant.

Taking equations (10.16a) and (10.16b), and introducing temporarily a new variable

$$d\alpha = \omega_z' dt = \frac{C-A}{A}\omega_z dt \tag{10.20}$$

one obtains

$$\frac{d\omega_x}{d\alpha} + \omega_y = 0 \tag{10.21a}$$

$$\frac{d\omega_y}{d\alpha} + \omega_x = 0. \tag{10.21b}$$

Differentiating equation (10.21a) once more with respect to α, and substituting $d\omega_y/d\alpha$ from equation (10.21b) leads to

$$\frac{d^2\omega_x}{d\alpha^2} - \omega_x = 0. \tag{10.22}$$

Equation (10.22) has the solution

$$\omega_x = C_1 \cos(\alpha + C_2). \tag{10.23}$$

The y-component is found by applying equation (10.21b)

$$\omega_y = C_1 \sin(\alpha + C_2). \tag{10.24}$$

In order to simplify further discussion, one can, without loss of generality, set

$$C_2 = 0. \tag{10.25}$$

The value of the integration constant C_1 is found by using equation (10.19), giving

$$C_1 = \omega_v. \tag{10.26}$$

Thus

$$\omega_x = \omega_v \cos\alpha \tag{10.27}$$

$$\omega_y = \omega_v \sin\alpha. \tag{10.28}$$

The variable α is obtained by integrating equation (10.20), with ω_z from equation (10.17),

$$\alpha = \frac{C-A}{A}(\omega_{z_0} t + \frac{M}{2C}t^2) + \alpha_0. \tag{10.29}$$

If we choose $\alpha_0 = ((C-A)/A)(M/2C)t_0^2$, with $t_0 = (2C/M)\omega_{z_0}$, then equations (10.27) and (10.28) become

$$\omega_x = \omega_v \cos\frac{C-A}{A}\frac{M}{2C}(t+t_0)^2 \tag{10.30}$$

$$\omega_y = \omega_v \sin\frac{C-A}{A}\frac{M}{2C}(t+t_0)^2. \tag{10.31}$$

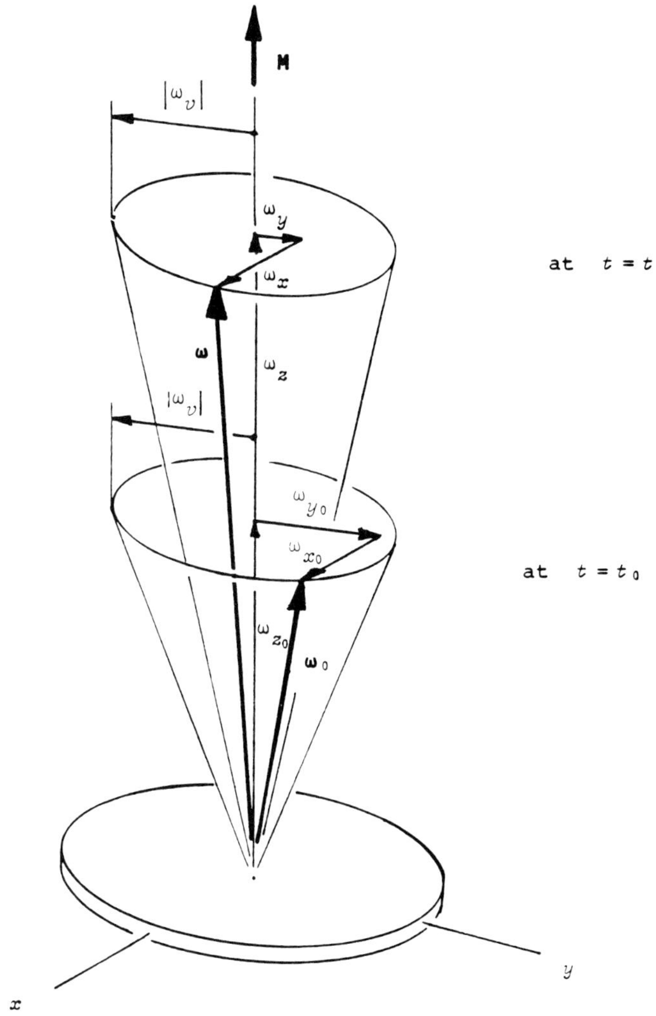

FIGURE 10.2. Oblate axisymmetric gyro with constant torque about the z-axis.

When equations (10.30) and (10.31) are compared with equations (6.21) and (6.22) for torquefree axisymmetric gyros, it is seen that

$$\int_{t_0}^{t} \omega'_z \, dt$$

has taken the place of $\omega'_z t$, when $t_0 = 0$.

Since ω_v = constant, merely the frequency of their cycle variation, not the amplitudes (i.e., the maximum values during a cycle) of ω_x and ω_y change, as an inspection of equations (10.19), (10.30), and (10.31) will indicate.

The velocity components of a flattened axisymmetric gyro, subjected to a constant torque M about the z-axis acting from a time t_0 to a time t are shown in Figure 10.2.

Suggested Reading

1. Grammel, R. *Der Kreisel, seine Theorie und seine Anwendungen*, Springer-Verlag, 1950, I, 281 pp.; II, 268 pp.

2. Jarmolow, K. "Dynamics of a Spinning Rocket with Varying Inertia and Applied Moment," *Journal of Applied Physics*, 28, 3, 1957, 308–313.

Problems

10.1. Given is a disk gyro with

$$[I] = \begin{bmatrix} 1000 & 0 & 0 \\ 0 & 1000 & 0 \\ 0 & 0 & 2000 \end{bmatrix} \text{m}^2\text{kg}.$$

At time $t < 0$, the gyro is torquefree, and has an angular momentum of $H_0 = 20\,000$ m kg/s, and a nutation angle between the gyro's z-axis and the $\mathbf{H}_0$-vector of $\nu_0 = 30°$.

Determine (a) the components H_{0_a} and H_{0_z}; (b) the angular velocity components ω_{0v} and ω_{0z}; (c) the angular velocity ω_0; (d) the relative angular velocity ω'_{0z} of motion of the v-axis with respect to the body-fixed x-axis; (e) the kinetic energy T_0 of the gyro. (f) Make a sketch.

At $t = 0$, and for a duration of 3 hours, a torque is applied about the gyro's z-axis, given by

$$\mathbf{M} = [\mathbf{e}_x \ \mathbf{e}_y \ \mathbf{e}_z] \begin{bmatrix} 0 \\ 0 \\ 0.94 \end{bmatrix} \text{Nm}.$$

For $t_f = 3$ h, determine (g) the impulse Mt_f; (h) the components H_{fv} and H_{fz} of the new angular momentum; (i) the new angular momentum $\mathbf{H}_f$; (j) the new angular velocity components ω_{fv} and ω_{fz}; (k) the new angular velocity $\boldsymbol{\omega}_f$; (l) the relative angular velocity ω'_{fz} at $t = t_f^-$ (i.e., when $M = M$); (m) the relative angular velocity ω_{fz} at $t = t_f^+$ (i.e., when $M = 0$); (n) the new kinetic energy T_f; (o) the work done by the applied torque during the 3 hours. (p) Complete the sketch.

10.2. Given is a rigid symmetrical gyro, which is torquefree for $t < 0$ and for $t > t_f$. The inertia tensor of the gyro is

$$[I] = \begin{bmatrix} 500 & 0 & 0 \\ 0 & 500 & 0 \\ 0 & 0 & 900 \end{bmatrix} \mathrm{m^2kg}.$$

The angular velocity for $t \leq 0$ is

$$\boldsymbol{\omega}_0 = [\mathbf{e}_u \ \mathbf{e}_v \ \mathbf{e}_z] \begin{bmatrix} 0 \\ 4 \\ 3 \end{bmatrix} \mathrm{rad/s}.$$

At $t = 0$ a set of minirockets is turned on (negligible mass exchange) which produce a constant torque

$$\mathbf{H} = [\mathbf{e}_x \ \mathbf{e}_y \ \mathbf{e}_z] \begin{bmatrix} 2400 \\ 0 \\ 0 \end{bmatrix} \mathrm{Nms}$$

for the period $0 < t < t_f$. At t_f the rockets are switched off, such that $\mathbf{M} = 0$ when $t > t_f$. (a) Determine ω_{vf} (with $\mathbf{M} = 0$) for $t_f = 5\pi/2\omega'_{z0}$, and make a sketch. (b) Determine ω_{vf} (with $\mathbf{M} = 0$) for $t_f = 2\pi/\omega'_{z0}$, and make a sketch.

10.3. Minirockets on an axisymmetric satellite, with $C = 9000\,\mathrm{m^2kg}$, and $A = 6000\,\mathrm{m^2kg}$, are installed such that they produce a torque M about the satellite's body-fixed x-axis. The satellite's mass loss due to the firing of the minirockets is negligible. Before the minirockets are fired, the satellite has an $\omega_z = 0.4$ rad/s about its symmetry axis and a nutation cone angle of $\nu = 0$. At $t = 0$ the minirockets are fired. They produce a constant torque of $M = 800$ Nm. At $t = 1.5\,\pi$ s they are switched off. Determine the new angular velocity components ω_x, ω_y, and ω_z just after the minirockets are switched off.

11

Rigid Gyrostats

A *gyrostat,* according to Kelvin, consists of two parts: a (typically small) axisymmetric *rotor* (or *momentum wheel,* or simply *wheel*) inside a (large) *platform* (or *carrier*). The rotor spins about its axis of symmetry with respect to the platform. The rotor spin consequently does not affect the mass distribution of the gyrostat. Thus the inertia tensor of the gyrostat remains constant, provided of course that a platform and rotor are both rigid bodies.

Gyrostat satellites are used as spacecraft in instances where the gyroscopic characteristics of a fast spinner are to be exploited, e.g., for stabilizing the spacecraft, while on the other hand, a platform pointing in a prescribed direction is needed.

For practical applications we find the platform either at rest or rotating very slowly, e.g., in locked rotation. The rotor spins rapidly and serves to stabilize the satellite's attitude. The stabilization is semi-active. It is achieved passively by the gyroscopic effect, aided if necessary by built-in nutation dampers. In order to maintain the spin of the rotor, however, a (small) power plant is needed, representing an active power input. The power required is very small, but must, on the other hand, be available throughout the duration of the serive life of the spacecraft.

For the subsequent investigation of the dynamic characteristics of gyrostats, we first discuss an arbitrary platform. Later we restrict ourselves to an axisymmetric platform, where the axes of symmetry of platform and rotor coincide.

It will be shown that a torquefree axisymmetric gyrostat of prolate shape ($C < A$) can be endowed with the characteristics of an oblate ($C > A$) single gyro, if the rotor is given sufficient spin. This attribute is the gyrostat's chief *raison d'etre.*

The first gyrostat satellite, OSO I (Orbiting Solar Observatory I), was launched in 1962. The first prolate gyrostat satellite was a synchronous satellite named MILCOMSAT (↑1969-February-09) circa 7 m high, of circa 2.5 m diameter, and of a mass of 725 kg, divided almost equally between platform and rotor. Immediately after insertion into its orbit it displayed a nutation angle of 1°, which died away to less than 0.05° within a month.

The platform of the first commercial prolate gyrostat satellite, INTELSAT IV (↑1971-January-26), has a mass of 180 kg and is about twice as large as its rotor mass.

In the meantime, numerous gyrostat satellites orbit the Earth. Space engineers, by the way, often refer to gyrostat satellites as ***dual-spin satellites*** or simply *dual spinners*.

While it is possible in principle, and sometimes advantageous, to have several rotors installed (Hagedorn, 1982), or one rotor on an axis whose direction is adjustable within the gyrostat, the present chapter is devoted solely to gyrostats with one single rotor on a fixed rigid axis within the platform.

11.1 Angular Momentum and Euler Equations

The subsequent discussion is for the practically important case of a platform P and a single axisymmetric rotor R, whose z-axes coincide. Point C is the common mass center. The $Cxyz$ coordinate system is platform-fixed, as shown in Figure 11.1.

One distinctive property of the gyrostat is that its inertia tensor remains constant with time. In other words, the spinning rotor inside must not cause a time-varying change of the inertia tensor. This can only be achieved by an *axisymmetric* rotor, whose inertia tensor with respect to the common mass center

$$[I^{R}] = \begin{bmatrix} A^{R} & 0 & 0 \\ 0 & A^{R} & 0 \\ 0 & 0 & C^{R} \end{bmatrix} \tag{11.1}$$

for the case where the rotor's z-axis is the symmetry axis (Figure 11.1). The platform, on the other hand, need not be symmetric. The principal inertia tensor of the entire gyrostat may then be written

$$[I] = \begin{bmatrix} A & 0 & 0 \\ 0 & B & 0 \\ 0 & 0 & C \end{bmatrix} \tag{11.2}$$

with

$$A = A^{P} + A^{R} \tag{11.3a}$$

$$B = B^{P} + A^{R} \tag{11.3b}$$

$$C = C^{P} + C^{R}. \tag{11.3c}$$

The angular velocity components are characterized by

$$\omega_x^{P} = \omega_x^{R} = \omega_x \tag{11.4}$$

$$\omega_y^{R} = \omega_y^{R} = \omega_y . \tag{11.5}$$

In the interest of notational brevity, we shall simply use ω_x and ω_y from now on. No such simplification is, however, possible for the z-components. We observe that for the platform one may still simplify

$$\omega_z^{P} = \omega_z \tag{11.6}$$

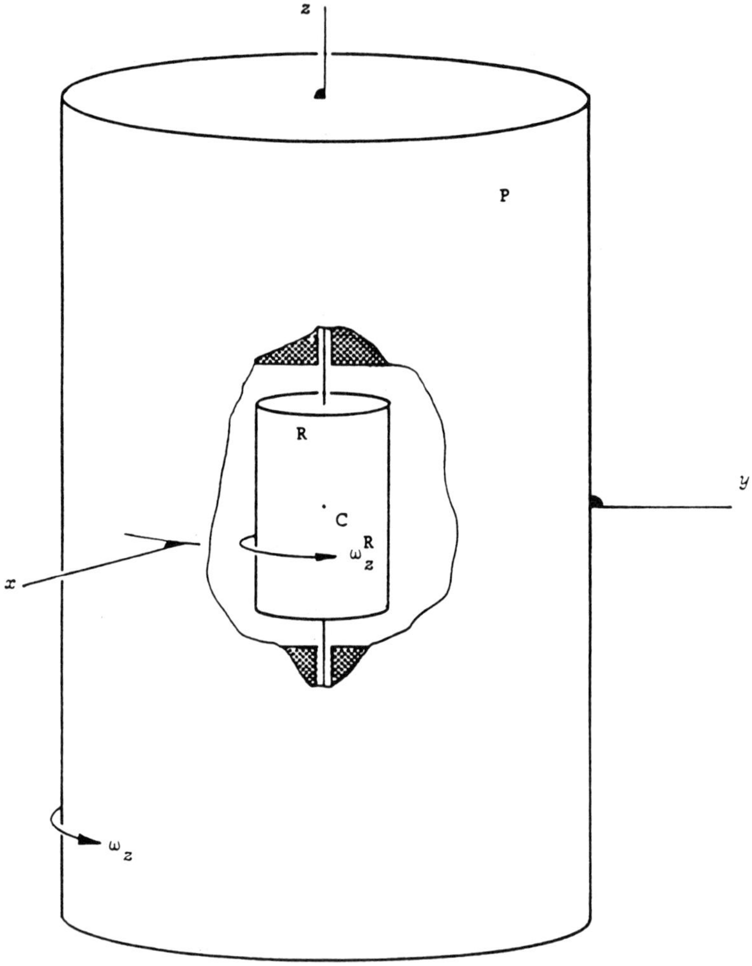

FIGURE 11.1. A simple gyrostat consisting of a platform and a single rotor.

but for the rotor one has

$$\omega_z^{\mathrm{R}} = \omega_z^{\mathrm{P}} + \omega_z^{\mathrm{E}} = \omega_z + \omega_z^{\mathrm{E}} \tag{11.7}$$

where $\omega_z^{\mathrm{E}} = \omega_z^{\mathrm{R}} - \omega_z$ is the *excess* angular velocity component, i.e., the amount by which the rotor spin speed exceeds the platform spin speed (and a measure of how much the angular momentum has been raised due to the installation of the rotor).

An expression for the angular momentum can now be established; it is

$$\mathbf{H} = [\mathbf{e}_x \ \mathbf{e}_y \ \mathbf{e}_z] \begin{bmatrix} A\omega_x \\ B\omega_y \\ C^{\mathrm{P}}\omega_z + C^{\mathrm{R}}\omega_z^{\mathrm{R}} \end{bmatrix} \tag{11.8}$$

where

$$C^{\mathrm{P}}\omega_z + C^{\mathrm{R}}\omega_z^{\mathrm{R}} = C\omega_z + C^{\mathrm{R}}\omega_z^{\mathrm{E}} = C'\omega_z \tag{11.9}$$

with a *reduced* inertia moment

$$C' = C\left(1 + \frac{C^{\mathrm{R}}}{C}\frac{\omega_z^{\mathrm{E}}}{\omega_z}\right) = C^{\mathrm{P}} + \frac{\omega_z^{\mathrm{R}}}{\omega_z}C^{\mathrm{R}}. \tag{11.10}$$

From $\dot{\mathbf{H}} = \overset{\circ}{\mathbf{H}} + \boldsymbol{\Omega} \times \mathbf{H}$, one obtains for the time derivative of the angular momentum

$$\dot{\mathbf{H}} = [\mathbf{e}_x \ \mathbf{e}_y \ \mathbf{e}_z] \begin{bmatrix} A\dot{\omega}_x - (B - C')\omega_z\omega_y \\ B\dot{\omega}_y - (C' - A)\omega_z\omega_x \\ C^{\mathrm{P}}\dot{\omega}_z + C^{\mathrm{R}}\dot{\omega}_z^{\mathrm{R}} - (A - B)\omega_x\omega_y \end{bmatrix}. \tag{11.11}$$

Using Euler's law $\mathbf{M} = \dot{\mathbf{H}}$, one can then write the Euler equations either as

$$M_x = A\dot{\omega}_x - (B - C')\omega_z\omega_y \tag{11.12a}$$

$$M_y = B\dot{\omega}_y - (C' - A)\omega_z\omega_x \tag{11.12b}$$

$$M_z = C^{\mathrm{P}}\dot{\omega}_z + C^{\mathrm{R}}\dot{\omega}_z^{\mathrm{R}} - (A - B)\omega_x\omega_y \tag{11.12c}$$

or as

$$M_x = A\dot{\omega}_x - (B - C^{\mathrm{P}})\omega_z\omega_y + C^{\mathrm{R}}\omega_z^{\mathrm{R}}\omega_y \tag{11.13a}$$

$$M_y = B\dot{\omega}_y - (C^{\mathrm{P}} - A)\omega_z\omega_x - C^{\mathrm{R}}\omega_z^{\mathrm{R}}\omega_x \tag{11.13b}$$

$$M_z = C^{\mathrm{P}}\dot{\omega}_z + C^{\mathrm{R}}\dot{\omega}_z^{\mathrm{R}} - (A - B)\omega_x\omega_y. \tag{11.13c}$$

11.2 Attitude Stability in the Short Run of Torquefree Gyrostat

For a torquefree gyrostat consisting of an absolutely rigid platform and an absolutely rigid rotor, the stability of rotation about the z-axis can be investigated by assuming the gyrostat initially spinning about the z-axis, with the other velocity components zero:

$$\boldsymbol{\omega} = [\mathbf{e}_x \ \mathbf{e}_y \ \mathbf{e}_z] \begin{bmatrix} 0 \\ 0 \\ \omega_0 \end{bmatrix}. \tag{11.14}$$

The initial steady motion is then perturbed, i.e., a small increase ϵ and correspondingly small components ω_x and ω_y are superimposed, such that

$$\boldsymbol{\omega} + \delta\boldsymbol{\omega} = [\mathbf{e}_x \ \mathbf{e}_y \ \mathbf{e}_z] \begin{bmatrix} \omega_x \\ \omega_y \\ \omega_0 + \epsilon \end{bmatrix}. \tag{11.15}$$

For a torquefree gyrostat, the Euler equations (11.12a–c) apply, with $M_x = M_y = M_z = 0$. Furthermore we arbitrarily prescribe that the z-component of the rotor's angular velocity remains constant, $\omega_z^R =$ constant. We neglect small terms of the second order and obtain, from equation (11.12c)

$$\dot{\omega}_z = 0 \tag{11.16}$$

leaving us with $\omega_z = \omega_0 + \epsilon =$ constant. From equations (11.12b) and (11.12c), we obtain

$$\dot{\omega}_x = \frac{B - C'}{A} \omega_0 \omega_y \tag{11.17a}$$

and

$$\dot{\omega}_y = \frac{C' - A}{B} \omega_0 \omega_x. \tag{11.17b}$$

Differentiation of equation (11.17b) with respect to time and neglect of small terms of second order, observing that $\dot{C}' = 0$, and replacement of $\dot{\omega}_x$ from equation (11.17a), yields

$$\ddot{\omega}_y + \frac{(C' - B)(C' - A)}{AB} \omega_0^2 \omega_y = 0. \tag{11.18}$$

Attitude stability (i.e., no increase in the amplitude of ω_y) is possibly only if the coefficient of ω_y is positive. This is the case if the reduced inertia moment C' is the largest of the three, i.e., if simultaneously

$$C' > B \tag{11.19a}$$

$$C' > A. \tag{11.19b}$$

For a gyrostat with given inertia moments, inequalities (11.19a and b) mean that

$$\omega_z^{\mathrm{R}} > \frac{B - C^{\mathrm{P}}}{C^{\mathrm{R}}}\omega_z \tag{11.20a}$$

and

$$\omega_z^{\mathrm{R}} > \frac{A - C^{\mathrm{P}}}{C^{\mathrm{R}}}\omega_z. \tag{11.20b}$$

An inspection of inequalities (11.20a and b) indicates that stability can always be achieved simply by increasing the spin speed of the rotor.

A stable solution is also obtained if the reduced inertia moment C' is the smallest of the three, i.e.,

$$C' < B \tag{11.21a}$$

$$C' < A. \tag{11.21b}$$

For a gyrostat with given inertia moments, inequalities (11.21a and b) mean that the rotor speed

$$\omega_z^{\mathrm{R}} < \frac{B - C^{\mathrm{P}}}{C^{\mathrm{R}}}\omega_z \tag{11.22a}$$

$$\omega_z^{\mathrm{R}} < \frac{A - C^{\mathrm{P}}}{C^{\mathrm{R}}}\omega_z. \tag{11.22b}$$

An inspection of inequalities (11.22a and b) indicates that stability can always be achieved by making the rotor's spin speed small enough, including negative (i.e., ω_z^{R} and ω_z with opposite signs) if necessary.

If, however,

$$B < C' < A \quad \text{for} \quad B < A \tag{11.23}$$

or

$$A < C' < B \quad \text{for} \quad A < B \tag{11.24}$$

then there is instability. The instability region disappears altogether for axisymmetric gyrostats, i.e., when $B = A$.

For a rotation about the gyrostat axis coinciding with the rotor axis to be stable, the reduced principal inertia moment about that gyrostat axis must either be the largest or the smallest. The preceding criterion has been derived by assuming absolute *rigidity*, a condition that can at best only be used to establish *stability in the short run*. Stability in the long run must take into consideration that real bodies do deform, at least slightly, during tumbling, that there is consequently internal energy dissipation, and thus a decline of kinetic energy and a drift of attitude.

A rigid gyrostat with an internal spring-mass-dashpot system has been used by Likins (1967) in a stability analysis, according to which a single-rotor gyrostat is stable if

$$\omega_z^{\mathrm{R}} > \frac{B - C^{\mathrm{P}}}{C^{\mathrm{R}}}\omega_z \tag{11.25a}$$

and

$$\omega_z^{\mathrm{R}} > \frac{A - C^{\mathrm{P}}}{C^{\mathrm{R}}}\omega_z \tag{11.25b}$$

i.e., the same as inequalities (11.20a and b).

11.3 Torquefree Axisymmetric Gyrostats

For a torquefree axisymmetric gyrostat with a single rotor, the Euler equations (11.13a–c) become

$$A\dot{\omega}_x + (C^{\mathrm{P}} - A)\omega_z\omega_y + C^{\mathrm{R}}\omega_z^{\mathrm{R}}\omega_y = 0 \tag{11.26a}$$

$$A\dot{\omega}_y - (C^{\mathrm{P}} - A)\omega_z\omega_x - C^{\mathrm{R}}\omega_z^{\mathrm{R}}\omega_x = 0 \tag{11.26b}$$

$$C^{\mathrm{P}}\dot{\omega}_z + C^{\mathrm{R}}\dot{\omega}_z^{\mathrm{R}} = 0. \tag{11.26c}$$

From equation (11.26c) one can immediately conclude that

$$C^{\mathrm{P}}\omega_z + C^{\mathrm{R}}\omega_z^{\mathrm{R}} = C'\omega_z = \text{constant.} \tag{11.27}$$

If the rotor has constant spin

$$\omega_z^{\mathrm{R}} = \text{constant} \tag{11.28}$$

then, as a consequence for the platform,

$$\omega_z = \text{constant.} \tag{11.29}$$

The case where ω_z and ω_z^{R} do change, but such that $C'\omega_z = $ constant, is discussed later, where bearing friction is introduced.

If we now introduce an auxiliary frequency defined by

$$\omega_z' = \frac{C' - A}{A}\omega_z \tag{11.30}$$

then we may write instead of equations (11.26a) and (11.26b),

$$\dot{\omega}_x + \omega_z'\omega_y = 0 \tag{11.31a}$$

$$\dot{\omega}_y - \omega_z'\omega_x = 0. \tag{11.31b}$$

Multiplication of equation (11.31a) with ω_x and of equation (11.31b) with ω_y, and a subsequent addition of the equations gives

$$\omega_x\dot{\omega}_x + \omega_y\dot{\omega}_y = 0 \tag{11.32}$$

representing a differential equation that can be solved immediately

$$\omega_x^2 + \omega_y^2 = \omega_v^2 = \text{constant} \tag{11.33}$$

where we have introduced ω_v^2 as an integration constant. We have thus found that the component ω_z and the combination $\sqrt{\omega_x^2 + \omega_y^2}$ of the other two components of the angular velocity of the platform are constants.

The magnitude of the platform's angular velocity is

$$\omega = \sqrt{\omega_x^2 + \omega_y^2 + \omega_z^2} = \sqrt{\omega_v^2 + \omega_z^2}. \tag{11.34}$$

The angular velocity vector of the platform may be expressed in different ways (Figure 11.2), e.g.,

$$\boldsymbol{\omega} = [\mathbf{e}_x \ \mathbf{e}_y \ \mathbf{e}_z] \begin{bmatrix} \omega_x \\ \omega_y \\ \omega_z \end{bmatrix} = [\mathbf{e}_u \ \mathbf{e}_v \ \mathbf{e}_z] \begin{bmatrix} 0 \\ \omega_v \\ \omega_z \end{bmatrix}. \tag{11.35}$$

The component ω_v rotates in the xy-plane. Its frequency can be obtained by combining equations (11.31a and b) into a linear differential equation of second order

$$\ddot{\omega}_x + {\omega'_z}^2 \omega_x = 0 \tag{11.36}$$

whose general solution is

$$\omega_x = A_1 \cos(\omega'_z t + A_2). \tag{11.37}$$

We set $A_2 = 0$, which does not affect the general applicability of the solution.

With the aid of Figure 11.2 we now determine the integration constant A_1. When $t = 0$ we have $\omega_x = \omega_v$ and thus obtain

$$A_1 = \omega_v. \tag{11.38}$$

Equation (11.45) is now

$$\omega_x = \omega_v \cos \omega'_z t. \tag{11.39}$$

With the aid of equations (11.31a and b) we obtain for the y-component

$$\omega_y = -\frac{\dot{\omega}_x}{\omega'_z} = \omega_v \sin \omega'_z t. \tag{11.40}$$

We conclude that the vector ω_v rotates in the xy-plane at a relative angular velocity ω'_z. The plane defined by the vectors ω_z and ω_v contains the vector $\boldsymbol{\omega}$ and the angular momentum vector $\mathbf{H}$, as can be seen by an inspection of Figure 11.2. This plane consequently also rotates about the z-axis at a relative frequency ω'_z, and at an absolute frequency of

$$\omega_z + \omega'_z = \frac{C'}{A}\omega_z.$$

The nutation angle ν is obtained from

$$\tan \nu = \frac{A\omega_v}{C'\omega_z}. \tag{11.41}$$

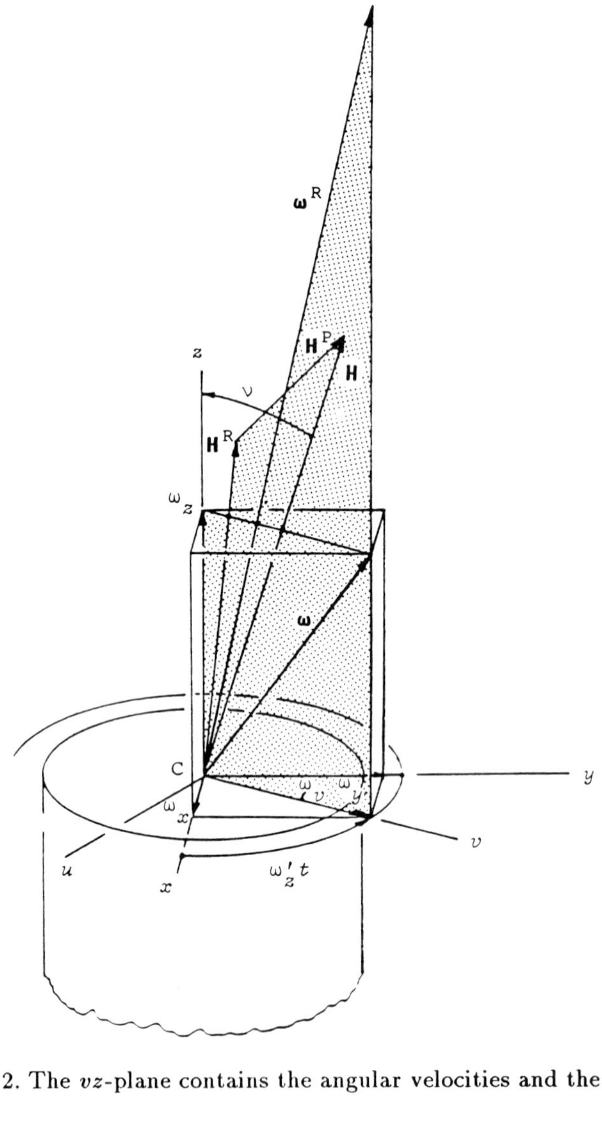

FIGURE 11.2. The vz-plane contains the angular velocities and the angular momenta.

Equation (11.41), solved for the angular velocity component ω_v, gives

$$\omega_v = \frac{C' \tan\nu}{A}\omega_z = \frac{H \sin\nu}{A}. \tag{11.42}$$

In reality axisymmetric gyrostats will not have a rotor installation that is perfectly axial, nor will the rotor or the platform be perfectly axisymmetric. For a discussion of the almost-axial, almost-axisymmetric gyrostat, the reader is referred to Roberson (1979).

The platform moves at an angular velocity given by equation (11.35). The rotor moves at an angular velocity:

$$\boldsymbol{\omega}^{\mathrm{R}} = [\mathbf{e}_x\ \mathbf{e}_y\ \mathbf{e}_z]\begin{bmatrix}\omega_x\\ \omega_y\\ \omega_z^{\mathrm{R}}\end{bmatrix} = [\mathbf{e}_u\ \mathbf{e}_v\ \mathbf{e}_z]\begin{bmatrix}0\\ \omega_v\\ \omega_z^{\mathrm{R}}\end{bmatrix}.$$

The floating $Cuvz$ coordinate system moves at an angular velocity

$$\boldsymbol{\Omega} = [\mathbf{e}_u\ \mathbf{e}_v\ \mathbf{e}_z]\begin{bmatrix}0\\ \omega_v\\ \frac{C'}{A}\omega\end{bmatrix}.$$

With $x = \omega_z^{\mathrm{R}}/\omega_z$, the torque transmitted from the platform to the rotor is

$$\mathbf{M}^{\mathrm{R}} = [\mathbf{e}_u\ \mathbf{e}_v \mathbf{e}_z]\begin{bmatrix}(C^{\mathrm{R}}x - \frac{A^{\mathrm{R}}}{A}C')\omega_v\omega_z\\ 0\\ 0\end{bmatrix}.$$

11.4 Bearing Friction

The rotor is necessarily borne within the platform in bearings that absorb friction energy. For a torquefree gyrostat, the angular momentum remains constant.

Bearing friction will lead to an internal torque about the z-axis that reduces the spin of the rotor, while the equal and opposite torque applied to the platform increases the platform's spin, both effects taking place while H_z remains constant, or, in other words, *bearing friction does not affect the attitude of the gyrostat* (Kane, 1970). From

$$H_z = C^{\mathrm{P}}\omega_z + C^{\mathrm{R}}\omega_z^{\mathrm{R}} = C\omega_z + C^{\mathrm{R}}\omega_z^{\mathrm{E}} = \text{constant} \tag{11.43}$$

we conclude that the spin speed changes are related by

$$\dot{\omega}_z = -\frac{C^{\mathrm{R}}}{C}\dot{\omega}_z^{\mathrm{E}} \tag{11.44}$$

that is, any reduction (through friction) of ω_z^{E} is compensated by an increase in ω_z.

When the relative motion of the rotor has eventually ceased altogether, the platform will have a somewhat higher angular velocity component along z, amounting to

$$\omega_{zf} = \omega_{zo} + \frac{C^{\mathrm{R}}}{C}\omega_{zo}^{\mathrm{E}} \tag{11.45}$$

where ω_{zf} is the final and ω_{zo} and ω_{zo}^{E} are the original angular velocity components. If ω_z^{R} = constant is a mission requirement, then a (small) motor must be installed, which keeps ω_z^{E} constant, or, in other words, replenishes the energy dissipated by friction in the bearings.

The kinetic energy of the rotor is

$$T^{\mathrm{R}} = \frac{1}{2}(A^{\mathrm{R}}\omega_v^2 + C^{\mathrm{R}}\omega_z^{\mathrm{R}^2}). \tag{11.46}$$

For ω_v = constant and $\dot{\omega}_z$ as given by equation (11.44), the rotor speed changes according to

$$\dot{\omega}_z^{\mathrm{R}} = \dot{\omega}_z^{\mathrm{R}} + \dot{\omega}_z = \left(1 - \frac{C^{\mathrm{R}}}{C}\right)\dot{\omega}_z^{\mathrm{E}} \tag{11.47}$$

and the rotor's kinetic energy changes by

$$\dot{T}^{\mathrm{R}} = C^{\mathrm{R}}\omega_z^{\mathrm{R}}\dot{\omega}_z^{\mathrm{R}} = C^{\mathrm{R}}\omega_z^{\mathrm{R}}\left(1 - \frac{C^{\mathrm{R}}}{C}\right)\dot{\omega}_z^{\mathrm{E}}. \tag{11.48}$$

At the same time, the kinetic energy of the platform

$$T^{\mathrm{P}} = \frac{1}{2}(A^{\mathrm{P}}\omega_z^2 + C^{\mathrm{P}}\omega_z^2) \tag{11.49}$$

changes by

$$\dot{T}^{\mathrm{P}} = C^{\mathrm{P}}\omega_z\dot{\omega}_z = -\frac{C^{\mathrm{P}}C^{\mathrm{R}}}{C}\omega_z\dot{\omega}_z^{\mathrm{E}}. \tag{11.50}$$

If the two are added, the total becomes

$$\dot{T} = \dot{T}^{\mathrm{R}} + \dot{T}^{\mathrm{P}} = C^{\mathrm{R}}\omega_z^{\mathrm{R}}\left(1 - \frac{C^{\mathrm{R}}}{C}\right)\dot{\omega}_z^{\mathrm{E}} - \frac{C^{\mathrm{P}}C^{\mathrm{R}}}{C}\omega_z\dot{\omega}_z^{\mathrm{E}}$$

which can be simplified to

$$\dot{T} = \frac{C^{\mathrm{P}}C^{\mathrm{R}}}{C}\omega_z^{\mathrm{E}}\dot{\omega}_z^{\mathrm{E}}. \tag{11.51}$$

Since $\dot{\omega}_z^{\mathrm{E}}$ is negative when the relative excess speed is declining, and since $C^{\mathrm{R}} < C$ by definition, the quantity $\dot{T}$ is negative, thus representing an energy *loss* of the gyrostat.

If originally the rotor spun with ω_{zo}^{E} and eventually this had died down to zero, then the gyrostat's kinetic energy has fallen from originally

$$T_0 = \frac{1}{2}(A\omega_v^2 + C^{\mathrm{P}}\omega_{zo}^2 + C^{\mathrm{R}}\omega_{zo}^{\mathrm{R}^2}) \tag{11.52}$$

to

$$T_f = \frac{1}{2}(A\omega_v^2 + C\omega_{zf}^2) \tag{11.53}$$

finally. With

$$\omega_{zf} = \omega_{zo} + \frac{C^{\mathrm{R}}}{C}\omega_{zo}^{\mathrm{E}} \tag{11.54}$$

the final kinetic energy becomes

$$T_f = \frac{1}{2}\left[A\omega_v^2 + C\left(\omega_{zo} + \frac{C^{\mathrm{R}}}{C}\omega_{zo}^{\mathrm{E}}\right)^2\right]. \tag{11.55}$$

In other words, the gyrostat has experienced an energy decline of

$$\Delta T = T_f - T_0 = -\frac{1}{2}\left(1 - \frac{C^{\mathrm{R}}}{C}\right)C^{\mathrm{R}}\omega_{zo}^{\mathrm{E}^2} \tag{11.56}$$

which is again seen to be negative, because $C^{\mathrm{R}} < C$ by definition. The reader should note that the gyrostat's angular momentum **H** as well as its attitude ν have remained the same in the process. Energy dissipation by itself, in other words, is not sufficient to bring about attitude drift.

When ω_z and ω_z^{R} are not constant, then equations (11.36)–(11.40) are no longer applicable. Using $d\alpha = (C' - A)\omega_z dt/A$, equation (11.36) becomes

$$\frac{d^2\omega_x}{d\alpha^2} + \omega_x = 0$$

giving

$$\omega_x = \omega_v \cos\alpha$$

$$\omega_y = \omega_v \sin\alpha.$$

Thus $\omega_x^2 + \omega_y^2 = \omega_v^2$ still applies, but the angle between ω_v and x is now

$$\alpha = \frac{1}{A}\int_0^t (C' - A)\omega_z\, dt$$

and no longer $(1/A)(C' - A)\omega_z t$.

11.5 Euler Angles for a Torquefree Axisymmetric Gyrostat

In order to express the motion of a gyrostat with respect to space-fixed coordinates $CXYZ$, one may, e.g., use Euler angles of the first kind (Appendix A).

The angular momentum **H** of a torquefree axisymmetric gyrostat is fixed in space. We select the $CXYZ$ coordinate system such that the **H**-vector is directed along the Z-axis. The components of the angular velocity $\boldsymbol{\omega}$ of the gyrostat (i.e., its platform) are then, with $\dot{\nu} = 0$,

$$\omega_x = \dot{\psi} \sin \nu \sin \sigma \tag{11.57a}$$

$$\omega_y = \dot{\psi} \sin \nu \cos \sigma \tag{11.57b}$$

$$\omega_z = \dot{\sigma} + \dot{\psi} \cos \nu. \tag{11.57c}$$

Since it can be shown that for a torquefree axisymmetric gyrostat, $\nu =$ constant, $\dot{\psi} =$ constant, and $\dot{\sigma} =$ constant, the time derivatives become

$$\dot{\omega}_x = \dot{\psi}\, \dot{\sigma} \sin \nu \cos \sigma \tag{11.58a}$$

$$\dot{\omega}_y = -\dot{\psi}\, \dot{\sigma} \sin \nu \sin \sigma \tag{11.58b}$$

$$\dot{\omega}_z = 0. \tag{11.58c}$$

We now enter the results of equations (11.58a) and (11.57b and c) into equation (11.12a)

$$A\dot{\sigma} + (C' - A)(\dot{\sigma} + \dot{\psi} \cos \nu) = 0 \tag{11.59}$$

from which the precession (also referred to as "nutation velocity") as a function of the spin $\dot{\sigma}$ is

$$\dot{\psi} = -\frac{C'}{(C' - A) \cos \nu} \dot{\sigma}. \tag{11.60}$$

Retrograde (pericycloidal) precession of the gyrostat prevails provided that

$$C' > A. \tag{11.61}$$

Solved for the rotor velocity, relation (11.61) gives the minimum value that must be exceeded in order to ensure retrograde precession of the entire gyrostat

$$\omega_z^{\mathrm{R}} > \frac{A - C^{\mathrm{P}}}{C^{\mathrm{R}}} \omega_z. \tag{11.62}$$

With the aid of equations (11.57c) and (11.60), one can obtain the spin of the gyrostat platform as a function of ω_z and ω_z^{R},

$$\dot{\sigma} = -\frac{C' - A}{A} \omega_z \tag{11.63}$$

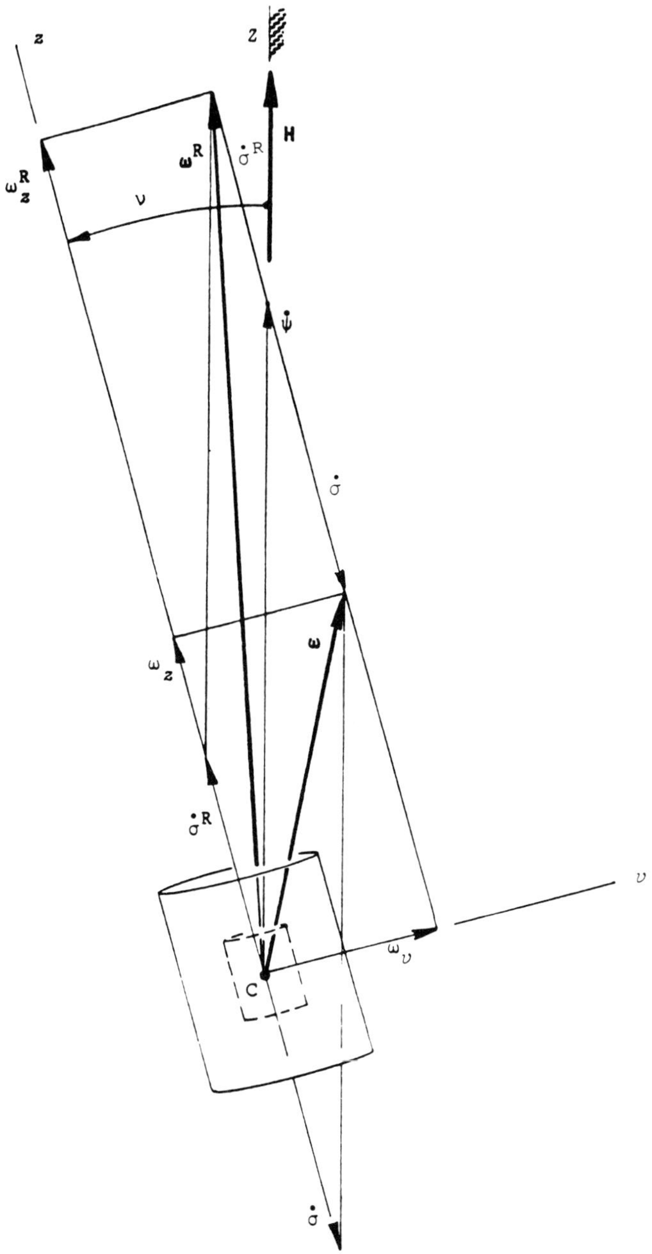

FIGURE 11.3. Angular velocities and their components.

and the precession of the gyrostat platform as

$$\dot{\psi} = \frac{C'}{A\cos\nu}\omega_z = \frac{H}{A} = \frac{\omega_v}{\sin\nu}. \tag{11.64}$$

Equation (11.64) applies to the rotor as well, that is, $\dot{\psi}^{\mathrm{R}} = \dot{\psi}$. The spin of the rotor is (from Figure 11.3)

$$\dot{\sigma}^{\mathrm{R}} = \frac{A - C^{\mathrm{R}}}{A}\omega_z^{\mathrm{R}} - \frac{C^{\mathrm{P}}}{A}\omega_z$$

$$\dot{\sigma}^{\mathrm{R}} = \frac{Ax - C'}{A}\omega_z \tag{11.65}$$

with $x = \omega_z^{\mathrm{R}}/\omega_z$. Note that

$$\omega_z^{\mathrm{R}} - \dot{\sigma}^{\mathrm{R}} = \omega_z - \dot{\sigma} = \frac{C'}{A}\omega_z.$$

The relation between precession and rotor spin is

$$\dot{\psi} = \frac{C'}{(A(\omega_z^{\mathrm{R}}/\omega_z) - C')\cos\nu}\dot{\sigma}^{\mathrm{R}}. \tag{11.66}$$

11.6 The Inertial Force Fields

Similar to the inertial force field for the single rigid gyro, the inertial force fields for the platform and for the rotor can be found.

The inertial force field for the platform can be established by choosing a platform-fixed point P^{P} of mass dm, and finding its acceleration. The resulting inertial force field of the platform is

$$d\mathbf{f}^{\mathrm{P}} = [\mathbf{e}_u\ \mathbf{e}_v\ \mathbf{e}_z]\begin{bmatrix} \omega^2 u\,dm \\ \omega_z^2 v\,dm - \frac{C'}{A}\omega_v\omega_z\,dm \\ \omega_v^2 z\,dm - (2 - \frac{C'}{A})\omega_v\omega_z v\,dm \end{bmatrix} \tag{11.67}$$

with $\omega^2 = \omega_v^2 + \omega_z^2$.

The inertial force field for the rotor can be established by choosing a rotor-fixed point P^{R} of mass dm, and determining its acceleration. The resulting inertial force field for the rotor is

$$d\mathbf{f}^{\mathrm{R}} = [\mathbf{e}_u\ \mathbf{e}_v\ \mathbf{e}_z]\begin{bmatrix} \omega^{\mathrm{R}^2} u\,dm \\ \omega_z^{\mathrm{R}^2} v\,dm - \frac{C'}{A}\omega_v\omega_z z\,dm \\ \omega_v^2 z\,dm - (2x - \frac{C'}{A})\omega_v\omega_z v\,dm \end{bmatrix} \tag{11.68}$$

with $x = \omega_z^{\mathrm{R}}/\omega_z$ and $\omega^{\mathrm{R}^2} = \omega_v^2 + \omega_z^{\mathrm{R}^2}$. Since

$$\int_m u\,dm = \int_m v\,dm = \int_m z\,dm = 0$$

because the origin of the $Cuvz$ coordinate system and the gyrostat's mass center C coincide,

$$\mathbf{f} = \mathbf{f}^{\mathrm{R}} + \mathbf{f}^{\mathrm{P}} = \int_{m^{\mathrm{R}}} d\mathbf{f}^{\mathrm{R}} + \int_{m^{\mathrm{P}}} d\mathbf{f}^{\mathrm{P}} = 0. \tag{11.69}$$

The resultant of the inertial force system of the gyro vanishes and

$$\mathbf{f}^{\mathrm{R}} = -\mathbf{f}^{\mathrm{P}}. \tag{11.70}$$

With

$$\mathbf{r} = [\mathbf{e}_u \ \mathbf{e}_v \ \mathbf{e}_z] \begin{bmatrix} u \\ v \\ z \end{bmatrix} \tag{11.71}$$

and $d\mathbf{f}^{\mathrm{R}}$ as given by equation (11.23), and considering that all inertia products vanish, the torque $\mathbf{M}^{\mathrm{R}}$ applied to the rotor from the platform is

$$\mathbf{M}^{\mathrm{R}} = -\int_{m^{\mathrm{R}}} \mathbf{r} \times d\mathbf{f}^{\mathrm{R}} = [\mathbf{e}_u \ \mathbf{e}_v \ \mathbf{e}_z] \begin{bmatrix} (C^{\mathrm{R}}\omega_z^{\mathrm{R}} - \frac{A^{\mathrm{R}}}{A} C'\omega_z)\omega_v \\ 0 \\ 0 \end{bmatrix}. \tag{11.72}$$

Similarly, the torque $\mathbf{M}^{\mathrm{P}}$ applied to the platform from the rotor is

$$\mathbf{M}^{\mathrm{P}} = -\int_{m^{\mathrm{P}}} \mathbf{r} \times d\mathbf{f}^{\mathrm{P}} = [\mathbf{e}_u \ \mathbf{e}_v \ \mathbf{e}_z] \begin{bmatrix} (C^{\mathrm{P}} - \frac{A^{\mathrm{P}}}{A} C')\omega_v\omega_z \\ 0 \\ 0 \end{bmatrix}. \tag{11.73}$$

The reader is invited to show that

$$\int_{m^{\mathrm{P}}} \mathbf{r} \times d\mathbf{f}^{\mathrm{P}} = -\int_{m^{\mathrm{R}}} \mathbf{r} \times d\mathbf{f}^{\mathrm{R}}. \tag{11.74}$$

It is important to realize that the platform's, as well as the rotor's, inertial force fields are constant with time in $Cuvz$ space. Expressed in other words, the inertial force fields rotate at the angular velocity $\boldsymbol{\Omega}$ of the $Cuvz$ floating coordinate system, and so do the torques $\mathbf{M}^{\mathrm{R}}$ and $\mathbf{M}^{\mathrm{P}}$.

Of some interest are two borderline cases of motion, namely, a nonspinning platform ($\dot{\sigma} = 0$), and a nonspinning rotor ($\dot{\sigma}^{\mathrm{R}} = 0$).

When $C' = A$, then $\dot{\sigma} = 0$ and the floating $Cuvz$ coordinate system becomes, in effect, rigidly attached to the platform. The inertial force fields become

$$d\mathbf{f}^{\mathrm{R}} = [\mathbf{e}_u \ \mathbf{e}_v \ \mathbf{e}_z] \begin{bmatrix} \omega^{\mathrm{R}^2} u\, dm \\ \omega_z^{\mathrm{R}^2} v\, dm - \omega_v\omega_z z\, dm \\ \omega_v^2 z\, dm - (2x - 1)\omega_v\omega_z v\, dm \end{bmatrix} \tag{11.75}$$

and

$$d\mathbf{f}^{\mathrm{P}} = [\mathbf{e}_u \ \mathbf{e}_v \ \mathbf{e}_z] \begin{bmatrix} \omega^2 u\, dm \\ \omega_z^2 v\, dm - \omega_v\omega_z z\, dm \\ \omega_v^2 z\, dm - \omega_v\omega_z v\, dm \end{bmatrix}. \tag{11.76}$$

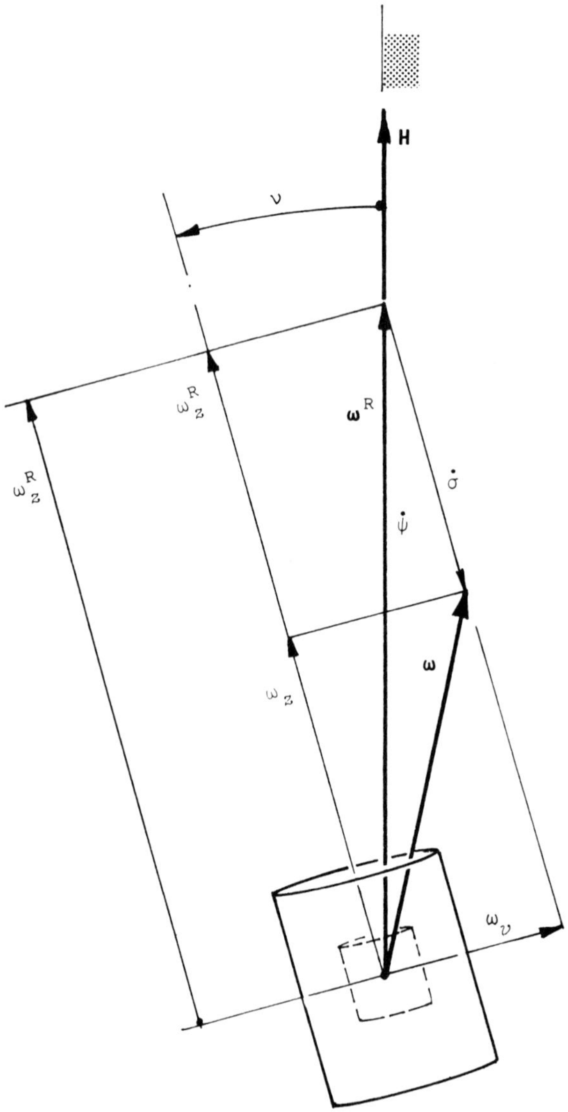

FIGURE 11.4. Angular velocities and components when $\dot{\sigma}^R = 0$.

When $C' = Ax$, then $\dot{\sigma}^{\mathrm{R}} = 0$ and the floating $Cuvz$ system becomes, in effect, rigidly attached to the rotor. The inertial force fields then become

$$d\mathbf{f}^{\mathrm{R}} = [\mathbf{e}_u\ \mathbf{e}_v\ \mathbf{e}_z] \begin{bmatrix} \omega^{\mathrm{R}^2} u\, dm \\ \omega_z^{\mathrm{R}^2} v\, dm - \omega_v \omega_z^{\mathrm{R}} z\, dm \\ \omega_v^2 z\, dm - \omega_v \omega_z^{\mathrm{R}} v\, dm \end{bmatrix} \tag{11.77}$$

$$d\mathbf{f}^{\mathrm{P}} = [\mathbf{e}_u\ \mathbf{e}_v\ \mathbf{e}_z] \begin{bmatrix} \omega^2 u\, dm \\ \omega_z^2 v\, dm - \omega_v \omega_z^{\mathrm{R}} z\, dm \\ \omega_v^2 z\, dm - (2-x)\omega_v \omega_z v\, dm \end{bmatrix}. \tag{11.78}$$

The case $C' = Ax$ is represented in Figure 11.4.

11.7 Spinning Axisymmetric Gyrostat in a Central Force Field

For an axisymmetric gyrostat orbiting the Earth sphere (Figure 11.5), or some other master, under conditions similar to those described in section 4.2 for a single gyro, equation (4.22) remains applicable, i.e., if ω_z^{R} = constant, then

$$\omega_z = \dot{\zeta} + \Omega = \omega_0 = \text{constant}. \tag{11.79}$$

Instead of equations (4.25) and (4.26), we now have

$$A\ddot{\xi} + (aC\omega_0 - 2A\Omega)\dot{\eta} + (aC\omega_0\Omega - A\Omega^2)\xi = 0 \tag{11.80a}$$

$$A\ddot{\eta} - (aC\omega_0 - 2A\Omega)\dot{\xi} + (aC\omega_0\Omega + 3C\Omega^2 - 4A\Omega^2)\eta = 0 \tag{11.80b}$$

and

$$p = \left(3\frac{C}{A} - 1\right)\Omega^2 - 2\frac{aC}{A}\omega_0\Omega + \left(\frac{aC}{A}\omega_0\right)^2 \tag{11.81}$$

$$q = \left(4 - 3\frac{C}{A}\right)\Omega^4 + \frac{aC}{A}\left(3\frac{C}{A} - 5\right)\omega_0\Omega^3 + \left(\frac{aC}{A}\omega_0\right)^2\Omega^2 \tag{11.82}$$

with

$$a = \frac{C'}{C} = \frac{C^{\mathrm{P}} + C^{\mathrm{R}}\frac{\omega_z^{\mathrm{R}}}{\omega_z}}{C^{\mathrm{P}} + C^{\mathrm{R}}}. \tag{11.83}$$

Attitude stability (i.e., no increase in the amplitude of ξ and η) is only possible if

$$p > 0 \tag{11.84a}$$

$$q > 0 \tag{11.84b}$$

$$p^2 - 4q > 0. \tag{11.84c}$$

The three curves delimitating the areas of attitude stability resulting from inequality (11.62) are plotted in Figures 11.6–11.9. Figure 11.6 is for

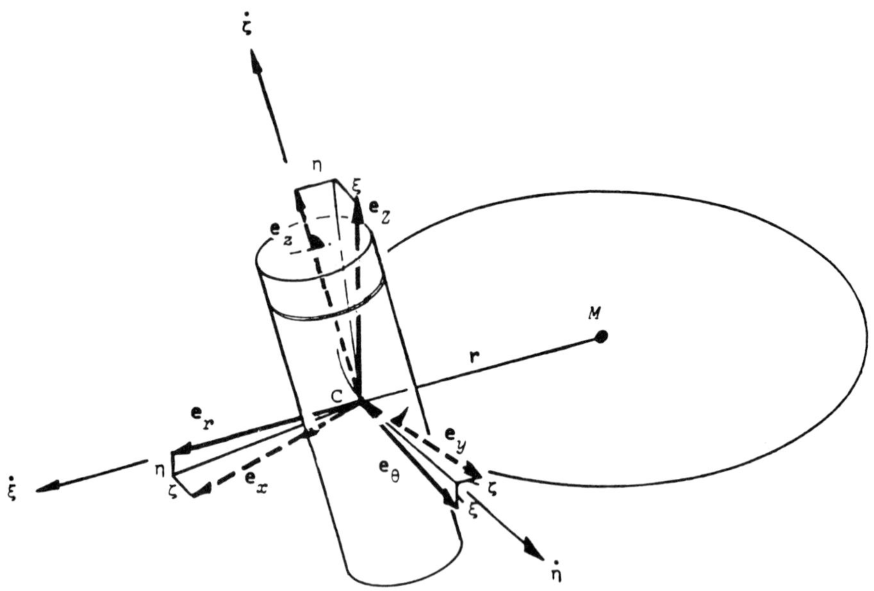

FIGURE 11.5. Orbital cylindrical coordinates r, θ, Z; Cardan angles ξ, η, ζ; and body-fixed $Cxyz$ coordinates.

the case of a gyrostat where $\omega_z^{\mathrm{R}}/\omega_z = 1$, i.e., there is no relative spin between rotor and platform, the gyrostat is equivalent to a single gyro, and Figure 11.6 is thus the same as Figure 4.3. Figures 11.7–11.9 show how the stability regions increase as the rotor spin increases. See also Vigneron (1971).

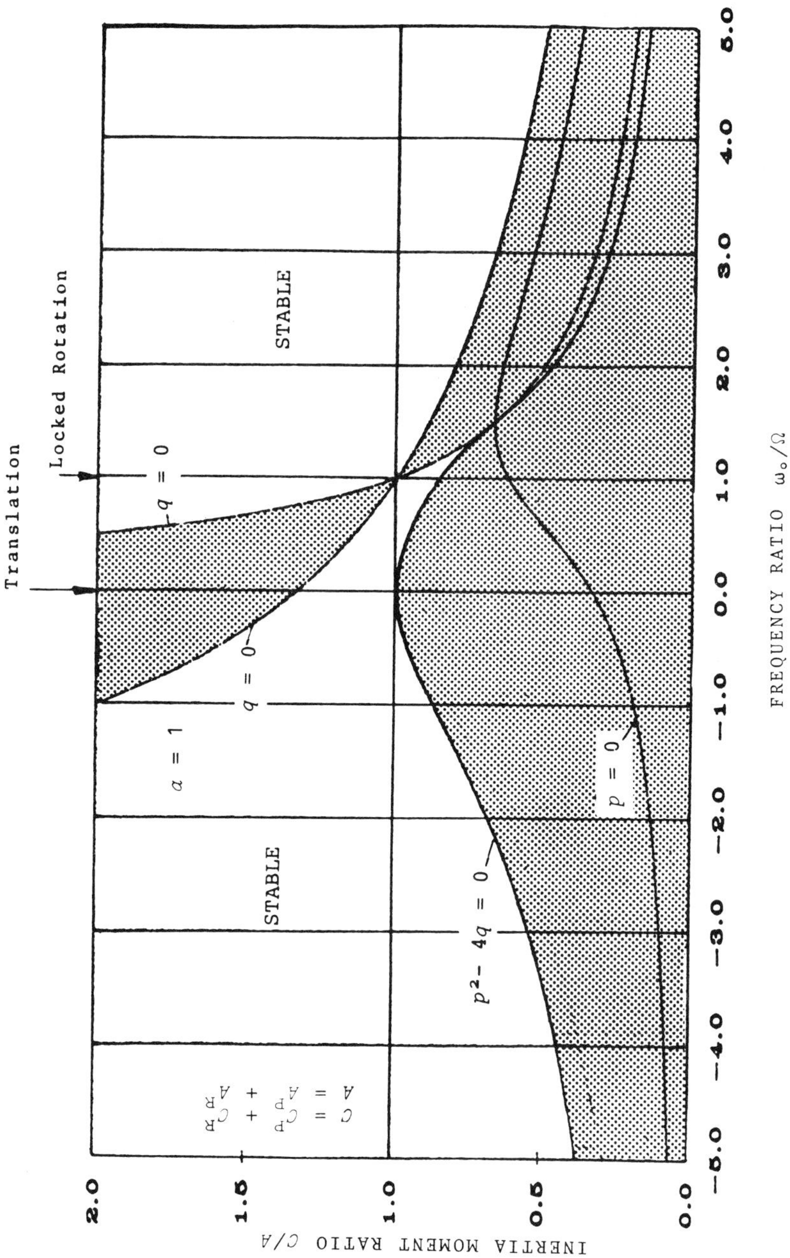

FIGURE 11.6. Short-term stability regions for a dual spinner $C'/C = 1$.

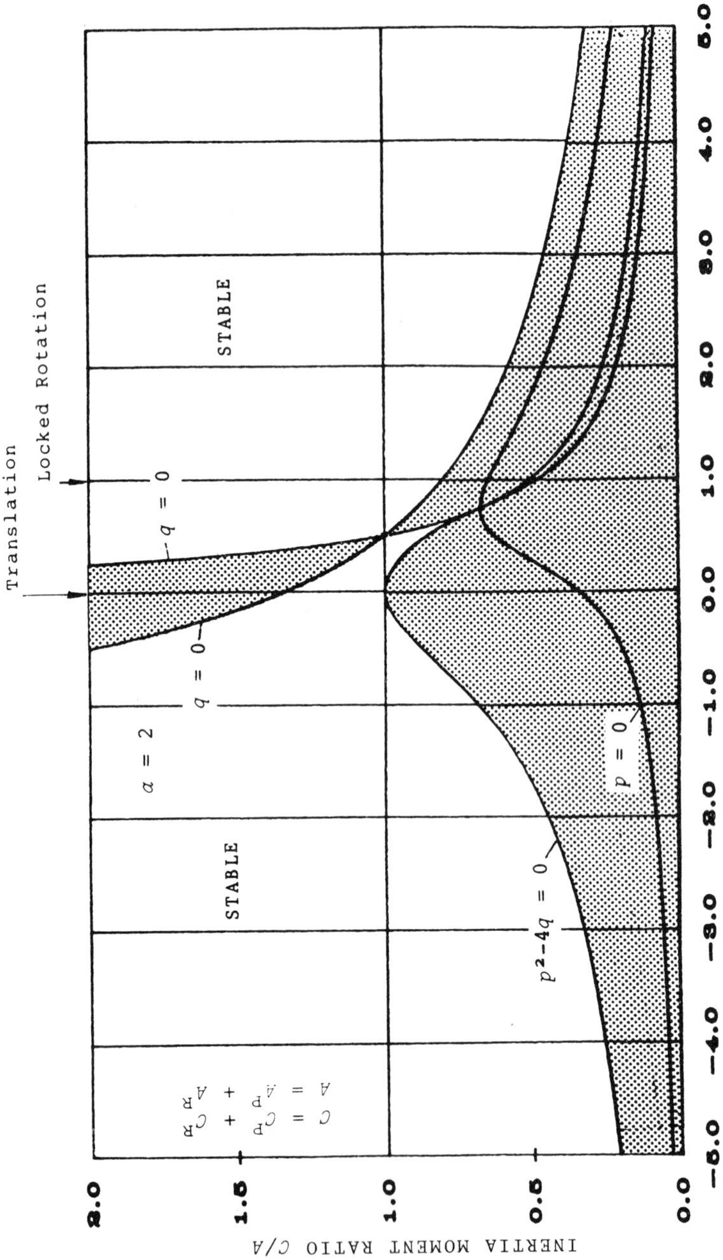

FIGURE 11.7. Short-term stability regions for a dual spinner $C'/C = 2$.

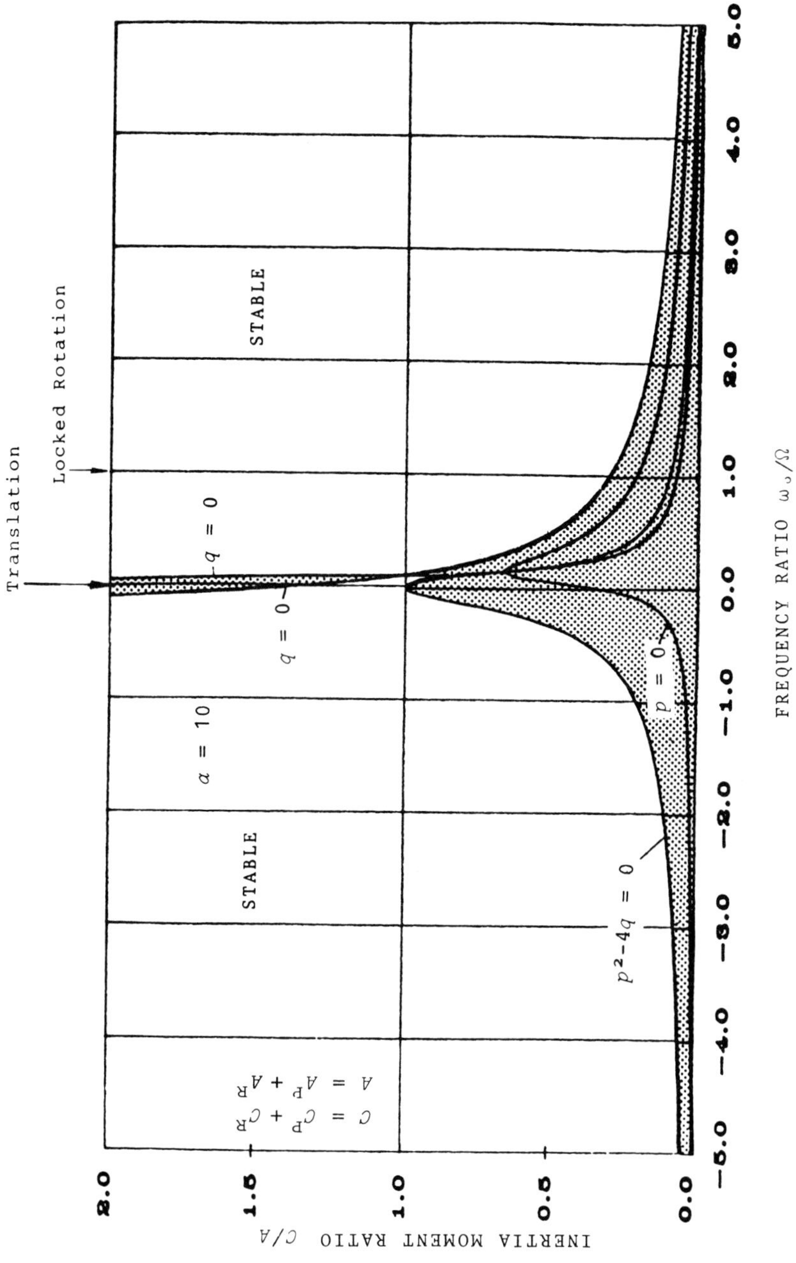

FIGURE 11.8. Short-term stability regions for a dual spinner $C'/C = 10$.

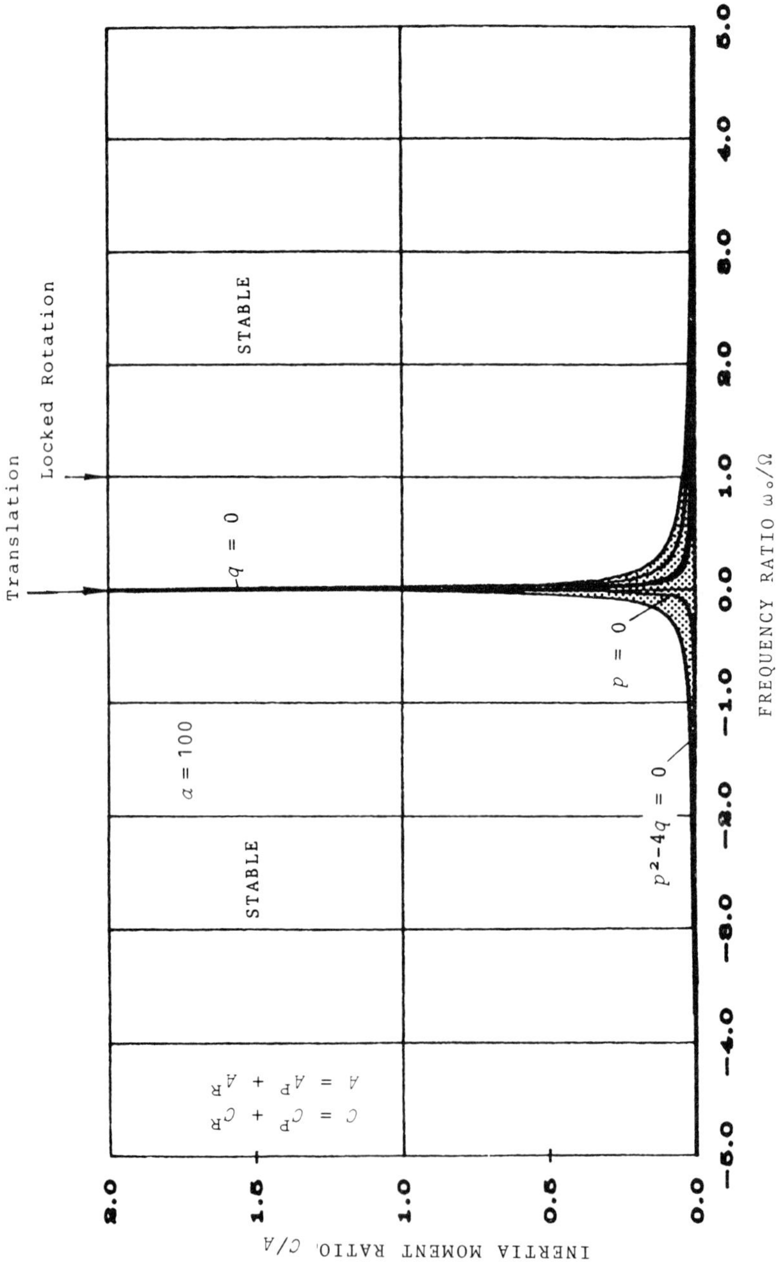

FIGURE 11.9. Short-term stability regions for a dual spinner $C'/C = 100$.

Suggested Reading

1. Hagedorn, P. "Steady Motions in Controlled Gyrostat Systems," *Ingenieur-Archiv,* 52, 1982, 183–204.

2. Kane, T.R. "Solution of the Equations of Rotational Motion for a Class of Torque-Free Gyrostats," *AIAA Journal,* 8, 6, 1970, 1141–1143.

3. Kaplan, M.H. *Modern Spacecraft Dynamics and Control,* John Wiley and Sons, 1976, 415 pp.

4. Likins, P.W. "Attitude Stability Criteria for Dual-Spin Spacecraft," *Journal of Spacecraft and Rockets,* 4, 12, 1967, 1638–1643.

5. Magnus, K. *Kreisel,* Springer-Verlag, 1971, 353 pp.

6. Magnus, K. *Dynamics of Satellites,* CISM, Udine, 56, 1970, 134 pp.

7. Magnus, K. "Der Gyrostat mit elastischer Welle," *Ingenieur-Archiv,* 48, 2, 1979, 97–111.

8. Roberson, R.E. "Spin of an Almost-Axial, Almost-Symmetric, Torque-Free Gyrostat near an Oscillation Condition," *Ingenieur-Archiv,* 48, 1, 1979, 27–33.

9. Schiehlen, W. *Dynamics of Satellites,* CISM, Udine, 56, 1970, 134 pp.

10. Vigneron, F. "Stability of a Dual-Spin Satellite in Circular Orbit," *AIAA Journal of Spacecraft and Rockets,* 8, 10, 1971, 1092–1094.

11. Wittenburg, J. "Beiträge zur Dynamik von Gyrostaten," *Acc. Naz. dei Lincei,* 217, 1975, 217–354.

12. Wittenburg, J. *Dynamics of Systems of Rigid Bodies,* B.G. Teubner, Stuttgart, 1977, 224 pp.

Problems

11.1. A symmetrical gyrostat has a platform of 300 kg mass and $i_x = i_y = 0.5$ m and $i_z = 0.1$ m. It precesses steadily at a nutation angle of 50°. The rotor has a mass of 100 kg, with gyration radii $i_x = i_y = 0.2$ m and $i_z = 0.1$ m. The rotor spins faster than the platform such that $\omega_z^{\mathrm{R}}/\omega_z = 100$. The angular velocity of the gyrostat has a magnitude of 0.9 rad/s. Determine (a) the total angular momentum; (b) the angular momentum of the platform alone; (c) the angular momentum of the rotor along; (d) the angular velocity of the gyrostat; (e) the angular velocity of the rotor; (f) the total kinetic energy; (g) the kinetic energy of the platform alone; (h) the kinetic energy of the rotor alone; (i) the spin of the gyrostat; (j) the spin of the rotor; (k) the precession of the gyrostat; and (l) the precession of the rotor. (m) Make a sketch.

11.2. A symmetrical gyrostat has a platform of 300 kg mass, $i_x = i_y = 0.5$ m and $i_z = 0.1$ m. The rotor has a mass of 100 kg, and $i_x = i_y = 0.2$ m and $i_z = 0.1$ m. The rotor spins faster than the platform, such that $\omega_z^{\mathrm{R}}/\omega_z = 100$. The angular velocity of the gyrostat has a magnitude of 0.9 rad/s. The nutation angle is 50°. Determine (a) the relative angular velocity of the v-direction as measured from the x-axis; (b) the absolute angular velocity of the v-direction; and (c) the force transmitted through each rotor bearing.

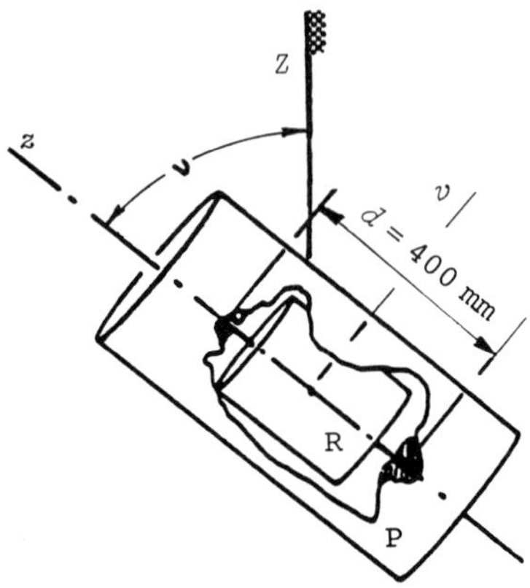

11.3. Given is a body-fixed point $\mathrm{P^P}$ of the rigid platform of a torquefree axisymmetric gyrostat. The platform's angular velocity is

$$\boldsymbol{\omega} = [\mathbf{e}_u \ \mathbf{e}_v \ \mathbf{e}_z] \begin{bmatrix} 0 \\ \omega_v \\ \omega_z \end{bmatrix}.$$

Use the floating $Cuvz$ coordinate system, and derive the acceleration of point $\mathrm{P^P}$.

11.4. Given is a body-fixed point $\mathrm{P^R}$ of the rigid rotor of a torquefree axisymmetric gyrostat. The rotor's angular velocity is

$$\boldsymbol{\omega}^{\mathrm{R}} = [\mathbf{e}_u \ \mathbf{e}_v \ \mathbf{e}_z] \begin{bmatrix} 0 \\ \omega_v \\ \omega_z^{\mathrm{R}} \end{bmatrix}.$$

Use the floating $Cuvz$ coordinate system, and derive the acceleration of point $\mathrm{P^R}$.

11.5. An axisymmetric gyrostat is torquefree and rigid. It rotates with $\omega_z^{\mathrm{R}} = 1{,}000$ rad/s, $\omega_z = 10$ rad/s, $\omega_v = 0.9$ rad/s. Furthermore $A = A^{\mathrm{R}} + A^{\mathrm{P}} = 900\,\mathrm{Ws}^3$, $C^{\mathrm{R}} = 15\,\mathrm{Ws}^3$, $C^{\mathrm{P}} = 90\,\mathrm{Ws}^3$. A point mass m, weighing 0.5 kg, is part of the platform and is mounted inside the platform at $x = 1$ m, $y = 0$, $z = 2$ m. Determine the $u-$, $v-$, and $z-$components of the inertial force acting on the point mass m, (a) at the moment when the x-axis coincides with the u-axis, and (b) at the moment when the x-axis coincides with the v-axis.

12

Torquefree Dissipative Axisymmetric Gyrostats

12.1 Near-Rigid Rotors and Platforms

For a near-rigid rotor in its undeformed state, using $Cuvz$ floating coordinates, the inertia tensor is

$$[I^{\mathrm{R}}]_{undef} = \begin{bmatrix} A^{\mathrm{R}} & 0 & 0 \\ 0 & A^{\mathrm{R}} & 0 \\ 0 & 0 & C^{\mathrm{R}} \end{bmatrix}. \tag{12.1}$$

Following equation (7.17), we conclude that when using the same $Cuvz$ coordinate system, the rotor's inertia tensor in the deformed state is

$$[I^{\mathrm{R}}] = \begin{bmatrix} A^{\mathrm{R}} & 0 & I^{\mathrm{R}}_{uz} \\ 0 & A^{\mathrm{R}} & I^{\mathrm{R}}_{vz} \\ I^{\mathrm{R}}_{uz} & I^{\mathrm{R}}_{vz} & C^{\mathrm{R}} \end{bmatrix}. \tag{12.2}$$

The inertia products I^{R}_{uz} and I^{R}_{vz} can be looked upon as including the effects of deformations of the rotor per se and of the bearings that carry the rotor.

The corresponding inertia tensors for the platform are, in its undeformed state,

$$[I^{\mathrm{P}}]_{undef} = \begin{bmatrix} A^{\mathrm{P}} & 0 & 0 \\ 0 & A^{\mathrm{P}} & 0 \\ 0 & 0 & C^{\mathrm{P}} \end{bmatrix} \tag{12.3}$$

and in its deformed state

$$[I^{\mathrm{P}}] = \begin{bmatrix} A^{\mathrm{P}} & 0 & I^{\mathrm{P}}_{uz} \\ 0 & A^{\mathrm{P}} & I^{\mathrm{P}}_{vz} \\ I^{\mathrm{P}}_{uz} & I^{\mathrm{P}}_{vz} & C^{\mathrm{P}} \end{bmatrix}. \tag{12.4}$$

As the deformations of near-rigid bodies are small by definition, the following inequalities hold

$$|I^{\mathrm{R}}_{uz}| << A^{\mathrm{R}}, C^{\mathrm{R}} \tag{12.5a}$$

$$|I^{\mathrm{P}}_{uz}| << A^{\mathrm{P}}, C^{\mathrm{P}} \tag{12.5b}$$

$$|I^{\mathrm{R}}_{vz}| << A^{\mathrm{R}}, C^{\mathrm{R}} \tag{12.5c}$$

$$|I^{\mathrm{P}}_{vz}| << A^{\mathrm{P}}, C^{\mathrm{P}}. \tag{12.5d}$$

Similar restrictions apply to partial and modified inertia products. In accordance with previous practice, i.e., equations (7.24), we introduce

$$I^{\mathrm{R}}_{uz} = J^{\mathrm{R}}_{uz} + J^{\mathrm{R}}_{zu} \tag{12.6a}$$

$$Y^{\mathrm{R}}_{uz} = \left(1 + \frac{\dot{\sigma}^{\mathrm{R}}}{\omega^{\mathrm{R}}_z}\right) J^{\mathrm{R}}_{uz} + \left(1 - \frac{\dot{\sigma}^{\mathrm{R}}}{\omega^{\mathrm{R}}_z}\right) J^{\mathrm{R}}_{zu} \tag{12.6b}$$

$$I^{\mathrm{P}}_{uz} = J^{\mathrm{P}}_{uz} + J^{\mathrm{P}}_{zu} \tag{12.7a}$$

$$Y^{\mathrm{P}}_{uz} = \left(1 + \frac{\dot{\sigma}}{\omega_z}\right) J^{\mathrm{P}}_{uz} + \left(1 - \frac{\dot{\sigma}}{\omega_z}\right) J^{\mathrm{P}}_{zu}. \tag{12.7b}$$

The total angular momentum is

$$\mathbf{H} = \mathbf{H}^{\mathrm{P}} + \mathbf{H}^{\mathrm{R}}. \tag{12.8}$$

In agreement with equation (7.37) the platform's angular momentum is

$$\mathbf{H}^{\mathrm{P}} = [\mathbf{e}_u\ \mathbf{e}_v\ \mathbf{e}_z] \begin{bmatrix} A^{\mathrm{P}}\omega_u + Y^{\mathrm{P}}_{uz}\omega_z \\ A^{\mathrm{P}}\omega_v + Y^{\mathrm{P}}_{vz}\omega_z \\ C^{\mathrm{P}}\omega_z + I^{\mathrm{P}}_{uz}\omega_u + I^{\mathrm{P}}_{vz}\omega_z \end{bmatrix} \tag{12.9}$$

and the rotor's angular momentum

$$\mathbf{H}^{\mathrm{R}} = [\mathbf{e}_u\ \mathbf{e}_v\ \mathbf{e}_z] \begin{bmatrix} A^{\mathrm{R}}\omega_u + Y^{\mathrm{R}}_{uz}\omega^{\mathrm{R}}_z \\ A^{\mathrm{R}}\omega_v + Y^{\mathrm{R}}_{vz}\omega^{\mathrm{R}}_z \\ C^{\mathrm{R}}\omega^{\mathrm{R}}_z + I^{\mathrm{R}}_{uz}\omega_u + I^{\mathrm{R}}_{vz}\omega^{\mathrm{R}}_z \end{bmatrix}. \tag{12.10}$$

The floating $Cuvz$ coordinate system is chosen such that $H_u = 0$. With $A = A^{\mathrm{P}} + A^{\mathrm{R}}$ and $\omega_u = \dot{\nu}$,

$$H_u = A\dot{\nu} + Y^{\mathrm{P}}_{uz}\omega_z + Y^{\mathrm{R}}_{uz}\omega^{\mathrm{R}}_z = 0 \tag{12.11}$$

and

$$\dot{\nu} = -\frac{Y^{\mathrm{P}}_{uz}}{A}\omega_z - \frac{Y^{\mathrm{R}}_{uz}}{A}\omega^{\mathrm{R}}_z \tag{12.12}$$

as fundamental relationship between attitude drift and platform deformation and rotor deformation. Since the rotor's modified inertia product Y^{R}_{uz} is typically negative, rotor and platform are found to be working at cross purposes, with a dissipative rotor working against stability and a dissipative platform working for stability ($\dot{\nu} < 0$).

Internal damping deformations are represented by a phase lag between load and deformation (Zhang and Ling, 1985), which is proportional to the spin (equation 7.106). For the rotor, one can consequently make an ansatz

$$Y^{\mathrm{R}}_{uz} = -\beta^{\mathrm{R}}\omega_v\omega^{\mathrm{R}}_z\dot{\sigma}^{\mathrm{R}} \tag{12.13a}$$

and for the platform

$$Y^{\mathrm{P}}_{uz} = -\beta^{\mathrm{P}}\omega_v\omega_z\dot{\sigma}. \tag{12.13b}$$

The internal damping deformation coefficients β^{R} and β^{P} which are positive and measured in Ws^6, are inter alia functions of $x = \omega_z^{\mathrm{R}}/\omega_z$, as an inspection of the consequences of the inertial loading as given by equations (11.67) and (11.68) will indicate.

When

$$J^{\mathrm{R}}_{uz} = -\beta^{\mathrm{R}}_1\left(2 - \frac{C'}{Ax}\right)\omega_v\omega_z^{\mathrm{R}}\dot{\sigma}^{\mathrm{R}}$$

and

$$J^{\mathrm{R}}_{zu} = -\beta^{\mathrm{R}}_2\frac{C'}{Ax}\omega_v\omega_z^{\mathrm{R}}\dot{\sigma}^{\mathrm{R}}$$

then

$$\beta^{\mathrm{R}} = \left(2 - \frac{C'}{Ax}\right)^2\beta^{\mathrm{R}}_1 + \left(\frac{C'}{Ax}\right)^2\beta^{\mathrm{R}}_2.$$

Similarly, for

$$J^{\mathrm{P}}_{uz} = -\beta^{\mathrm{P}}_1\left(2 - \frac{C'}{A}\right)\omega_v\omega_z\dot{\sigma}$$

and

$$J^{\mathrm{P}}_{zu} = -\beta^{\mathrm{P}}_2\frac{C'}{A}\omega_v\omega_z\dot{\sigma}$$

one obtains

$$\beta^{\mathrm{P}} = \left(2 - \frac{C'}{A}\right)^2\beta^{\mathrm{P}}_1 + \left(\frac{C'}{A}\right)^2\beta^{\mathrm{P}}_2.$$

12.2 Secular Attitude Drift for a Torquefree Axisymmetric Gyrostat with Dissipative Platform

Similar to the case of the single gyro, we search again for a condition for stability in the long run, i.e., for which the kinetic energy is a minimum. We observe that the angular momentum remains constant, i.e.,

$$\mathbf{H} = \text{constant}.$$

The kinetic energy is, according to equation (7.45),

$$T = \frac{1}{2}\left(A\omega_v^2 + C^{\mathrm{P}}\omega_z^2 + C^{\mathrm{R}}\omega_z^{\mathrm{R}^2}\right) \tag{12.14}$$

which can be brought into the form

$$T = \frac{1}{2}\left(\frac{H^2\sin^2\nu}{A} + \frac{(H\cos\nu - C^{\mathrm{R}}\omega_z^{\mathrm{R}})^2}{C^{\mathrm{P}}} + C^{\mathrm{R}}\omega_z^{\mathrm{R}^2}\right). \tag{12.15}$$

If there is an attitude drift $d\nu$, then the change dT of kinetic energy associated with this attitude drift is

$$dT = \left(\frac{H^2 \sin\nu \cos\nu}{A} - \frac{(H\cos\nu - C^{\mathrm{R}}\omega_z^{\mathrm{R}})}{C^{\mathrm{P}}} \left(H\sin\nu + C^{\mathrm{R}} \frac{\partial \omega_z^{\mathrm{R}}}{\partial \nu} \right) \right.$$
$$\left. + C^{\mathrm{R}}\omega_z^{\mathrm{R}} \frac{\partial \omega_z^{\mathrm{R}}}{\partial \nu} \right) d\nu. \tag{12.16}$$

Following Landon and Stewart (1964), we first assume the rotor R to be a rigid body, which has to obey Euler's equations for rigid-body motion. Euler's equations indicate that, no matter how the nutation angle ν and the spin rate ω_z of the platform vary because of damping in the platform, the spin ω_z^{R} of the rigid rotor must remain constant, i.e., $\partial\omega_z^{\mathrm{R}}/\partial\nu = 0$ (Figure 12.1). The change of kinetic energy consequently is

$$dT = \left(\frac{H^2 \sin\nu \cos\nu}{A} - \frac{(H\cos\nu - C^{\mathrm{R}}\omega_z^{\mathrm{R}})H\sin\nu}{C^{\mathrm{P}}} \right) d\nu. \tag{12.17}$$

If stabilization about the Z-axis is sought, which is equivalent to saying that $d\nu < 0$, then a loss of kinetic energy, i.e., $dT < 0$, is assured if the expression in brackets is positive, i.e., if

$$\frac{H^2 \sin\nu \cos\nu}{A} > \frac{(H\cos\nu - C^{\mathrm{R}}\omega_z^{\mathrm{R}})H\sin\nu}{C^{\mathrm{P}}} \tag{12.18}$$

which can be arranged to give

$$C' > A \tag{12.19}$$

or

$$\omega_z^{\mathrm{R}} > \frac{A - C^{\mathrm{P}}}{C^{\mathrm{R}}}\omega_z \tag{12.20}$$

or

$$C^{\mathrm{R}} > \frac{A - C^{\mathrm{P}}}{x} \tag{12.21}$$

with $x = \omega_z^{\mathrm{R}}/\omega_z \geq 1$ for simplicity. For $x < 0$ see Annett et al. (1980).

Inequality (12.19) is the same as that obtained by Likins (1967), equations (11.25), for $B = A$.

The torque transmitted from a dissipative platform to a rigid rotor is

$$\mathbf{M}^{\mathrm{R}} = [\mathbf{e}_u \ \mathbf{e}_v \ \mathbf{e}_z] \begin{bmatrix} (C^{\mathrm{R}}\omega_z^{\mathrm{R}} - \frac{A^{\mathrm{R}}}{A}C'\omega_z)\omega_v \\ (2\frac{A^{\mathrm{R}}}{A}C'\omega_z - C^{\mathrm{R}}\omega_z^{\mathrm{R}})\dot{\nu} \\ 0 \end{bmatrix}. \tag{12.22}$$

Inequality (12.19) must remain valid throughout the attitude drift, during which the value of the angular velocity component ω_z is found to increase (slightly). Inequality (12.19) must thus still hold, when $\nu = 0$ and

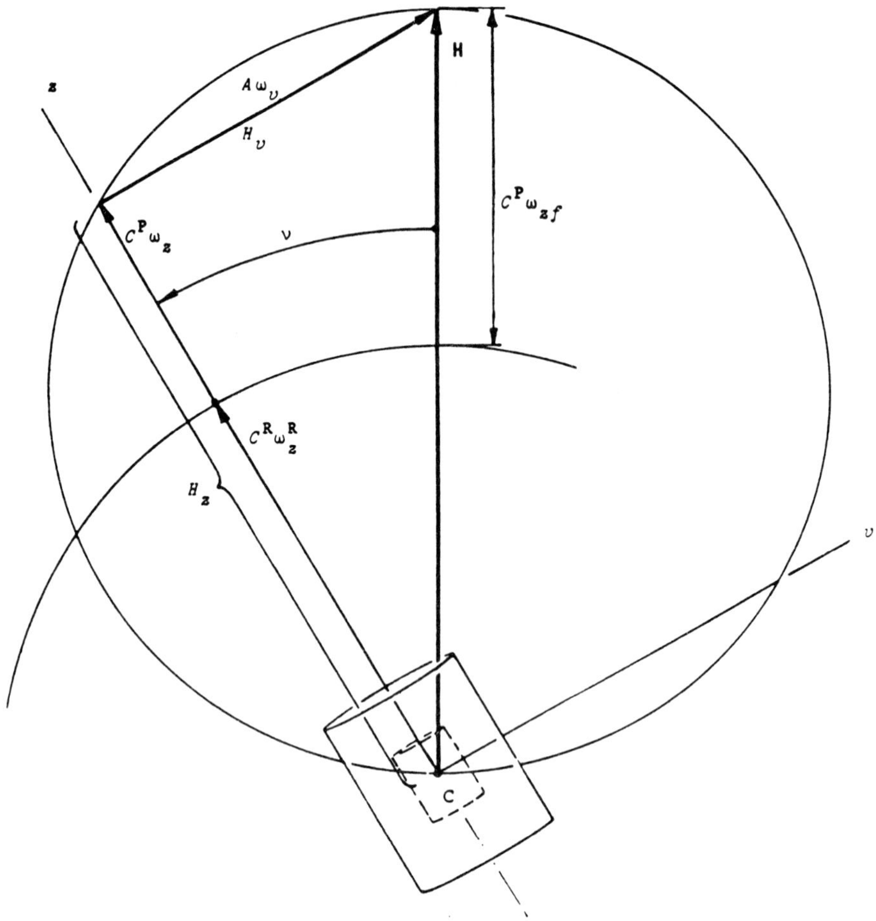

FIGURE 12.1. The constant angular momentum and its components for the case ω_z^R = constant.

$\omega_z = \omega_{zf}$. It must then satisfy

$$\omega_z^{\mathrm{R}} > \frac{A - C^{\mathrm{P}}}{C^{\mathrm{R}}}\omega_{zf}.$$

Since ω_{zf} can readily be shown to be (Figure 12.1)

$$\omega_{zf} = \omega_z + \frac{H}{C^{\mathrm{P}}}(1 - \cos\nu) \tag{12.23}$$

the rotor speed for stability (i.e., $\dot{\nu} < 0$) throughout the attitude drift must fulfill

$$\omega_z^{\mathrm{R}} > \frac{A - C^{\mathrm{R}}}{C^{\mathrm{R}}}\left(\omega_z + \frac{H}{C^{\mathrm{P}}}(1 - \cos\nu)\right) \tag{12.24}$$

with

$$H = \sqrt{H^2\omega_v^2 + C'^2\omega_z^2}. \tag{12.25}$$

For all practical cases, inequality (12.24) will lead only to marginally greater values for the rotor speed than inequality (12.19).

By making use of the constancy of the total angular momentum, it can be shown that for the assumption $d\omega_z^{\mathrm{R}} = 0$ the changes of angular velocity components are

$$d\omega_v = \frac{H\cos\nu}{A}d\nu \tag{12.26}$$

and

$$d\omega_z = -\frac{H\sin\nu}{C^{\mathrm{P}}}d\nu. \tag{12.27}$$

As the gyrostat drifts, by a small amount $|d\nu|$, closer to its stable attitude, the rotor's kinetic energy

$$T^{\mathrm{R}} = \frac{1}{2}\left(A^{\mathrm{R}}\frac{H^2\sin^2\nu}{A^2} + C^{\mathrm{R}}\omega_z^{\mathrm{R}^2}\right) \tag{12.28}$$

changes by

$$dT^{\mathrm{R}} = A^{\mathrm{R}}\frac{H^2\sin\nu\cos\nu}{A^2}d\nu \tag{12.29}$$

which is negative as long as $0 < \nu < 90°$ and $d\nu < 0$.

The platform's kinetic energy

$$T^{\mathrm{P}} = \frac{1}{2}\left(A^{\mathrm{P}}\frac{H^2\sin^2\nu}{A^2} + \frac{(H\cos\nu - C^{\mathrm{R}}\omega_z^{\mathrm{R}})^2}{C^{\mathrm{P}}}\right) \tag{12.30}$$

changes by an amount

$$dT^{\mathrm{P}} = \left(A^{\mathrm{P}}\frac{H^2\sin\nu\cos\nu}{A^2} - \frac{(H\cos\nu - C^{\mathrm{R}}\omega_z^{\mathrm{R}})H\sin\nu}{C^{\mathrm{P}}}\right)d\nu. \tag{12.31}$$

Since, for convenience $0 \leq \nu \leq 90°$, and since always $C^{\mathrm{R}}\omega_z^{\mathrm{R}} < H\cos\nu$, the energy change is negative as long as $d\nu < 0$ and inequality (12.19) is satisfied. There is an intriguing narrow range where $dT < 0$, $dT^{\mathrm{R}} < 0$, but $dT^{\mathrm{P}} > 0$, as an inspection of equations (12.16), (12.30), and (12.31) will indicate, or, in other words, where the gyrostat is stable, i.e., its attitude drifts towards $\nu = 0$ and the gyrostat as a whole loses kinetic energy, but the platform gains kinetic energy. This range is bounded by

$$\frac{A - C^{\mathrm{P}}}{C^{\mathrm{R}}}\omega_z < \omega_z^{\mathrm{R}} < \frac{A^2/A^{\mathrm{P}} - C^{\mathrm{P}}}{C^{\mathrm{R}}}\omega_z. \tag{12.32}$$

The rotor, being absolutely rigid, is characterized by no deformation, thus

$$I_{uz}^{\mathrm{R}} = Y_{uz}^{\mathrm{R}} = 0. \tag{12.33}$$

The platform is deforming and exhibits a

$$Y_{uz}^{\mathrm{P}} = -\beta^{\mathrm{P}}\omega_v\omega_z\dot{\sigma} = \frac{(C' - A)H^2}{2A^2C'^2}\beta^{\mathrm{P}}\cos\nu\sin 2\nu \tag{12.34}$$

where β^{P} is the platform's dissipation coefficient, measured in Ws^6.

The floating $Cuvz$ coordinate is chosen such that always $H_u = 0$, giving

$$H_u = A\dot{\nu} + Y_{uz}^{\mathrm{P}}\omega_z = 0 \tag{12.35}$$

or

$$\dot{\nu} = -\frac{Y_{uz}^{\mathrm{P}}}{A}\omega_z. \tag{12.36}$$

Combining equations (12.34) and (12.36), the drift rate becomes

$$\dot{\nu} = \frac{\beta^{\mathrm{P}}}{A}\omega_v\omega_z^2\dot{\sigma} = -\frac{(C' - A)H^4}{2A^3C'^3}\beta^{\mathrm{P}}\cos^2\nu\sin 2\nu. \tag{12.37}$$

The drift rate is seen to be negative in the first quadrant, i.e., for $0 < \nu < 90°$, thus the attitude drifts back to the reference attitude $\nu = 0°$, and one can conclude that a gyrostat with a dissipative platform, and a nondissipative rotor, is stable, provided $C' > A$.

Equation (12.36), can be used in two ways. If the modified inertia product Y_{uz}^{P} of the platform is known, then the drift rate can be obtained from equation (12.36):

$$\dot{\nu} = -\frac{Y_{uz}^{\mathrm{P}}}{A}\omega_z.$$

If the drift rate $\dot{\nu}$ is known, then the modified inertia product can be found from

$$Y_{uz}^{\mathrm{P}} = -\frac{\dot{\nu}}{\omega_z}A. \tag{12.38}$$

The energy dissipation rate $\dot{D}^{\mathrm{P}}$ of the platform is, for the present case, the negative of the total kinetic energy change rate

$$\dot{D}^{\mathrm{P}} = -\dot{T} = -(C' - A)\omega_v \omega_z \dot{\nu} \tag{12.39a}$$

or

$$\dot{D}^{\mathrm{P}} = -Y^{\mathrm{P}}_{uz}\omega_v \omega_z \dot{\sigma} \tag{12.39b}$$

or

$$\dot{D}^{\mathrm{P}} = \beta^{\mathrm{P}} \omega_v^2 \omega_z^2 \dot{\sigma}^2 \tag{12.39c}$$

or

$$\dot{D}^{\mathrm{P}} = \frac{(C' - A)^2 H^6}{2A^4 C'^4} \beta^{\mathrm{P}} \cos^2 \nu \sin^2 2\nu. \tag{12.39d}$$

The energy dissipation rate is defined such that it must always be positive, for physical reasons. The reader should note that the relatively simple representation of the attitude drift process due to internal energy dissipation shown here has been possible because the deformations leading to a Y^{P}_{vz} were ignored.

It is also of significance to realize that both energy dissipation (as expressed by $\dot{D}$) and an associated deformation (as expressed by Y^{P}_{uz}) are necessary and sufficient for attitude drift.

With Y^{P}_{uz} known from equation (12.34), the torque (12.22) transmitted to the rotor from the platform is

$$\mathbf{M}^{\mathrm{R}} = [\mathbf{e}_u \ \mathbf{e}_v \ \mathbf{e}_z] \begin{bmatrix} (C^{\mathrm{R}} x - \frac{A^{\mathrm{R}}}{A} C')\omega_v \omega_z \\ \frac{Y^{\mathrm{P}}_{uz}}{A}(C^{\mathrm{R}} x - 2\frac{A^{\mathrm{R}}}{A} C')\omega_z^2 \\ 0 \end{bmatrix}. \tag{12.40}$$

12.3 Secular Attitude Drift for a Torquefree Axisymmetric Gyrostat with Dissipative Rotor

Following equation (7.45), one may express the gyrostat's kinetic energy also in the form

$$T = \frac{1}{2}\left(\frac{H^2 \sin^2 \nu}{A} + \frac{(H \cos \nu - C^{\mathrm{P}} \omega_z)^2}{C^{\mathrm{R}}} + C^{\mathrm{P}} \omega_z^2\right). \tag{12.41}$$

If one now assumes the platform P to be a rigid body obeying Euler's equations for rigid-body motion, then the spin rate ω_z of the platform must remain constant, if there is no applied torque about the spin axis, no matter how the nutation angle ν and the spin ω_z^{R} of the rotor may vary due to internal damping in the rotor. With ω_z = constant, the kinetic energy change of the gyrostat, as a result of an attitude drift $d\nu$, becomes

$$dT = \left(\frac{H^2 \sin\nu \cos\nu}{A} - \frac{H\cos\nu - C^{\mathrm{P}}\omega_z}{C^{\mathrm{R}}} H \sin\nu \right) d\nu. \tag{12.42}$$

For stability $d\nu < 0$ and $dT < 0$, thus the expression in brackets must be positive, leading to

$$C' > Ax \tag{12.43}$$

or

$$\omega_z^{\mathrm{R}} < \frac{C^{\mathrm{P}}}{A - C^{\mathrm{R}}}\omega_z \tag{12.44}$$

or

$$C^{\mathrm{R}} > A - \frac{C^{\mathrm{P}}}{x} \tag{12.45}$$

with $x = \omega_z^{\mathrm{R}}/\omega_z$. Inequality (12.45) shows that attitude stability can only be achieved with very oblate systems, which are of little practical interest, because in practical applications one typically encounters

$$Ax > C'. \tag{12.46}$$

The torque transmitted from a rigid platform to a dissipative rotor is

$$\mathbf{M}^{\mathrm{R}} = [\mathbf{e}_u\ \mathbf{e}_v\ \mathbf{e}_z] \begin{bmatrix} (C^{\mathrm{R}}\omega_z^{\mathrm{R}} - \frac{A^{\mathrm{R}}}{A}C'\omega_z)\omega_v \\ (2\frac{A^{\mathrm{R}}}{A}C'\omega_z - C'\omega_z - C^{\mathrm{R}}\omega_z^{\mathrm{R}})\dot{\nu} \\ 0 \end{bmatrix}. \tag{12.47}$$

Since the rotor is deforming it exhibits a

$$Y_{uz}^{\mathrm{R}} = -\beta^{\mathrm{R}}\omega_v\omega_z^{\mathrm{R}}\dot{\sigma}^{\mathrm{R}} = -\frac{(Ax - C')H^3}{2A^2C'^2}x\beta^{\mathrm{R}}\cos\nu\sin 2\nu. \tag{12.48}$$

The platform is rigid, hence

$$Y_{uz}^{\mathrm{P}} = I_{uz}^{\mathrm{P}} = 0. \tag{12.49}$$

Since the floating $Cuvz$ coordinate system is chosen such that always $H_u = 0$,

$$H_u = A\dot{\nu} + Y_{vz}^{\mathrm{R}}\omega_z^{\mathrm{R}} = 0 \tag{12.50}$$

or

$$\dot{\nu} = -\frac{Y_{uz}^{\mathrm{R}}}{A}\omega_z^{\mathrm{R}}. \tag{12.51}$$

Combining equations (12.48) and (12.51), the drift rate becomes

$$\dot{\nu} = -\frac{\beta^{\mathrm{R}}}{A}\omega_v\omega_z^{\mathrm{R}^2}\dot{\sigma}^{\mathrm{R}} = \frac{Ax - C'}{2A^3C'^3}x^2\beta^{\mathrm{R}}\cos^2\nu\sin 2\nu \tag{12.52}$$

and is seen to be positive in the first quadrant, i.e., for $0 < \nu < 90°$, thus the attitude drifts away from the reference attitude $\nu = 0°$, and one can conclude that a gyrostat with a rigid platform and a dissipative rotor is unstable, when $Ax > C'$.

The energy dissipation rate is

$$\dot{D}^{\mathrm{R}} = (Ax - C')\omega_v \omega_z^{\mathrm{R}} \dot{\nu} \tag{12.53a}$$

or

$$\dot{D}^{\mathrm{R}} = -Y_{uz}^{\mathrm{P}} \omega_v \omega_z^{\mathrm{R}} \dot{\sigma}^{\mathrm{R}} \tag{12.53b}$$

or

$$\dot{D}^{\mathrm{R}} = \beta^{\mathrm{R}} \omega_v^2 \omega_z^{\mathrm{R}^2} \dot{\sigma}^{\mathrm{R}^2} \tag{12.53c}$$

or

$$\dot{D}^{\mathrm{R}} = \frac{(Ax - C')H^6}{2A^4 C'^4} x^2 \beta^{\mathrm{R}} \cos^2 \nu \sin^2 2\nu. \tag{12.53d}$$

With the help of equation (12.51), the torque (12.47) transmitted to the rotor from the platform can also be expressed by

$$\mathbf{M}^{\mathrm{R}} = [\mathbf{e}_u \ \mathbf{e}_v \ \mathbf{e}_z] \begin{bmatrix} (xC^{\mathrm{R}} - \frac{A^{\mathrm{R}}}{A} C')\omega_v \omega_z \\ \frac{Y_{uz}^{\mathrm{R}}}{A}(xC^{\mathrm{R}} + (1 - 2\frac{A^{\mathrm{R}}}{A})C')x\omega_z^2 \\ 0 \end{bmatrix}. \tag{12.54}$$

12.4 Secular Attitude Drift for a Torquefree Axisymmetric Gyrostat with Dissipative Rotor and Dissipative Platform

When there is energy dissipation in both rotor and platform, then, according to Rimrott and Li (1982), no torque transmitted between rotor and platform about the spin axis means that the rotor spin changes during an attitude drift $d\nu$ by

$$C^{\mathrm{R}} d\omega_z^{\mathrm{R}} = -\eta A \omega_v \, d\nu \tag{12.55}$$

and the platform spin by

$$C^{\mathrm{P}} d\omega_z = -(1 - \eta) A \omega_v \, d\nu. \tag{12.56}$$

Incorporating the *spin change allotment factor* η, the rotor's kinetic energy changes during an attitude drift $d\nu$ by

$$dT^{\mathrm{R}} = \left(\frac{A^{\mathrm{R}}}{A} C' - \eta A x \right) \omega_v \omega_z \, d\nu \tag{12.57}$$

with $x = \omega_z^{\mathrm{R}} / \omega_z$ and $C' = C^{\mathrm{P}} + C^{\mathrm{R}} x$. The platform's kinetic energy changes by

$$dT^{\mathrm{P}} = \left(\frac{A^{\mathrm{P}}}{A} C' - (1 - \eta) A \right) \omega_v \omega_z \, d\nu. \tag{12.58}$$

Together

$$dT = dT^{\mathrm{R}} + dT^{\mathrm{P}} = (C' - (1-\eta)A - \eta Ax)\omega_v\omega_z d\nu. \tag{12.59}$$

The energy dissipated by the rotor can be shown to amount to

$$dD^{\mathrm{R}} = -\eta(C' - Ax)\omega_v\omega_z d\nu \tag{12.60}$$

while the energy dissipated by the platform amounts to

$$dD^{\mathrm{P}} = -(1-\eta)(C' - A)\omega_v\omega_z d\nu. \tag{12.61}$$

The torque transmitted from the platform to the rotor can be shown (Rimrott and Li, 1982) to be

$$\mathbf{M}^{\mathrm{R}} = [\mathbf{e}_u\ \mathbf{e}_v\ \mathbf{e}_z]\begin{bmatrix} (C^{\mathrm{R}}\omega_z^{\mathrm{R}} - \frac{A^{\mathrm{R}}}{A}C'\omega_z)\omega_v \\ (2\frac{A^{\mathrm{R}}}{A}C'\omega_z - \eta C'\omega_z - C^{\mathrm{R}}\omega_z^{\mathrm{R}})\dot{\nu} \\ 0 \end{bmatrix}. \tag{12.62}$$

Note that its z-component is zero.

For stability, which is characterized by $d\nu < 0$ representing a return to the reference attitude $\nu = 0$, the following three conditions must be satisfied simultaneously:

(1) The total kinetic energy (12.59) must decline

$$dT < 0. \tag{12.63}$$

(2) The energy (12.60) dissipated by the rotor must satisfy

$$dD^{\mathrm{R}} \geq 0. \tag{12.64}$$

(3) The energy (12.61) dissipated by the platform must satisfy

$$dD^{\mathrm{P}} \geq 0. \tag{12.65}$$

For $\eta \leq 0$ criteria (12.63)–(12.65) are satisfied simultaneously if

$$A < C' < Ax. \tag{12.66a}$$

For $0 < \eta \leq 1$ criteria (12.63)–(12.65) are satisfied simultaneously if

$$Ax < C'. \tag{12.66b}$$

For $1 < \eta$ criteria (12.63)–(12.65) cannot be satisfied simultaneously. Summing up, we arrive at the following stability criterion:

An axisymmetric dual spinner is stable, if and only if either simultaneously

(1)

$$A < C' < Ax \quad \text{and} \quad \eta \leq 0 \tag{12.67a}$$

or

(2)

$$Ax < C'. \tag{12.67b}$$

For criterion (12.67b), which is of little practical interest, the spin factor is bounded by $0 \leq \eta \leq 1$. Values outside this region lead to inconsistencies.

For dual spinners in the $A < C' < Ax$ range, the range $0 < \eta \leq 1$ leads to inconsistencies, but the range $1 < \eta$ is possible. In this range dual spinners with $A < C' < Ax$ are unstable.

Whether a dual spinner with $A < C' < Ax$ is in the stable or unstable range depends on the relative energy absorption capacities of rotor and platform with respect to each other, and is based upon the well-known rule that a comparatively nondissipative rotor combined with a dissipative platform leads to $\eta < 0$ and stability, while a dissipative rotor combined with a comparatively nondissipative platform leads to $\eta > 1$ and instability.

The rotor's internal damping deformation makes itself felt by a modified inertia product (see Rimrott, 1986)

$$\begin{aligned} Y_{uz}^{\mathrm{R}} &= -\beta^{\mathrm{R}}\omega_v\omega_z^{\mathrm{R}}\dot{\sigma}^{\mathrm{R}} = -\frac{(Ax - C')H^3}{A^2C'^2}x\beta^{\mathrm{R}}\cos\nu\sin 2\nu \\ &= -\eta\frac{A\dot{\nu}}{x\omega_z} \end{aligned} \tag{12.68}$$

while the platform's internal damping deformation leads to a

$$\begin{aligned} Y_{uz}^{\mathrm{P}} &= -\beta^{\mathrm{P}}\omega_v\omega_z\dot{\sigma} = \frac{(C' - A)H^3}{A^2C'^2}\beta^{\mathrm{P}}\cos\nu\sin 2\nu \\ &= -(1-\eta)\frac{A\dot{\nu}}{\omega_z}. \end{aligned} \tag{12.69}$$

Since the following $Cuvz$ coordinate system is selected such that H_u always vanishes, one has

$$H_u = A\dot{\nu} + Y_{uz}^{\mathrm{R}}\omega_z^{\mathrm{R}} + Y_{uz}^{\mathrm{P}}\omega_z = 0 \tag{12.70}$$

leading to

$$\dot{\nu} = -\frac{1}{A}(Y_{uz}^{\mathrm{P}}\omega_z + Y_{uz}^{\mathrm{R}}\omega_z^{\mathrm{R}}) \tag{12.71}$$

and when combining equations (12.27), (12.65), (12.68), (12.69), and (12.71), the drift rate becomes

$$\dot{\nu} = -\frac{(C' - A)H^4}{2A^3C'^3}\left(\beta^{\mathrm{P}} - \frac{Ax - C'}{C' - A}x^2\beta^{\mathrm{R}}\right)\cos^2\nu\sin 2\nu. \tag{12.72a}$$

The terms in the second bracket may be combined to an equivalent damping coefficient

$$\beta_e = \beta^{\mathrm{P}} - \frac{Ax - C'}{C' - A} x^2 \beta^{\mathrm{R}}. \tag{12.72b}$$

The attitude drift is negative, i.e., the attitude returns to the reference attitude $\nu = 0°$, when $\beta_e > 0$, or when

$$\beta^{\mathrm{P}} > \frac{Ax - C'}{C' - A} x^2 \beta^{\mathrm{R}}. \tag{12.73}$$

The spin change allotment factor η can be shown to be

$$\eta = -\frac{(Ax - C')x^2\beta^{\mathrm{R}}}{(C' - A)\beta^{\mathrm{P}} - (Ax - C')x^2\beta^{\mathrm{R}}}. \tag{12.74}$$

It must be negative for stable gyrostats, i.e., gyrostats with $\dot{\nu} < 0$.

The total energy dissipation rate is (see Rimrott, 1986)

$$\dot{D} = -\dot{T} = -(C' - (1-\eta)A - \eta x A)\omega_v \omega_z \dot{\nu} \tag{12.75}$$

or

$$\dot{D} = -\left(Y^{\mathrm{P}}_{uz} - \frac{Ax - C'}{C' - A} x Y^{\mathrm{R}}_{uz}\right) \omega_v \omega_z \dot{\sigma} \tag{12.76}$$

or

$$\dot{D} = \left(\beta^{\mathrm{P}} - \frac{Ax - C'}{C' - A} x^2 \beta^{\mathrm{R}}\right) \omega_v^2 \omega_z^2 \dot{\sigma}^2. \tag{12.77}$$

The rotor's energy dissipation rate is

$$\dot{D}^{\mathrm{R}} = \eta(Ax - C')\omega_v \omega_z \dot{\nu}. \tag{12.78}$$

The platform's energy dissipation rate is

$$\dot{D}^{\mathrm{P}} = -(1-\eta)(C' - A)\omega_v \omega_z \dot{\nu}. \tag{12.79}$$

$\eta = 0$ means a nondissipative rotor; $\eta = 1$ means a nondissipative platform; $\eta = -\infty$ means $\dot{\nu} = 0$.

The kinetic energy rates are (from equations (12.57) and (12.58)):

$$\dot{T}^{\mathrm{R}} = \left(\frac{A^{\mathrm{R}}}{A} C' - \eta A x\right) \omega_v \omega_z \dot{\nu}$$

$$\dot{T}^{\mathrm{P}} = \left(\frac{A^{\mathrm{P}}}{A} C' - (1-\eta) A\right) \omega_v \omega_z \dot{\nu}.$$

Instead of the spin change allotment factor η, one can also use an *effective dissipation ratio* Q of rotor to platform given by (Rimrott, 1986)

$$Q = \frac{(C' - A)\dot{D}^{\mathrm{R}}}{(Ax - C')\dot{D}^{\mathrm{P}}} = \frac{\eta}{\eta - 1} = -\frac{Y^{\mathrm{R}}_{uz} x^2}{Y^{\mathrm{P}}_{uz}} = \frac{(Ax - C')x^2\beta^{\mathrm{R}}}{(C' - A)\beta^{\mathrm{P}}} = -\frac{C^{\mathrm{R}} d\omega_z^{\mathrm{R}}}{C^{\mathrm{P}} d\omega_z}. \tag{12.80}$$

Stability, i.e., drifting towards $\nu = 0$, for a gyrostat with $A < C' < Ax$ is thus assured, if

$$0 \leq Q < 1. \tag{12.81}$$

With the help of equation (12.71) the torque (12.62) transmitted to the rotor from the platform can be written

$$\mathbf{M}^{\mathrm{R}} = [\mathbf{e}_u \ \mathbf{e}_v \ \mathbf{e}_z] \begin{bmatrix} (C^{\mathrm{R}}x - \frac{A^{\mathrm{R}}}{A}C')\omega_v\omega_z \\ \frac{Y'_{uz}}{A}(C^{\mathrm{R}}x + (\eta - 2\frac{A^{\mathrm{R}}}{A})C')\omega_z^2 \\ 0 \end{bmatrix} \tag{12.82}$$

with

$$Y'_{uz} = Y^{\mathrm{P}}_{uz} + Y^{\mathrm{R}}_{uz}x. \tag{12.83}$$

It is very difficult to gain an impression of the magnitude and effects of the various quantities appearing in the theory. Let us therefore select a reasonable numerical example, having the following data:

$$m = 500\,\mathrm{kg}$$

$$H = 25\,\mathrm{Ws}^2$$

$$A^{\mathrm{R}} = B^{\mathrm{R}} = 1\,\mathrm{Ws}^3$$

$$\beta^{\mathrm{P}} = 5\,\mathrm{Ws}^6$$

$$A^{\mathrm{P}} = B^{\mathrm{P}} = 999\,\mathrm{Ws}^3$$

$$\beta^{\mathrm{R}} = 1.5(10^{-10})\,\mathrm{Ws}^6$$

$$C^{\mathrm{R}} = 1\,\mathrm{Ws}^3$$

$$\nu = 20^\circ$$

$$C^{\mathrm{P}} = 99\,\mathrm{Ws}^3$$

$$x = \omega_z^{\mathrm{R}}/\omega_z = 1000.$$

One obtains

$$A = 999 + 1 = 1000\,\mathrm{Ws}^3$$

$$C' = 99 + 1(1000) = 1099\,\mathrm{Ws}^3$$

$$C' - A = 1099 - 1000 = 99\,\mathrm{Ws}^3$$

$$Ax - C' = 1000(1000) - 1099 = 998\,901\,\mathrm{Ws}^3$$

$$\beta_e = 5 - \frac{998\,901}{99}1000^2(1.5)10^{-10} = 3.486\,\mathrm{Ws}^6$$

$$\omega_v = \frac{25\sin 20^\circ}{1000} = 0.008\,551\,\mathrm{rad/s}$$

$$\omega_z = \frac{25\cos 20^\circ}{1099} = 0.021\,376\,\mathrm{rad/s}$$

$$\omega_z^{\mathrm{R}} = 1000(0.021376) = 21.376\,\mathrm{rad/s}$$

$$\dot{\nu} = -\frac{99(24^4)}{1000^3\,1099^3}3.486\cos^2 20^\circ \sin 40^\circ = -57.646(10^{-12})\,\mathrm{rad/s}.$$

Note that $|\dot{\nu}| << \omega_z$ and $|\dot{\nu}| << \omega_v$. Also that $\dot{\nu} < 0$, thus the gyrostat drifts towards $\nu = 0^\circ$.

$$\dot{\sigma} = -\frac{C' - A}{A}\omega_z = -\frac{99}{1000}0.021\,376 = -0.001\,116\,\mathrm{rad/s}$$

$$\dot{\sigma}^{\mathrm{R}} = \frac{Ax - C'}{A}\omega_z = \frac{998\,901}{1000}0.021\,376 = 21.352\,\mathrm{rad/s}$$

$$\dot{\psi} = \frac{25}{1000} = 0.025\,\mathrm{rad/s}$$

$$\eta = -\frac{(Ax - C')x^2\beta^{\mathrm{R}}}{(C' - A)\beta^{\mathrm{R}} - (Ax - C')x^2\beta^{\mathrm{R}}} = -\frac{998\,901(1000^2)1.5(10^{-10})}{99(5) - 998\,901(1000^2)1.5(10^{-10})}$$

$$\eta = -0.434\,097\,3$$

$$Q = 0.302697$$

$$\dot{\omega}_z = -(1-\eta)\frac{A}{C^{\mathrm{P}}}\omega_v\dot{\nu} = (1.434\,097\,3)\frac{1000}{99}0.008\,551(-57.646)10^{-12}$$

$$= 7.141(10^{-12})\,\mathrm{rad/s}^2$$

$$\dot{\omega}_z^{\mathrm{R}} = -\eta\frac{A}{C^{\mathrm{R}}}\omega_v\dot{\nu} = -(-0.434)\frac{1000}{1}0.008\,551(-57.646)10^{-12}$$

$$= -213.980(10^{-12})\,\mathrm{rad/s}^2$$

$$\dot{H}_z = C^{\mathrm{P}}\dot{\omega}_z + C^{\mathrm{R}}\dot{\omega}_z^{\mathrm{R}} = 99(7.141)10^{-12} + 1(-213.980)10^{-12}$$

$$= 492.979(10^{-12})\,\mathrm{Ws}$$

$$\dot{\omega}_v = \frac{H\cos\nu}{A}\dot{\nu} = \frac{25\cos 20^\circ}{1000}(-57.646)10^{-12} = -1.354(10^{-12})\,\mathrm{rad/s}^2$$

$$\dot{H}_v = 1000(-1.354)(10^{-12}) = -1354.238(10^{-12})\,\mathrm{Ws}.$$

Note that the rotor spin decreases, and the platform spin increases, as the attitude drifts towards 0°.

$$\dot{D}^{\mathrm{R}} = \eta(Ax - C')\omega_v\omega_z\dot{\nu}$$

$$= -434(998\,901)0.008\,551(0.021\,376)(-57.646)10^{-12}$$

$$= 4569.43\,\mathrm{nW}$$

$$\dot{D}^{\mathrm{P}} = -(1-\eta)(C' - A)\omega_v\omega_z\dot{\nu}$$

$$= -1.434(99)0.008\,551(0.021\,376)(-37.054)10^{-12}$$

$$\dot{D}^{\mathrm{P}} = 1.50\,\mathrm{nW}$$
$$\dot{D} = 4570.93\,\mathrm{nW}.$$

Note that the (small!) rotor's energy dissipation is 3054 times as great as the platform's!

$$\dot{T}^{\mathrm{R}} = \left(\frac{1}{1000}1099 + 0.434(1000)1000\right)\frac{25^2\sin 40^\circ}{2(1000)1099}(-57.464)10^{-12}\,\mathrm{W}$$
$$= -4574.47\,\mathrm{nW}$$
$$\dot{T}^{\mathrm{P}} = \left(\frac{999}{1000}1099 - 1.434(1000)\right)\frac{25^2\sin 40^\circ}{2(1000)1099}(-57.646)10^{-12}\,\mathrm{W}$$
$$= 3.44\,\mathrm{nW}$$
$$\dot{T} = -4570.93\,\mathrm{nW}.$$

Note that $\dot{T} = -\dot{D}$, and that the rotor loses kinetic energy, and the platform gains kinetic energy, while the gyrostat attitude drifts towards 0°.

$$Y^{\mathrm{R}}_{uz} = -\frac{(998\,901)25^3}{1000^3(1099^2)}1000(1.5)10^{-10}\cos 20^\circ\sin 40^\circ\,\mathrm{Ws}^3$$
$$= -1.1708\,\mathrm{nWs}^3.$$

Note that $|Y^{\mathrm{R}}_{uz}| << A^{\mathrm{R}}$ and $|Y^{\mathrm{R}}_{uz}| << C^{\mathrm{R}}$

$$Y^{\mathrm{P}}_{uz} = \frac{(99)25^3}{1000^2(1099^2)}5\cos 20^\circ\sin 40^\circ\,\mathrm{Ws}^3 = 3867.9705\,\mathrm{nWs}^3.$$

Note that $|Y^{\mathrm{P}}_{uz}| << A^{\mathrm{P}}$ and $|Y^{\mathrm{P}}_{uz}| << C^{\mathrm{P}}$.

Suggested Reading

1. Annett, R.; W.J. Shackcloth, S.W. Tonkin. "Theory, Simulation, and Practical Tests of a Counterspun Nutation Damper for Prolate Spinners," *Journal of Guidance and Control,* 3, 4, 1980, 290–297.

2. Cochran, J.E. "Nonlinear Resonances in the Attitude Motion of Dual-Spin Spacecraft," *Journal of Spacecraft and Rockets,* 14, 9, 1977, 562–572.

3. Iorillo, A.J. "Nutation Damping Dynamics of Axisymmetric Rotor Stabilized Satellites," *Hughes SSD 70 538 B,* 1965.

4. Landon, V.; B. Stewart, "Nutational Stability of an Axisymmetric Body Containing a Rotor," *Journal of Spacecraft and Rockets,* 1, 6, 1964, 682–684.

5. Likins, P.W. "Attitude Stability Criteria for Dual Spin Spacecraft," *AIAA Journal of Spacecraft and Rockets,* 4, 12, 1967, 1638–1643.

6. Rimrott, F.P.J. "Global Treatment of Energy Dissipation Effects for Multibody Satellites," *Proceedings of IUTAM-IFToMM Symposium on Dynamics of Multibody Systems* (Ed. G. Bianchi and W. Schiehlen), Udine, 1985, Springer-Verlag, 1986, 213–225.

7. Rimrott, F.P.J.; Y.-S. Li. "Attitude Stability of Torque-free Axisymmetric Dual Spinners with Energy Sinks, Considering Spin Change Allotment between Rotor and Platform," *Mechanics Research Communications,* 9, 6, 1982, 403–410.

8. Rimrott, F.P.J.; Y. Yu. "The Near-Rigid Solid in Gyrodynamics," *Proceedings* (Ed. G. Faulkner), CANCAM 87, Edmonton, 1987, CSME Montreal, A78–A79.

9. Roberson, R.E. "Torques on a Satellite Vehicle from Internal Moving Parts," *ASME Journal of Applied Mechanics,* 25, 3, 1958, 196–200.

10. Zhang, W.; F.H. Ling. "Dynamic Stability of the Rotating Shaft Made of Boltzman Viscoelastic Solid," *ASME Journal of Applied Mechanics,* 53, 1985, 424–429.

Problems

12.1. A torquefree axisymmetric gyrostat consists of a rigid rotor in frictionfree bearings and a dissipative platform. At a certain moment in time, the attitude angle ν equals 30°. After the elapse of a time interval $\Delta t = 172\,800$ s, the attitude angle amounts to 29°. Calculate the changes during this time interval experienced by: (a) the ω_v-component; (b) the ω_z-component; (c) the ω_z^{R}-component; (d) the rotor's kinetic energy; (e) the platform's kinetic energy; and (f) the dissipated energy. Calculate also: (g) the attitude drift rate $\dot{\nu}$; (h) the dissipation inertia product Y_{uz}^{P}; (i) the torque transmitted to the rotor; (j) the dissipation coefficient β^{P}; and (k) the gyrostat's kinetic energy. The physical data are $H = 25\,\mathrm{Ws}^2$, $A^{\mathrm{R}} = 10\,\mathrm{Ws}^3$, $C^{\mathrm{R}} = 5\,\mathrm{Ws}^3$, $C^{\mathrm{P}} = 100\,\mathrm{Ws}^3$, $x = \omega_z^{\mathrm{R}}/\omega_z = 800$, $A^{\mathrm{P}} = 990\,\mathrm{Ws}^3$. (l) Repeat the problem with $x = \omega_z^{\mathrm{R}}/\omega_z = 181$.

12.2. A torquefree axisymmetric gyrostat consists of a dissipative rotor in frictionfree bearings and a rigid platform. At a certain moment in time, the attitude angle ν equals 30°. After the elapse of a time interval $\Delta t = 172\,800$ s, the attitude angle amounts to 31°. Calculate the changes during this time interval experienced by: (a) the ω_v-component; (b) the ω_z-component; (c) the ω_z^{R}-component; (d) the rotor's kinetic energy; (e) the platform's kinetic energy; and (f) the dissipated energy. Calculate also: (g) the attitude drift rate $\dot{\nu}$; (h) the dissipation inertia product Y_{uz}^{R}; (i) the torque transmitted to the rotor; (j) the dissipation coefficient β^{R}; and (k) the gyrostat's kinetic energy. The physical data are $H = 25\,\mathrm{Ws}^2$, $A^{\mathrm{R}} = 10\,\mathrm{Ws}^3$, $A^{\mathrm{P}} = 900\,\mathrm{Ws}^3$, $C^{\mathrm{R}} = 5\,\mathrm{Ws}^3$, $C^{\mathrm{P}} = 100\,\mathrm{Ws}^3$, $x = \omega_z^{\mathrm{R}}/\omega_z = 800$.

12.3. A torquefree axisymmetric gyrostat consists of a dissipative rotor in frictionfree bearings and a dissipative platform. At a certain moment in time, the attitude angle ν equals 30°. After the elapse of a time interval $\Delta t = 172\,800$ s, the attitude angle amounts to 29.5°. Calculate the changes during this time interval experienced by: (a) the ω_v-component; (b) the ω_z-component; (c) the ω_z^{R}-component; (d) the rotor's kinetic energy; (e) the platform's kinetic energy; (f) the rotor's dissipated energy; and (g) the platform's dissipated energy. Calculate also: (h) the attitude drift rate $\dot{\nu}$, (i) the dissipation inertia product Y_{uz}^{R}; (j) the dissipation inertia product Y_{uz}^{P}; (k) the torque transmitted to the rotor; (l) the dissipation coefficient β^{R}; (m) the dissipation coefficient β^{P}; and (n) the gyrostat's kinetic energy. The physical data are $H = 25\,\mathrm{Ws}^2$, $A^{\mathrm{R}} = 10\,\mathrm{Ws}^3$, $A^{\mathrm{P}} = 990\,\mathrm{Ws}^3$, $C^{\mathrm{R}} = 5\,\mathrm{Ws}^3$, $C^{\mathrm{P}} = 100\,\mathrm{Ws}^3$, $x = \omega_z^{\mathrm{R}}/\omega_z = 800$, $\eta = -2$.

12.4. A dual spinner has $A^{\mathrm{P}} = 500\,\mathrm{m}^2\mathrm{kg}$, $A^{\mathrm{R}} = 2\,\mathrm{m}^2\mathrm{kg}$, $C^{\mathrm{P}} = 200\,\mathrm{m}^2\mathrm{kg}$, $C^{\mathrm{R}} = 3\,\mathrm{m}^2\mathrm{kg}$, $\omega_z = 1$ rad/s, $\omega_z^{\mathrm{R}} = 1000$ rad/s, $\nu = 30°$, $\eta = -0.01$. (a) Determine ω_v. (b) Will the dual spinner be stable, i.e., tend to move towards $\nu = 0$? (c) For $\dot{\nu} = -0.001$ rad/s, determine the rotor's kinetic energy change rate $\dot{T}^{\mathrm{R}}$; (d) the rotor's energy absorption rate $\dot{D}^{\mathrm{R}}$, (e) the platform's kinetic energy change rate $\dot{T}^{\mathrm{P}}$; (f) the platform's energy absorption rate $\dot{D}^{\mathrm{P}}$; and (g) the energy transmission rate $\dot{E}^{\mathrm{T}}$ from platform to rotor.

12.5. Two equal, very thin rigid disks ($C = 2A$) 1 and 2 of equal size and mass are mounted in frictionfree bearings on a massless structure S as shown in the first figure. Both disks rotate initially at the same angular speed ω while the structure S is at rest. There is no external torque acting on the system. After a long period of time the attitude of the system has changed to that shown in the second figure. Disk 1 is an absolute rigid and has kept its angular speed ω. Disk 2 is a dissipative rigid. Determine (a) the new angular speed ω_2 of disk 2, (b) the old and new kinetic energy of disk 2, and (c) the energy loss (in percent) of the whole system.

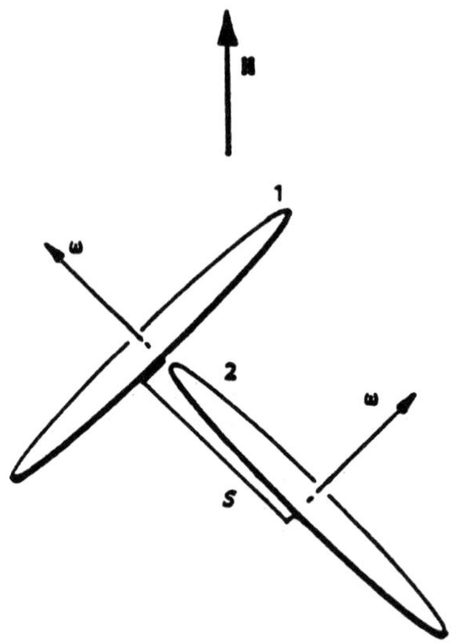

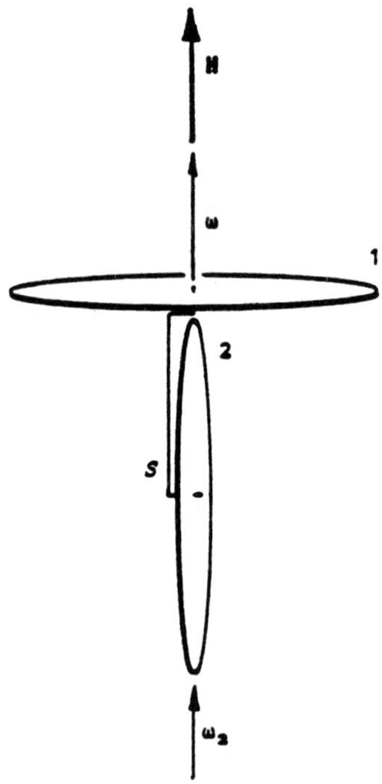

12.6. The dual spinner assembly shown consists of two equal very thin ($C = 2A$), rigid disks of radius r, connected by frictionfree bearings to a massless structure S which separates the disk centroids by a distance of $\sqrt{2}r$. The structure S does not rotate initially, when each disk has an angular speed of ω as shown. Disk 1 is a dissipative rigid, disk 2 is an absolute rigid and retains its angular velocity ω during any attitude drift. There is no external force or torque acting. Eventually the dual spinner assumes the attitude shown in the second figure. (a) Calculate the angular speed ω_1 of disk 1, and (b) give the energy loss (in percent) the system has suffered during the 45° attitude drift.

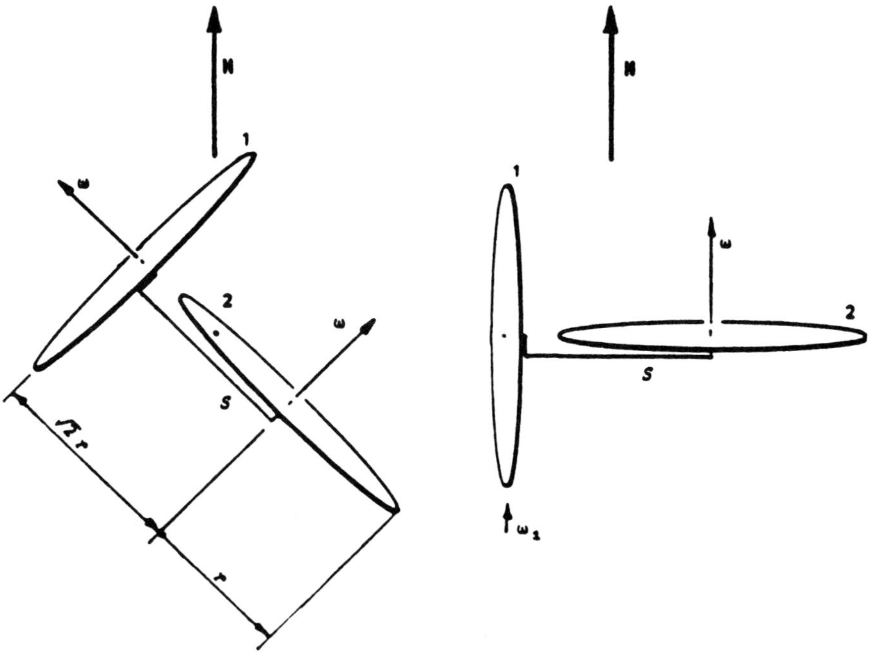

12.7. The two equal spinners, each of mass m and inertia moments A, C, of the gyrostat shown are connected by a massfree shaft and run in frictionfree bearings. Spinner 1 is an absolute rigid and has an angular velocity of

$$\boldsymbol{\omega}_1 = [\mathbf{e}_u \ \mathbf{e}_v \ \mathbf{e}_z] \begin{bmatrix} \frac{C \sin \nu}{2A} \omega \\ 0 \\ \omega \end{bmatrix}.$$

Spinner 2 is dissipative and has an angular velocity of

$$\boldsymbol{\omega}_2 = [\mathbf{e}_u \ \mathbf{e}_v \ \mathbf{e}_z] \begin{bmatrix} \frac{C \sin \nu}{2A} \omega \\ 0 \\ -(1 - \cos \nu)\omega \end{bmatrix}.$$

Determine (a) the angular momentum $\mathbf{H}$ of the system, and (b) the kinetic energy $T + \Delta T$. After some time the system is found to be aligned with the angular momentum vector $\mathbf{H}$. Determine (c) the new angular velocity of spinner 1, and (d) of spinner 2, (e) the new kinetic energy T of the system, and (f) the change $\Delta T/T$. (g) If $C = 2A$ and $\nu = 30°$, what is $\Delta T/T$ (in percent)? (h) If $C = A/2$ and $\nu = 30°$, what is $\Delta T/T$ (in percent)? (i) If $C = 0$ and $\nu = 30°$, what is $\Delta T/T$ (in percent)?

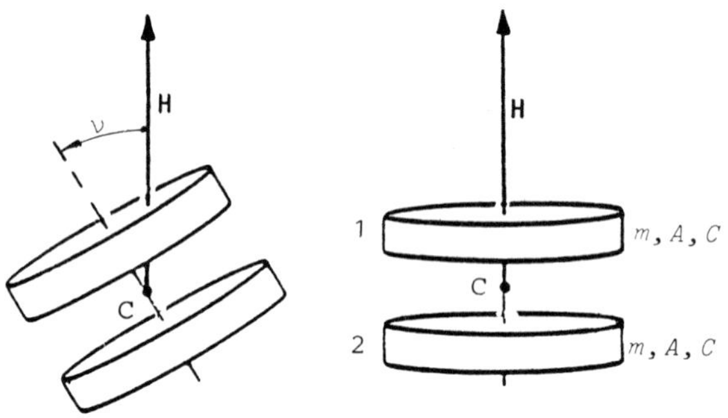

Appendix A

Euler Angles

Euler Angles

Three angles are required to describe the attitude of a (body-fixed) $Cxyz$ coordinate system with repsect to a (space-fixed) $CXYZ$ coordinate system. The sequence of angular rotations is significant. Euler angles involve one axis twice (once in its original attitude, then in its carried attitude) in this rotation process, one axis once, and one axis not at all. Altogether six kinds of Euler angles can be distinguished in this fashion. Since axes are, however, cyclically interchangeable, it suffices to investigate two kinds only. Selecting the z-axis for two rotations, they are shown in the following table.

Rotation Sequence	First Kind (Figure A.1)	Second Kind (Figure A.2)
1.	ψ about z_0	ψ about z_0
2.	ν about x_1	μ about y_1
3.	σ about z_2	σ about z_2

Euler frequencies are integrable, thus

$$\psi = \psi_0 + \int \dot{\psi}\, dt$$

$$\nu = \nu_0 + \int \dot{\nu}\, dt \quad \text{or} \quad \mu = \mu_0 + \int \dot{\mu}\, dt.$$

$$\sigma = \sigma_0 + \int \dot{\sigma}\, dt.$$

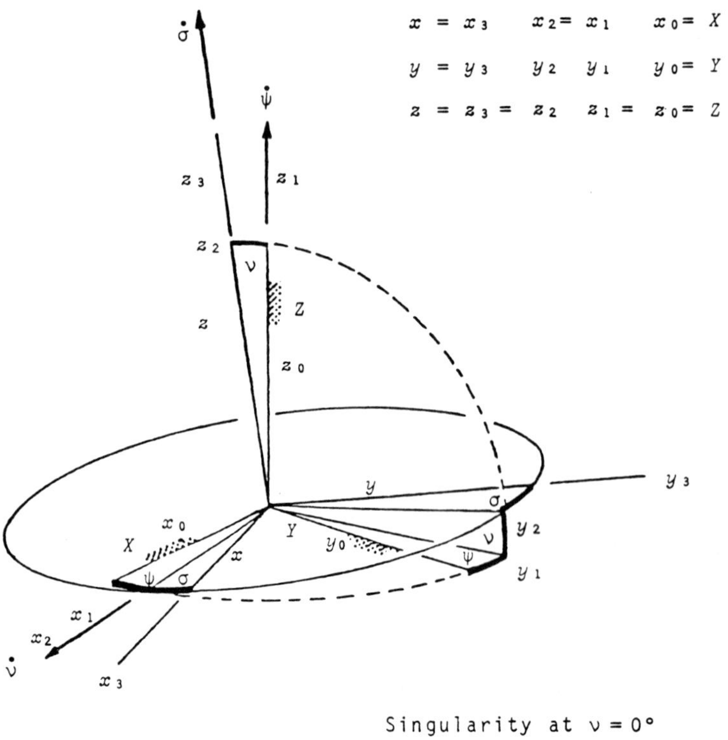

FIGURE A.1. Euler angles of the first kind.

Relationships

$$\mathbf{r} = [\mathbf{e}_{x_3}\ \mathbf{e}_{y_3}\ \mathbf{e}_{z_3}] \begin{bmatrix} x_3 \\ y_3 \\ z_3 \end{bmatrix} = [\mathbf{e}_{x_2}\ \mathbf{e}_{y_2}\ \mathbf{e}_{z_2}] \begin{bmatrix} x_2 \\ y_2 \\ z_2 \end{bmatrix} = [\mathbf{e}_{x_1}\ \mathbf{e}_{y_1}\ \mathbf{e}_{z_1}] \begin{bmatrix} x_1 \\ y_1 \\ z_1 \end{bmatrix}$$

$$= [\mathbf{e}_{x_0}\ \mathbf{e}_{y_0}\ \mathbf{e}_{z_0}] \begin{bmatrix} x_0 \\ y_0 \\ z_0 \end{bmatrix}$$

$$\begin{bmatrix} x_3 \\ y_3 \\ z_3 \end{bmatrix} = [T]_{3\leftarrow 2} \begin{bmatrix} x_2 \\ y_2 \\ z_2 \end{bmatrix} \qquad \begin{bmatrix} x_2 \\ y_2 \\ z_2 \end{bmatrix} = [T]_{2\leftarrow 1} \begin{bmatrix} x_1 \\ y_1 \\ z_1 \end{bmatrix}$$

$$\begin{bmatrix} x_1 \\ y_1 \\ z_1 \end{bmatrix} = [T]_{1\leftarrow 0} \begin{bmatrix} x_0 \\ y_0 \\ z_0 \end{bmatrix}$$

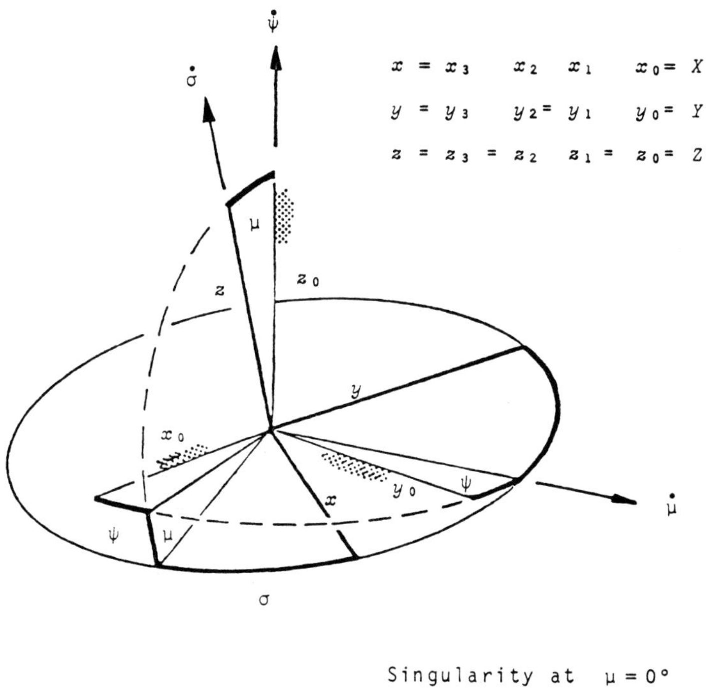

FIGURE A.2. Euler angles of the second kind.

$$\begin{bmatrix} x_3 \\ y_3 \\ z_3 \end{bmatrix} = [T]_{3\leftarrow 0} \begin{bmatrix} x_0 \\ y_0 \\ z_0 \end{bmatrix} \qquad [T]_{3\leftarrow 0} = [T]_{3\leftarrow 2}[T]_{2\leftarrow 1}[T]_{1\leftarrow 0}$$

$$\begin{bmatrix} x \\ y \\ z \end{bmatrix} \equiv \begin{bmatrix} x_3 \\ y_3 \\ z_3 \end{bmatrix} \qquad \begin{bmatrix} x_0 \\ y_0 \\ z_0 \end{bmatrix} \equiv \begin{bmatrix} X \\ Y \\ Z \end{bmatrix}.$$

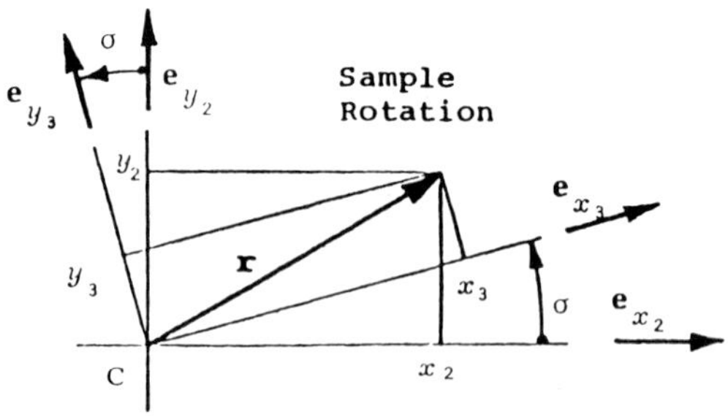

Transition Matrices

First Kind	Second Kind
$[T]_{1\leftarrow 0} = \begin{bmatrix} \cos\psi & \sin\psi & 0 \\ -\sin\psi & \cos\psi & 0 \\ 0 & 0 & 1 \end{bmatrix}$	$[T]_{1\leftarrow 0} = \begin{bmatrix} \cos\psi & \sin\psi & 0 \\ -\sin\psi & \cos\psi & 0 \\ 0 & 0 & 1 \end{bmatrix}$
$[T]_{2\leftarrow 1} = \begin{bmatrix} 1 & 0 & 0 \\ 0 & \cos\nu & \sin\nu \\ 0 & -\sin\nu & \cos\nu \end{bmatrix}$	$[T]_{2\leftarrow 1} = \begin{bmatrix} \cos\mu & 0 & -\sin\mu \\ 0 & 1 & 0 \\ \sin\mu & 0 & \cos\mu \end{bmatrix}$
$[T]_{3\leftarrow 2} = \begin{bmatrix} \cos\sigma & \sin\sigma & 0 \\ -\sin\sigma & \cos\sigma & 0 \\ 0 & 0 & 1 \end{bmatrix}$	$[T]_{3\leftarrow 2} = \begin{bmatrix} \cos\sigma & \sin\sigma & 0 \\ -\sin\sigma & \cos\sigma & 0 \\ 0 & 0 & 1 \end{bmatrix}$

Transition Matrix for the First Kind

$$[T]_{3\leftarrow 0}=\begin{bmatrix} \cos\psi\cos\sigma-\sin\psi\cos\nu\sin\sigma & \sin\psi\cos\sigma+\cos\psi\cos\nu\sin\sigma & \sin\nu\sin\sigma \\ -\cos\psi\sin\sigma-\sin\psi\cos\nu\cos\sigma & -\sin\psi\sin\sigma+\cos\psi\cos\nu\cos\sigma & \sin\nu\cos\sigma \\ \sin\psi\sin\nu & -\cos\psi\sin\nu & \cos\nu \end{bmatrix}$$

Transition Matrix for the Second Kind

$$[T]_{3\leftarrow 0}=\begin{bmatrix} \cos\psi\cos\mu\cos\sigma-\sin\psi\sin\sigma & \sin\psi\cos\mu\cos\sigma+\cos\psi\sin\sigma & -\sin\mu\cos\sigma \\ -\cos\psi\cos\mu\sin\sigma-\sin\psi\cos\sigma & -\sin\psi\cos\mu\sin\sigma+\cos\psi\cos\sigma & \sin\mu\sin\sigma \\ \cos\psi\sin\mu & \sin\psi\sin\mu & \cos\mu \end{bmatrix}$$

The angular velocity vector $\boldsymbol{\omega}$ may be expressed in several different coordinates, such as body-fixed or space-fixed coordinates:

$$\boldsymbol{\omega} = [\mathbf{e}_x \ \mathbf{e}_y \ \mathbf{e}_z] \begin{bmatrix} \omega_x \\ \omega_y \\ \omega_z \end{bmatrix} = [\mathbf{e}_X \ \mathbf{e}_Y \ \mathbf{e}_Z] \begin{bmatrix} \omega_X \\ \omega_Y \\ \omega_Z \end{bmatrix}.$$

It may also be expressed in (nonorthogonal) carried coordinates, i.e.,

$$\boldsymbol{\omega} = [\mathbf{e}_{z_0} \ \mathbf{e}_{x_1} \ \mathbf{e}_{z_2}] \begin{bmatrix} \dot{\psi} \\ \dot{\nu} \\ \dot{\sigma} \end{bmatrix} \quad \text{for Euler angles of the first kind}$$

and

$$\boldsymbol{\omega} = [\mathbf{e}_{z_0} \ \mathbf{e}_{y_1} \ \mathbf{e}_{z_2}] \begin{bmatrix} \dot{\psi} \\ \dot{\mu} \\ \dot{\sigma} \end{bmatrix} \quad \text{for Euler angles of the second kind.}$$

Transition Matrices for Velocity Components

First Kind	Second Kind
$\begin{bmatrix} \omega_x \\ \omega_y \\ \omega_z \end{bmatrix} = \begin{bmatrix} \sin\nu\sin\sigma & \cos\sigma & 0 \\ \sin\nu\cos\sigma & -\sin\sigma & 0 \\ \cos\nu & 0 & 1 \end{bmatrix} \begin{bmatrix} \dot\psi \\ \dot\nu \\ \dot\sigma \end{bmatrix}$	$\begin{bmatrix} \omega_x \\ \omega_y \\ \omega_z \end{bmatrix} = \begin{bmatrix} -\sin\mu\cos\sigma & \cos\sigma & 0 \\ \sin\mu\cos\sigma & -\sin\sigma & 0 \\ \cos\mu & 0 & 1 \end{bmatrix} \begin{bmatrix} \dot\psi \\ \dot\mu \\ \dot\sigma \end{bmatrix}$
$\begin{bmatrix} \dot\psi \\ \dot\nu \\ \dot\sigma \end{bmatrix} = \begin{bmatrix} \frac{\sin\sigma}{\sin\nu} & \frac{\cos\sigma}{\sin\nu} & 0 \\ \cos\sigma & -\sin\sigma & 0 \\ -\frac{\sin\sigma}{\tan\nu} & -\frac{\cos\sigma}{\tan\nu} & 1 \end{bmatrix} \begin{bmatrix} \omega_x \\ \omega_y \\ \omega_z \end{bmatrix}$	$\begin{bmatrix} \dot\psi \\ \dot\mu \\ \dot\sigma \end{bmatrix} = \begin{bmatrix} -\frac{\cos\sigma}{\sin\mu} & \frac{\sin\sigma}{\sin\mu} & 0 \\ \sin\sigma & \cos\sigma & 0 \\ \frac{\cos\sigma}{\tan\mu} & -\frac{\sin\sigma}{\tan\mu} & 1 \end{bmatrix} \begin{bmatrix} \omega_x \\ \omega_y \\ \omega_z \end{bmatrix}$
$\begin{bmatrix} \omega_X \\ \omega_Y \\ \omega_Z \end{bmatrix} = \begin{bmatrix} 0 & \cos\psi & \sin\psi\sin\nu \\ 0 & \sin\psi & -\cos\psi\sin\nu \\ 1 & 0 & \cos\nu \end{bmatrix} \begin{bmatrix} \dot\psi \\ \dot\nu \\ \dot\sigma \end{bmatrix}$	$\begin{bmatrix} \omega_X \\ \omega_Y \\ \omega_Z \end{bmatrix} = \begin{bmatrix} 0 & -\sin\psi & \cos\psi\sin\mu \\ 0 & \cos\psi & \sin\psi\sin\mu \\ 1 & 0 & \cos\mu \end{bmatrix} \begin{bmatrix} \dot\psi \\ \dot\mu \\ \dot\sigma \end{bmatrix}$
$\begin{bmatrix} \dot\psi \\ \dot\nu \\ \dot\sigma \end{bmatrix} = \begin{bmatrix} -\frac{\sin\psi}{\tan\nu} & \frac{\cos\psi}{\tan\nu} & 1 \\ \cos\psi & \sin\psi & 0 \\ \frac{\sin\psi}{\sin\nu} & -\frac{\cos\psi}{\sin\nu} & 0 \end{bmatrix} \begin{bmatrix} \omega_X \\ \omega_Y \\ \omega_Z \end{bmatrix}$	$\begin{bmatrix} \dot\psi \\ \dot\mu \\ \dot\sigma \end{bmatrix} = \begin{bmatrix} -\frac{\cos\psi}{\tan\mu} & \frac{\sin\psi}{\tan\mu} & 1 \\ -\sin\psi & \cos\psi & 0 \\ \frac{\cos\psi}{\sin\mu} & \frac{\sin\psi}{\sin\mu} & 0 \end{bmatrix} \begin{bmatrix} \omega_X \\ \omega_Y \\ \omega_Z \end{bmatrix}$

Transition and Rotation

While both transition and rotation involve a transformation, there is a fundamental difference in the concepts of *transition* and *rotation,* and the experienced engineer takes great care in clearly stating from the outset which of the two applies in a given transformation.

Transition is the process of changing the attitude of a coordinate system (for one and the same vector in absolute space). Rotation is the process of changing the attitude of a vector (in one and the same coordinate system).

The issue is so confusing because of the similar appearance of the transformation matrices containing the direction cosines, since one matrix is the transpose of the other.

In Figure A.3 a *transition* is shown from a coordinate system 0 to a coordinate system 1. One and the same vector $\mathbf{r}$ can be expressed in either system

$$\mathbf{r} = [\mathbf{e}_{x_1}\ \mathbf{e}_{y_1}\ \mathbf{e}_{z_1}] \begin{bmatrix} x_1 \\ y_1 \\ z_1 \end{bmatrix} = [\mathbf{e}_{x_0}\ \mathbf{e}_{y_0}\ \mathbf{e}_{z_0}] \begin{bmatrix} x_0 \\ y_0 \\ z_0 \end{bmatrix}.$$

In order to transit from coordinate sytem 0 to coordinate system 1, we write

$$\begin{bmatrix} x_1 \\ y_1 \\ z_1 \end{bmatrix} = [T] \begin{bmatrix} x_0 \\ y_0 \\ z_0 \end{bmatrix}.$$

An inspection of Figure A.3 shows that the transition matrix is

$$[T] = \begin{bmatrix} \cos\theta & \sin\theta & 0 \\ -\sin\theta & \cos\theta & 0 \\ 0 & 0 & 1 \end{bmatrix}.$$

In Figure A.4 the *rotation* of a vector $\mathbf{r}$ through an angle θ is shown. After rotation, the vector is

$$\mathbf{r}_1 = [\mathbf{e}_x\ \mathbf{e}_y\ \mathbf{e}_z] \begin{bmatrix} x_1 \\ y_1 \\ z_1 \end{bmatrix}.$$

Prior to rotation the vector was

$$\mathbf{r}_0 = [\mathbf{e}_x\ \mathbf{e}_y\ \mathbf{e}_z] \begin{bmatrix} x_0 \\ y_0 \\ z_0 \end{bmatrix}.$$

In order to express the (new) coordinates 1 of the rotated vector in terms of the (old) coordinates 0 of the vector prior to rotation, we write

$$\begin{bmatrix} x_1 \\ y_1 \\ z_1 \end{bmatrix} = [R] \begin{bmatrix} x_0 \\ y_0 \\ z_0 \end{bmatrix}.$$

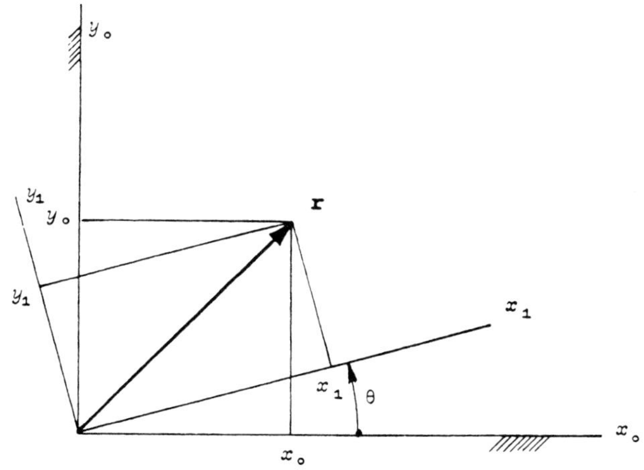

FIGURE A.3. Transition.

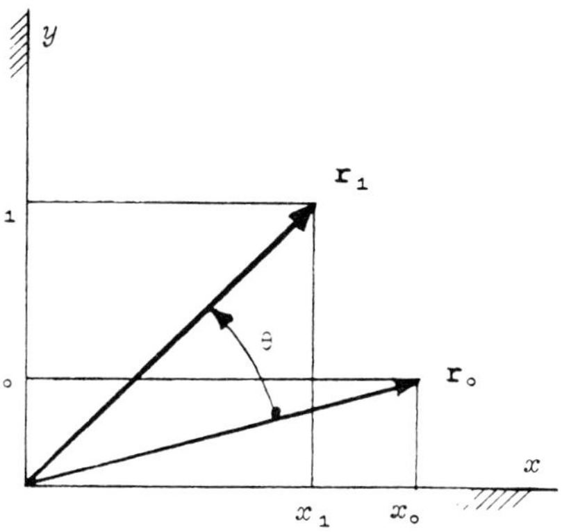

FIGURE A.4. Rotation.

An inspection of Figure A.4 shows that the rotation matrix is

$$[R] = \begin{bmatrix} \cos\theta & -\sin\theta & 0 \\ \sin\theta & \cos\theta & 0 \\ 0 & 0 & 1 \end{bmatrix}.$$

The reader is invited to compare the transformation matrices $[T]$ and $[R]$, and to confirm that

$$[R] = [T]^T.$$

The transition matrix $[T]$ is also known as the *direction cosine matrix,* and is often given the symbol $[C]$. Thus

$$[T_{ij}] = [R_{ij}]^T = [C_{ij}] = [\cos(ij)].$$

Appendix B

Euler Parameters

Euler Parameters

As Euler has shown, any attitude of one coordinate system with respect to another can be expressed by a *rotation axis* and a *rotation angle* about this axis. The rotation axis is identified by a unit vector $\boldsymbol{\lambda}$ along (or parallel to) the rotation axis; the rotation angle is given the symbol θ. To fully specify rotation axis and rotation angle, four items of information are required: the three components of the vector $\boldsymbol{\lambda}$, plus the rotation angle θ. Due to Euler is the computationally convenient concept of using four parameters. The first three ϵ_X, ϵ_Y, and ϵ_Z, are the components of the *Euler vector* $\boldsymbol{\epsilon}$, while the fourth, ϵ_θ, is directly related to the rotation angle.

$$\boldsymbol{\epsilon} = \boldsymbol{\lambda} \sin \frac{\theta}{2}$$

with

$$\epsilon_X = \boldsymbol{\epsilon} \cdot \mathbf{e}_X = \epsilon_x = \boldsymbol{\epsilon} \cdot \mathbf{e}_x = \epsilon_1$$

$$\epsilon_Y = \boldsymbol{\epsilon} \cdot \mathbf{e}_Y = \epsilon_y = \boldsymbol{\epsilon} \cdot \mathbf{e}_y = \epsilon_2$$

$$\epsilon_Z = \boldsymbol{\epsilon} \cdot \mathbf{e}_Z = \epsilon_z = \boldsymbol{\epsilon} \cdot \mathbf{e}_z = \epsilon_3$$

and

$$\epsilon_\theta = \cos \frac{\theta}{2} = \epsilon_4.$$

The Euler parameters are not independent, but are related by

$$\epsilon_1^2 + \epsilon_2^2 + \epsilon_3^2 + \epsilon_4^2 = 1.$$

The unit rotation axis vector $\boldsymbol{\lambda}$ and the rotation angle θ in terms of Euler parameters are

$$\boldsymbol{\lambda} = \{\mathbf{e}_I\}^T \{\epsilon_I\} \frac{1}{\sqrt{\epsilon_X^2 + \epsilon_Y^2 + \epsilon_Z^2}} = \frac{\boldsymbol{\epsilon}}{|\boldsymbol{\epsilon}|}$$

$$\boldsymbol{\lambda} = [\mathbf{e}_X \ \mathbf{e}_Y \ \mathbf{e}_z] \begin{bmatrix} \dfrac{\epsilon_X}{\sqrt{\epsilon_X^2 + \epsilon_Y^2 + \epsilon_Z^2}} \\ \dfrac{\epsilon_Y}{\sqrt{\epsilon_X^2 + \epsilon_Y^2 + \epsilon_Z^2}} \\ \dfrac{\epsilon_Z}{\sqrt{\epsilon_X^2 + \epsilon_Y^2 + \epsilon_Z^2}} \end{bmatrix}$$

$$\theta = 2\arccos \epsilon_\theta \quad 0 \leq \theta \leq 360^\circ.$$

The entries T_{ij} of the transition matrix $[T_{Ij}]$ are the direction cosines

$$T_{Ij} = \mathbf{e}_I \cdot \mathbf{e}_j = \cos(Ij)$$

with $I = X, Y, Z$, and $j = x, y, z$. The following relations also apply

$$\{\mathbf{e}_I\} = [T_{Ij}]\{\mathbf{e}_j\}$$
$$\{\mathbf{e}_I\}^T = \{\mathbf{e}_j\}^T[T_{Ij}]^T$$
$$[T_{Ij}] = \{\mathbf{e}_I\} \cdot \{\mathbf{e}_j\}^T$$
$$\{\mathbf{e}_j\}^T = \{\mathbf{e}_I\}^T[T_{Ij}].$$

Direction cosines and Euler parameters are related by

$$T_{Xx} = \epsilon_1^2 - \epsilon_2^2 - \epsilon_3^2 + \epsilon_4^2 = 1 - 2\epsilon_2^2 - 2\epsilon_3^2$$
$$T_{Xy} = 2(\epsilon_1\epsilon_2 - \epsilon_3\epsilon_4)$$
$$T_{Xz} = 2(\epsilon_3\epsilon_1 + \epsilon_2\epsilon_4)$$

$$T_{Yx} = 2(\epsilon_1\epsilon_2 + \epsilon_3\epsilon_4)$$
$$T_{Yy} = \epsilon_2^2 - \epsilon_3^2 - \epsilon_1^2 + \epsilon_4^2 = 1 - 2\epsilon_3^2 - 2\epsilon_1^2$$
$$T_{Yz} = 2(\epsilon_2\epsilon_3 - \epsilon_1\epsilon_4)$$

$$T_{Zx} = 2(\epsilon_3\epsilon_1 - \epsilon_2\epsilon_4)$$
$$T_{Zy} = 2(\epsilon_2\epsilon_3 + \epsilon_1\epsilon_4)$$
$$T_{Zz} = \epsilon_3^2 - \epsilon_1^2 - \epsilon_2^2 + \epsilon_4^2 = 1 - 2\epsilon_1^2 - \epsilon_2^2.$$

Euler parameters in terms of direction cosines are given by

$$\epsilon_1 = -\frac{T_{Yz} - T_{Zy}}{2\sqrt{1 + T_{Xx} + T_{Yy} + T_{Zz}}}$$

$$\epsilon_2 = -\frac{T_{Zx} - T_{Xz}}{2\sqrt{1 + T_{Xx} + T_{Yy} + T_{Zz}}}$$

$$\epsilon_3 = -\frac{T_{Xy} - T_{Yx}}{2\sqrt{1 + T_{Xx} + T_{Yy} + T_{Zz}}}$$

$$\epsilon_4 = \frac{1}{2}\sqrt{1 + T_{Xx} + T_{Yy} + T_{Zz}}.$$

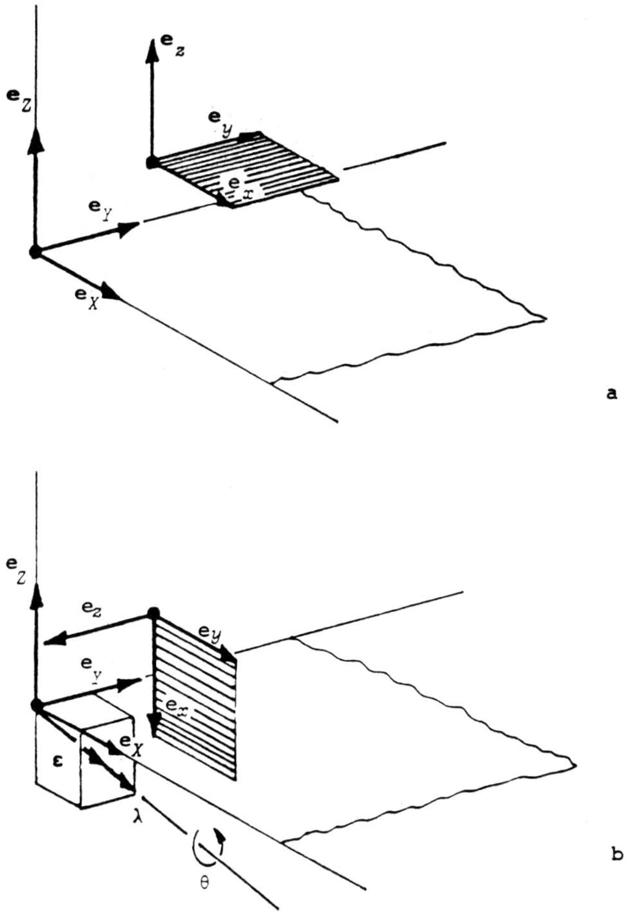

FIGURE B.1. Attitude of xyz coordinate system prior to and after rotation.

Using Euler parameters, the direction cosine matrix $[T_{Ij}]$, which is also called the *attitude matrix*, becomes

$$[T_{Ij}] = \begin{bmatrix} 1-2(\epsilon_2^2+\epsilon_3^2) & 2(\epsilon_1\epsilon_2-\epsilon_3\epsilon_4) & 2(\epsilon_1\epsilon_3+\epsilon_2\epsilon_4) \\ 2(\epsilon_1\epsilon_2+\epsilon_3\epsilon_4) & 1-2(\epsilon_3^2+\epsilon_1^2) & 2(\epsilon_2\epsilon_3-\epsilon_1\epsilon_4) \\ 2(\epsilon_1\epsilon_3-\epsilon_2\epsilon_4) & 2(\epsilon_2\epsilon_3+\epsilon_1\epsilon_4) & 1-2(\epsilon_1^2+\epsilon_2^2) \end{bmatrix}.$$

As an example, take the xyz system shown in Figure B.1. Prior to rotation xyz and XYZ are parallel, as shown in Figure B.1a. After rotation the xyz system has an attitude as shown in Figure B.1b. The direction cosine matrix is

$$[T_{ij}] = \begin{bmatrix} T_{Xx} & T_{Xy} & T_{Xz} \\ T_{Yx} & T_{Yy} & T_{Yz} \\ T_{Zx} & T_{Zy} & T_{Zz} \end{bmatrix} = \begin{bmatrix} 0 & 1 & 0 \\ 0 & 0 & -1 \\ -1 & 0 & 0 \end{bmatrix}.$$

The Euler parameters are

$$\epsilon_1 = -\frac{-1-0}{2\sqrt{1+0+0+0}} = \frac{1}{2}$$

$$\epsilon_2 = -\frac{-1-0}{2\sqrt{1+0+0+0}} = \frac{1}{2}$$

$$\epsilon_3 = -\frac{1-0}{2\sqrt{1+0+0+0}} = -\frac{1}{2}$$

$$\epsilon_4 = \frac{1}{2}\sqrt{1+0+0+0} = \frac{1}{2}.$$

The Euler vector is

$$\boldsymbol{\epsilon} = [\mathbf{e}_X \ \mathbf{e}_Y \ \mathbf{e}_Z] \begin{bmatrix} \frac{1}{2} \\ \frac{1}{2} \\ -\frac{1}{2} \end{bmatrix}.$$

The unit rotation axis vector is

$$\boldsymbol{\lambda} = \frac{1}{\sqrt{\epsilon_1^2+\epsilon_2^2+\epsilon_3^2}}\boldsymbol{\epsilon} = [\mathbf{e}_X \ \mathbf{e}_Y \ \mathbf{e}_Z] \begin{bmatrix} \frac{1}{\sqrt{3}} \\ \frac{1}{\sqrt{3}} \\ -\frac{1}{\sqrt{3}} \end{bmatrix}.$$

The rotation angle is $\theta = 2\arccos 1/2 = 120°$.

The *Euler rotation theorem* states that the attitude of a vector $\mathbf{r}$ fixed in the xyz coordinate system, can be expressed by

$$\mathbf{r}_1 = \mathbf{r}_0 + 2(\epsilon_\theta \boldsymbol{\epsilon} \ \times \mathbf{r}_0 + \ \boldsymbol{\epsilon} \ \times (\boldsymbol{\epsilon} \ \times \mathbf{r}_0))$$

with

$$\mathbf{r}_1 = [\mathbf{e}_X \ \mathbf{e}_Y \ \mathbf{e}_Z] \begin{bmatrix} X \\ Y \\ Z \end{bmatrix}$$

representing the vector $\mathbf{r}$ after rotation, and

$$\mathbf{r}_0 = [\mathbf{e}_X \ \mathbf{e}_Y \ \mathbf{e}_Z] \begin{bmatrix} x \\ y \\ z \end{bmatrix}$$

representing the vector $\mathbf{r}$ prior to rotation, i.e., when the xyz system and the XYZ system were parallel (Figure B.2).

Take, as an example, the vector

$$\mathbf{r} = [\mathbf{e}_x \ \mathbf{e}_y \ \mathbf{e}_z] \begin{bmatrix} 1 \\ 2 \\ 2 \end{bmatrix}$$

shown in Figure B.2. Before rotation of the xyz system, the vector has an attitude (with respect to the XYZ system) given by

$$\mathbf{r}_0 = [\mathbf{e}_X \ \mathbf{e}_Y \ \mathbf{e}_Z] \begin{bmatrix} 1 \\ 2 \\ 2 \end{bmatrix}.$$

After rotation of the xyz system given by the Euler vector

$$\boldsymbol{\epsilon} = [\mathbf{e}_X \ \mathbf{e}_Y \ \mathbf{e}_Z] \begin{bmatrix} \frac{1}{2} \\ \frac{1}{2} \\ -\frac{1}{2} \end{bmatrix}$$

and the Euler parameter $\epsilon_\theta = 1/2$, the vector's new attitude (with respect to the XYZ system) is given by the Euler rotation theorem. The cross products occuring therein are

$$\boldsymbol{\epsilon} \times \mathbf{r}_0 = \begin{vmatrix} \mathbf{e}_X & \mathbf{e}_Y & \mathbf{e}_Z \\ \frac{1}{2} & \frac{1}{2} & -\frac{1}{2} \\ 1 & 2 & 2 \end{vmatrix} = [\mathbf{e}_X \ \mathbf{e}_Y \ \mathbf{e}_Z] \begin{bmatrix} 2 \\ -\frac{3}{2} \\ \frac{1}{2} \end{bmatrix}$$

and

$$\boldsymbol{\epsilon} \times (\boldsymbol{\epsilon} \times \mathbf{r}_0) = \begin{vmatrix} \mathbf{e}_X & \mathbf{e}_Y & \mathbf{e}_Z \\ \frac{1}{2} & \frac{1}{2} & -\frac{1}{2} \\ 2 & -\frac{3}{2} & \frac{1}{2} \end{vmatrix} = [\mathbf{e}_X \ \mathbf{e}_Y \ \mathbf{e}_Z] \begin{bmatrix} -\frac{1}{2} \\ -\frac{5}{4} \\ -\frac{7}{4} \end{bmatrix}$$

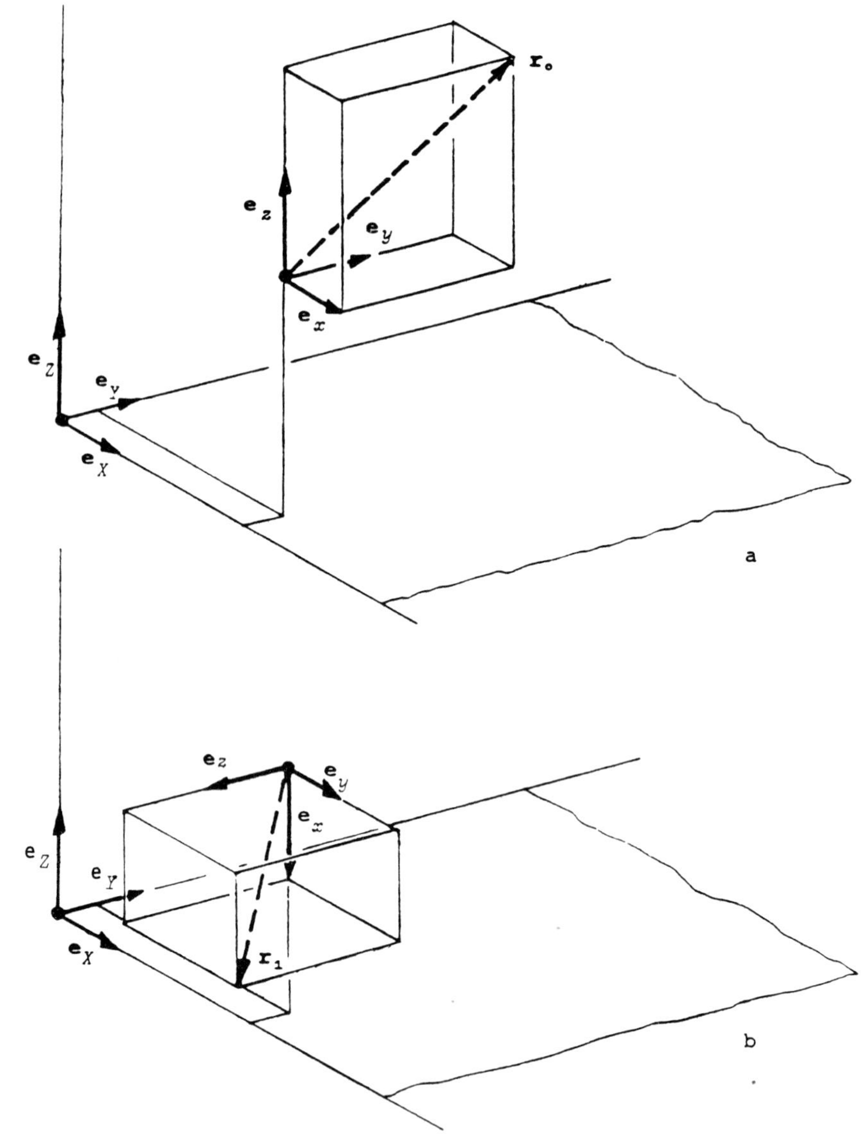

FIGURE B.2. Attitude of **r** vector prior to and after rotation of xyz system.

such that

$$\begin{bmatrix} X \\ Y \\ Z \end{bmatrix} = \begin{bmatrix} 1 \\ 2 \\ 2 \end{bmatrix} + 2\left(\frac{1}{2}\begin{bmatrix} 2 \\ -\frac{3}{2} \\ \frac{1}{2} \end{bmatrix} + \begin{bmatrix} -\frac{1}{2} \\ -\frac{3}{2} \\ -\frac{7}{4} \end{bmatrix}\right) = \begin{bmatrix} 2 \\ -2 \\ -1 \end{bmatrix}$$

that is

$$\mathbf{r}_1 = [\mathbf{e}_X \ \mathbf{e}_Y \ \mathbf{e}_Z]\begin{bmatrix} 2 \\ -2 \\ -1 \end{bmatrix}$$

as can be seen in Figure B.2.

The angular velocity $\boldsymbol{\omega}$ of the xyz system with respect to the XYZ system is given by

$$\boldsymbol{\omega} = [\mathbf{e}_x \ \mathbf{e}_y \ \mathbf{e}_z]\begin{bmatrix} \omega_x \\ \omega_y \\ \omega_z \end{bmatrix} = \{\mathbf{e}_j\}^T\{\omega_j\}.$$

It can be shown that

$$\omega_x = T_{Xz}\dot{T}_{Xy} + T_{Yz}\dot{T}_{Yy} + T_{Zz}\dot{T}_{Zy}$$

$$\omega_y = T_{Yx}\dot{T}_{Yz} + T_{Zx}\dot{T}_{Zz} + T_{Xx}\dot{T}_{Xz}$$

$$\omega_z = T_{Zy}\dot{T}_{Zx} + T_{Xy}\dot{T}_{Xx} + T_{Yy}\dot{T}_{Yx}.$$

The *Poisson kinematic equations* are

$$\dot{T}_{Xx} = T_{Xy}\omega_z - T_{Xz}\omega_y$$

$$\dot{T}_{Xy} = T_{Xz}\omega_x - T_{Xx}\omega_z$$

$$\dot{T}_{Xz} = T_{Xx}\omega_y - T_{Xy}\omega_x$$

$$\dot{T}_{Yx} = T_{Yy}\omega_z - T_{Yz}\omega_y$$

$$\dot{T}_{Yy} = T_{Yz}\omega_x - T_{Yx}\omega_z$$

$$\dot{T}_{Yz} = T_{Yx}\omega_y - T_{Yy}\omega_x$$

$$\dot{T}_{Zx} = T_{Zy}\omega_z - T_{Zz}\omega_y$$

$$\dot{T}_{Zy} = T_{Zz}\omega_x - T_{Zx}\omega_z$$

$$\dot{T}_{Zz} = T_{Zx}\omega_y - T_{Zy}\omega_x.$$

The angular velocity in terms of Euler parameters is

$$\boldsymbol{\omega} = 2[\epsilon_4 \overset{\circ}{\boldsymbol{\epsilon}} - \dot{\epsilon}_4 \boldsymbol{\epsilon} - \boldsymbol{\epsilon} \times \overset{\circ}{\boldsymbol{\epsilon}}]$$

with

$$\boldsymbol{\epsilon} = [\mathbf{e}_x \ \mathbf{e}_y \ \mathbf{e}_z] \begin{bmatrix} \epsilon_1 \\ \epsilon_2 \\ \epsilon_3 \end{bmatrix}$$

$$\overset{\circ}{\boldsymbol{\epsilon}} = [\mathbf{e}_x \ \mathbf{e}_y \ \mathbf{e}_z] \begin{bmatrix} \dot{\epsilon}_1 \\ \dot{\epsilon}_2 \\ \dot{\epsilon}_3 \end{bmatrix}.$$

The Euler parameter representation has as an advantage the fact that the absolute value of any of the four Euler parameters does not exceed unity. A disadvantage is that one of the four is redundant. By forming a so-called *Rodrigues vector* this disadvantage is removed. The Rodrigues vector, in terms of unit rotation axis vector $\boldsymbol{\lambda}$ and rotation angle θ is defined as

$$\boldsymbol{\rho} = \boldsymbol{\lambda} \tan \frac{\theta}{2}$$

which, in terms of Euler parameters is

$$\boldsymbol{\rho} = \frac{1}{\epsilon_\theta} \boldsymbol{\epsilon} = \frac{1}{\epsilon_4} \boldsymbol{\epsilon}.$$

The *Rodrigues parameters,* i.e., the components of the Rodrigues vector, are

$$\rho_X = \frac{1}{\epsilon_\theta} \epsilon_X = \frac{1}{\epsilon_\theta} \epsilon_x = \rho_x = \frac{1}{\epsilon_4} \epsilon_1 = \rho_1$$

$$\rho_Y = \frac{1}{\epsilon_\theta} \epsilon_Y = \frac{1}{\epsilon_\theta} \epsilon_y = \rho_y = \frac{1}{\epsilon_4} \epsilon_2 = \rho_2$$

$$\rho_Z = \frac{1}{\epsilon_\theta} \epsilon_Z = \frac{1}{\epsilon_\theta} \epsilon_z = \rho_z = \frac{1}{\epsilon_4} \epsilon_3 = \rho_3.$$

Rodrigues parameters can assume any value, including infinity.

Problem

B.1. An $Oxyz$ coordinate system's attitude with respect to a (fixed) $OXYZ$ coordinate system is obtained by rotating it through the Euler angles $\psi = 0°$, $\nu = 4°$, $\sigma = 10°$. (a) Determine the direction cosines; (b) find the four Euler parameters; (c) find and sketch the Euler rotation axis vector $\boldsymbol{\lambda}$; and (d) compute the Euler rotation angle θ.

Appendix C

Cardan Angles

Depending upon the sequence of rotations, six kinds of Cardan angles can be defined (Figure C.1).

Rotation Sequence	First Kind (Figure C.2)	Second Kind (Figure C.3)	Third Kind (Figure C.4)
1.	ξ about x_0	η about y_0	ζ about z_0
2.	η about y_1	ζ about z_1	ξ about x_1
3.	ζ about z_2	ξ about x_2	η about y_2

	Fourth Kind (Figure C.5)	Fifth Kind (Figure C.6)	Sixth Kind (Figure C.7)
1.	ξ about x_0	ζ about z_0	η about y_0
2.	ζ about z_1	η about y_1	ξ about x_1
3.	η about y_2	ξ about x_2	ζ about z_2

Cardan frequencies can be integrated:

$$\xi = \xi_0 + \int \dot{\xi}\, dt$$

$$\eta = \eta_0 + \int \dot{\eta}\, dt$$

$$\zeta = \zeta_0 + \int \dot{\zeta}\, dt.$$

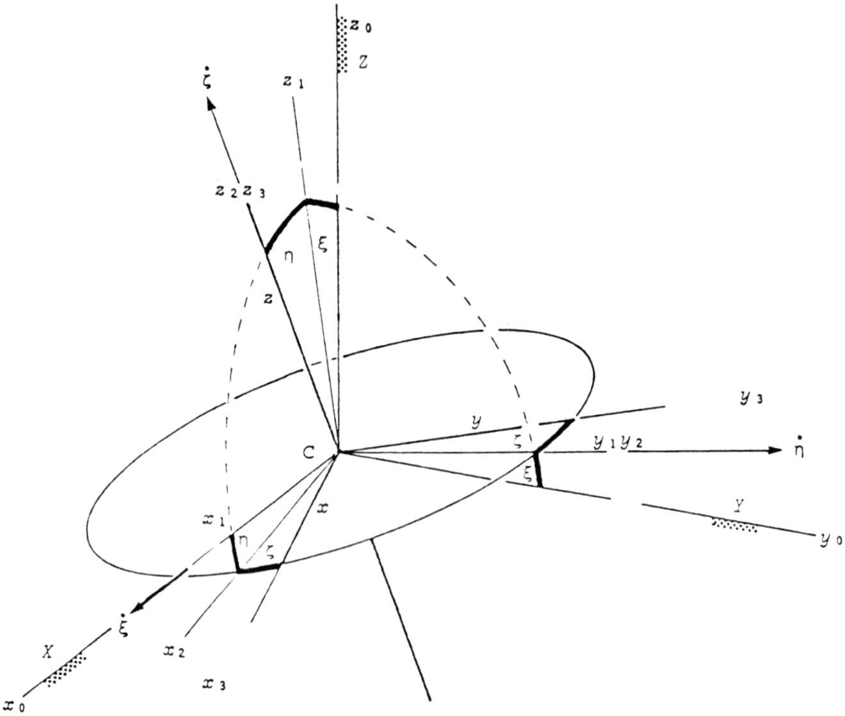

FIGURE C.1. Gyro axes $Cxyz$ rotated through Cardan angles of the first kind.

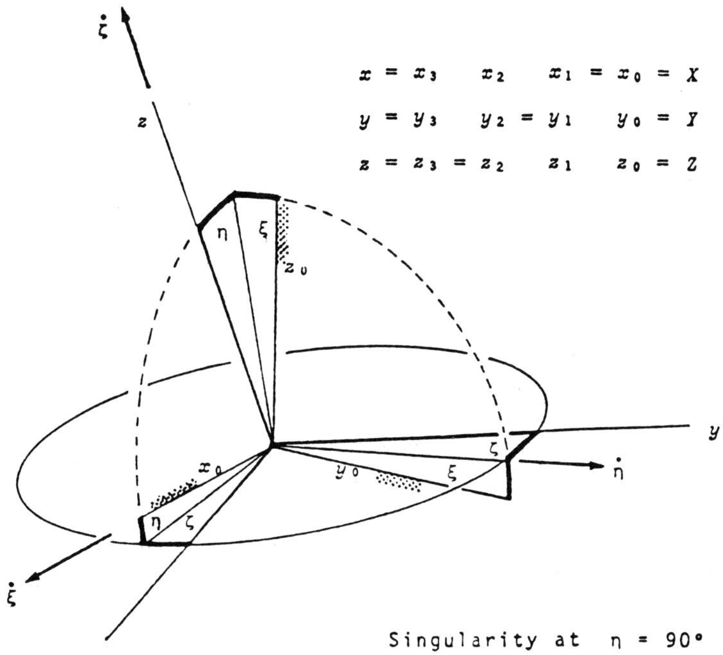

FIGURE C.2. Cardan angles of the first kind.

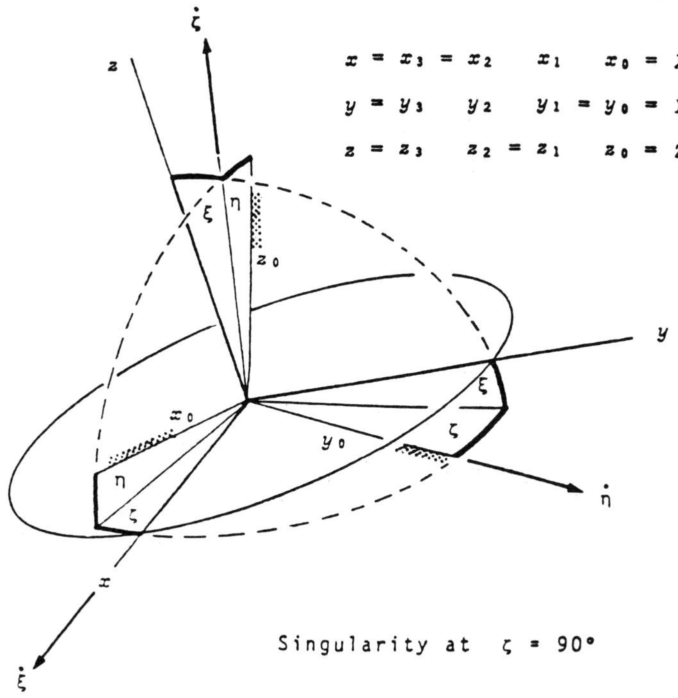

FIGURE C.3. Cardan angles of the second kind.

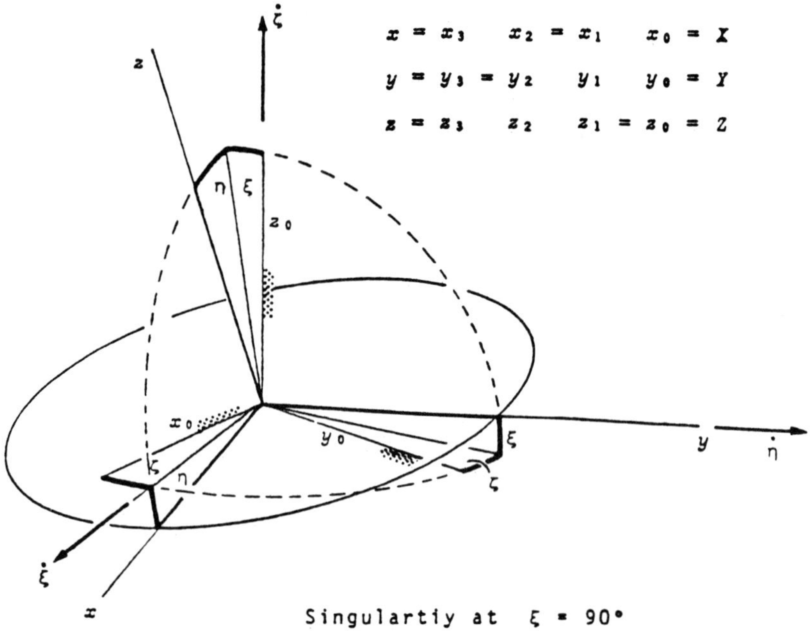

FIGURE C.4. Cardan angles of the third kind.

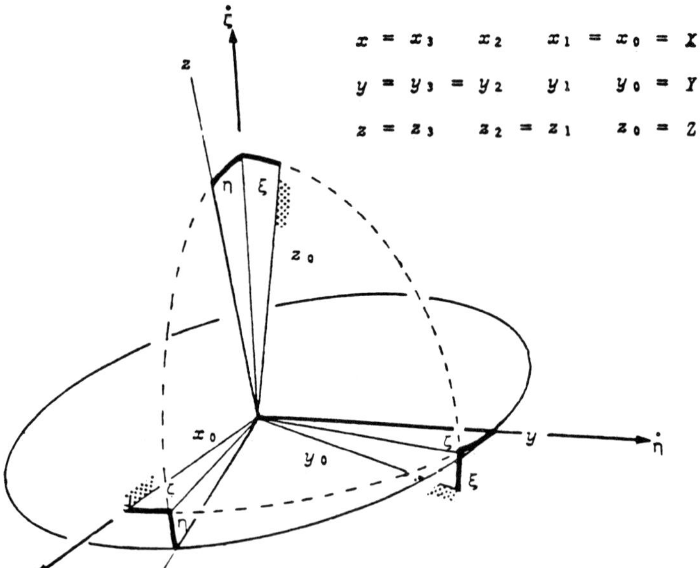

FIGURE C.5. Cardan angles of the fourth kind.

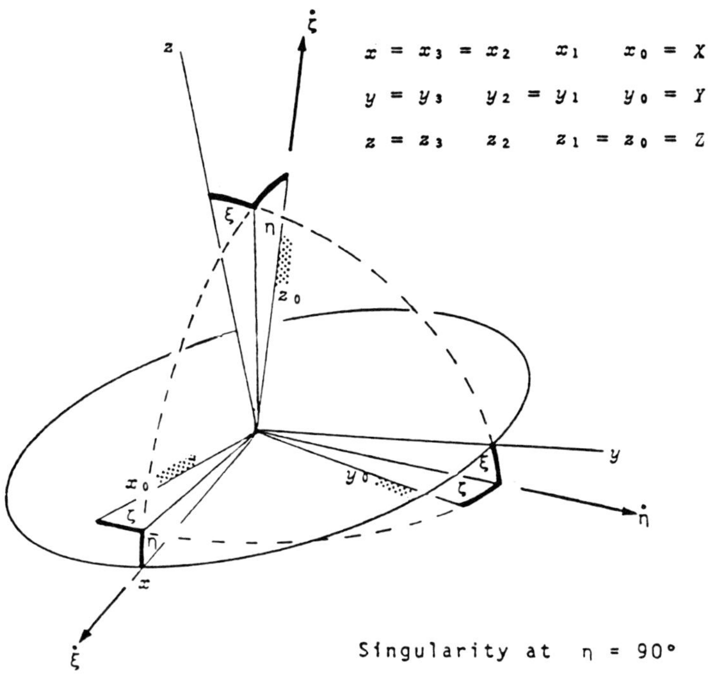

FIGURE C.6. Cardan angles of the fifth kind.

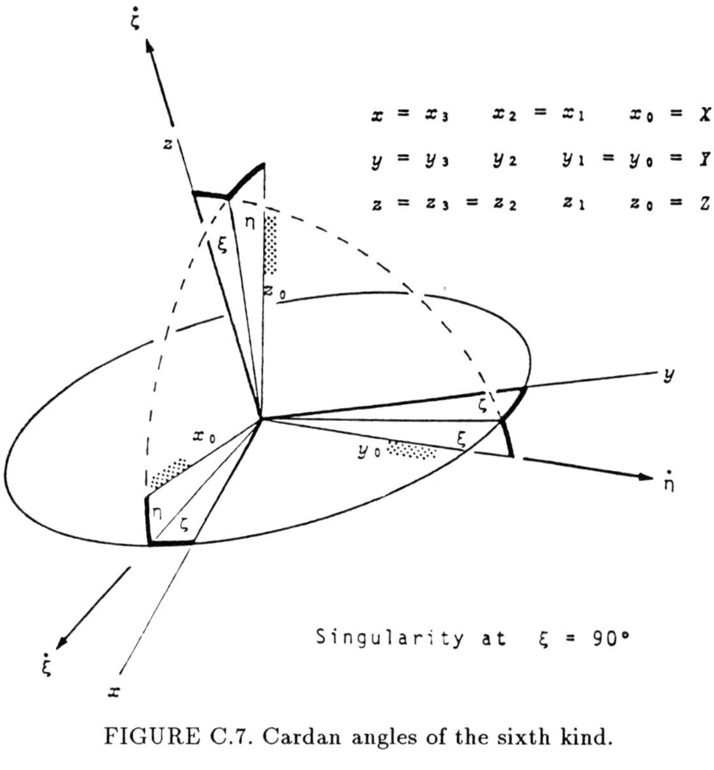

FIGURE C.7. Cardan angles of the sixth kind.

Relationships

$$\mathbf{r} = [\mathbf{e}_{x_3}\ \mathbf{e}_{y_3}\ \mathbf{e}_{z_3}]\begin{bmatrix} x_3 \\ y_3 \\ z_3 \end{bmatrix} = [\mathbf{e}_{x_2}\ \mathbf{e}_{y_2}\ \mathbf{e}_{z_2}]\begin{bmatrix} x_2 \\ y_2 \\ z_2 \end{bmatrix} = [\mathbf{e}_{x_1}\ \mathbf{e}_{y_1}\ \mathbf{e}_{z_1}]\begin{bmatrix} x_1 \\ y_1 \\ z_1 \end{bmatrix}$$

$$= [\mathbf{e}_{x_0}\ \mathbf{e}_{y_0}\ \mathbf{e}_{z_0}]\begin{bmatrix} x_0 \\ y_0 \\ z_0 \end{bmatrix}$$

$$\begin{bmatrix} x_3 \\ y_3 \\ z_3 \end{bmatrix} = [T]_{3\leftarrow 2}\begin{bmatrix} x_2 \\ y_2 \\ z_2 \end{bmatrix} \qquad \begin{bmatrix} x_2 \\ y_2 \\ z_2 \end{bmatrix} = [T]_{2\leftarrow 1}\begin{bmatrix} x_1 \\ y_1 \\ z_1 \end{bmatrix}$$

$$\begin{bmatrix} x_1 \\ y_1 \\ z_1 \end{bmatrix} = [T]_{1\leftarrow 0}\begin{bmatrix} x_0 \\ y_0 \\ z_0 \end{bmatrix}$$

$$\begin{bmatrix} x_3 \\ y_3 \\ z_3 \end{bmatrix} = [T]_{3\leftarrow 0}\begin{bmatrix} x_0 \\ y_0 \\ z_0 \end{bmatrix}$$

$$[T]_{3\leftarrow 0} = [T]_{3\leftarrow 2}[T]_{2\leftarrow 1}[T]_{1\leftarrow 0}$$

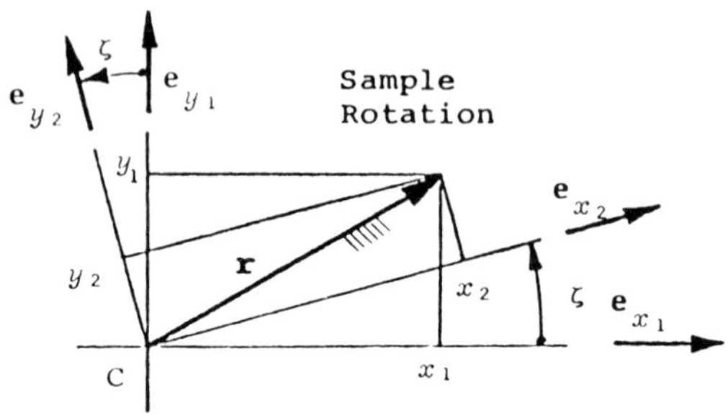

$$\begin{bmatrix} x \\ y \\ z \end{bmatrix} \equiv \begin{bmatrix} x_3 \\ y_3 \\ z_3 \end{bmatrix} \qquad \begin{bmatrix} x_0 \\ y_0 \\ z_0 \end{bmatrix} \equiv \begin{bmatrix} X \\ Y \\ Z \end{bmatrix}.$$

	First Kind	Second Kind	Third Kind
$[T]_{1\leftarrow 0}$	$\begin{bmatrix} 1 & 0 & 0 \\ 0 & \cos\xi & \sin\xi \\ 0 & -\sin\xi & \cos\xi \end{bmatrix}$	$\begin{bmatrix} \cos\eta & 0 & -\sin\eta \\ 0 & 1 & 0 \\ \sin\eta & 0 & \cos\eta \end{bmatrix}$	$\begin{bmatrix} \cos\zeta & \sin\zeta & 0 \\ -\sin\zeta & \cos\zeta & 0 \\ 0 & 0 & 1 \end{bmatrix}$
$[T]_{2\leftarrow 1}$	$\begin{bmatrix} \cos\eta & 0 & -\sin\eta \\ 0 & 1 & 0 \\ \sin\eta & 0 & \cos\eta \end{bmatrix}$	$\begin{bmatrix} \cos\zeta & \sin\zeta & 0 \\ -\sin\zeta & \cos\zeta & 0 \\ 0 & 0 & 1 \end{bmatrix}$	$\begin{bmatrix} 1 & 0 & 0 \\ 0 & \cos\xi & \sin\xi \\ 0 & -\sin\xi & \cos\xi \end{bmatrix}$
$[T]_{3\leftarrow 2}$	$\begin{bmatrix} \cos\zeta & \sin\zeta & 0 \\ -\sin\zeta & \cos\zeta & 0 \\ 0 & 0 & 1 \end{bmatrix}$	$\begin{bmatrix} 1 & 0 & 0 \\ 0 & \cos\xi & \sin\xi \\ 0 & -\sin\xi & \cos\xi \end{bmatrix}$	$\begin{bmatrix} \cos\eta & 0 & -\sin\eta \\ 0 & 1 & 0 \\ \sin\eta & 0 & \cos\eta \end{bmatrix}$
	Fourth Kind	Fifth Kind	Sixth Kind
$[T]_{1\leftarrow 0}$	$\begin{bmatrix} 1 & 0 & 0 \\ 0 & \cos\xi & \sin\xi \\ 0 & -\sin\xi & \cos\xi \end{bmatrix}$	$\begin{bmatrix} \cos\zeta & \sin\zeta & 0 \\ -\sin\zeta & \cos\zeta & 0 \\ 0 & 0 & 1 \end{bmatrix}$	$\begin{bmatrix} \cos\eta & 0 & -\sin\eta \\ 0 & 1 & 0 \\ \sin\eta & 0 & \cos\eta \end{bmatrix}$
$[T]_{2\leftarrow 1}$	$\begin{bmatrix} \cos\zeta & \sin\zeta & 0 \\ -\sin\zeta & \cos\zeta & 0 \\ 0 & 0 & 1 \end{bmatrix}$	$\begin{bmatrix} \cos\eta & 0 & -\sin\eta \\ 0 & 1 & 0 \\ \sin\eta & 0 & \cos\eta \end{bmatrix}$	$\begin{bmatrix} 1 & 0 & 0 \\ 0 & \cos\xi & \sin\xi \\ 0 & -\sin\xi & \cos\xi \end{bmatrix}$
$[T]_{3\leftarrow 2}$	$\begin{bmatrix} \cos\eta & 0 & -\sin\eta \\ 0 & 1 & 0 \\ \sin\eta & 0 & \cos\eta \end{bmatrix}$	$\begin{bmatrix} 1 & 0 & 0 \\ 0 & \cos\xi & \sin\xi \\ 0 & -\sin\xi & \cos\xi \end{bmatrix}$	$\begin{bmatrix} \cos\zeta & \sin\zeta & 0 \\ -\sin\zeta & \cos\zeta & 0 \\ 0 & 0 & 1 \end{bmatrix}$

First kind:

$$[T]_{3\leftarrow 0} = \begin{bmatrix} \cos\eta\cos\xi & \cos\xi\sin\eta + \sin\xi\sin\eta\cos\xi & \sin\xi\sin\zeta - \cos\xi\sin\eta\cos\zeta \\ -\cos\eta\sin\zeta & \cos\xi\cos\zeta - \sin\xi\sin\eta\sin\zeta & \sin\xi\cos\zeta + \cos\xi\sin\eta\sin\zeta \\ \sin\eta & -\sin\xi\cos\eta & \cos\xi\cos\eta \end{bmatrix}$$

Second kind:

$$[T]_{3\leftarrow 0} = \begin{bmatrix} \cos\eta\cos\zeta & \sin\zeta & -\sin\eta\cos\zeta \\ -\cos\xi\cos\eta\sin\zeta + \sin\xi\sin\eta & \cos\xi\cos\zeta & \cos\xi\sin\eta\sin\zeta + \sin\xi\cos\eta \\ \sin\xi\cos\eta\sin\zeta + \cos\xi\sin\eta & -\sin\xi\cos\zeta & -\sin\xi\sin\eta\sin\zeta + \cos\xi\cos\eta \end{bmatrix}$$

Third kind:

$$[T]_{3\leftarrow 0} = \begin{bmatrix} \cos\eta\cos\zeta - \sin\xi\sin\eta\sin\zeta & \cos\eta\sin\zeta + \sin\xi\sin\eta\cos\zeta & -\cos\xi\sin\eta \\ -\cos\xi\sin\zeta & \cos\xi\cos\zeta & \sin\xi \\ \sin\eta\cos\zeta + \sin\xi\cos\eta\sin\zeta & \sin\eta\sin\zeta - \sin\xi\cos\eta\cos\xi & \cos\xi\cos\eta \end{bmatrix}$$

Fourth kind:

$$[T]_{3\leftarrow 0} = \begin{bmatrix} \cos\eta\sin\zeta & \cos\xi\cos\eta\sin\zeta + \sin\xi\sin\eta & \sin\xi\cos\eta\sin\zeta - \cos\xi\sin\eta \\ -\sin\zeta & \cos\xi\cos\zeta & \sin\xi\cos\zeta \\ \sin\eta\cos\zeta & \cos\xi\sin\eta\sin\zeta - \sin\xi\cos\eta & \sin\xi\sin\eta\sin\zeta - \cos\xi\cos\eta \end{bmatrix}$$

Fifth kind:

$$[T]_{3\leftarrow 0} = \begin{bmatrix} \cos\eta\cos\zeta & \cos\eta\sin\zeta & -\sin\xi \\ -\cos\xi\sin\zeta + \sin\xi\sin\eta\cos\zeta & \cos\xi\cos\zeta + \sin\xi\sin\eta\sin\zeta & \sin\xi\cos\eta \\ \sin\xi\sin\zeta + \cos\xi\sin\eta\cos\zeta & -\sin\xi\cos\zeta + \cos\xi\sin\eta\sin\zeta & \cos\xi\cos\eta \end{bmatrix}$$

Sixth kind:

$$[T]_{3\leftarrow 0} = \begin{bmatrix} \cos\eta\cos\zeta + \sin\xi\sin\eta\sin\zeta & \cos\xi\sin\zeta & -\sin\eta\cos\zeta + \sin\xi\cos\eta\sin\zeta \\ -\cos\eta\sin\zeta & \cos\xi\cos\zeta & \sin\eta\sin\zeta + \sin\xi\cos\eta\cos\zeta \\ \cos\xi\sin\eta & -\sin\xi & \cos\xi\cos\eta \end{bmatrix}$$

For very small angles, i.e., when each Cardan angle becomes so small that one can set $\cos\xi = \cos\eta = \cos\zeta = 1$, $\sin\xi = \xi$, $\sin\eta = \eta$, $\sin\zeta = \zeta$, and $\xi\eta = \eta\zeta = \zeta\xi = 0$, the transition matrices of all six kinds become the same, i.e.,

$$[T]_{3\leftarrow 0} = \begin{bmatrix} 1 & \zeta & -\eta \\ -\zeta & 1 & \xi \\ \eta & -\xi & 1 \end{bmatrix}.$$

The vector $\boldsymbol{\omega}$ can also be expressed in terms of Cardan frequencies (which are not orthogonal):
First kind:

$$\boldsymbol{\omega} = [\mathbf{e}_{x_0}\ \mathbf{e}_{y_1}\ \mathbf{e}_{z_2}] \begin{bmatrix} \dot\xi \\ \dot\eta \\ \dot\zeta \end{bmatrix}$$

Second kind:

$$\boldsymbol{\omega} = [\mathbf{e}_{x_2}\ \mathbf{e}_{y_0}\ \mathbf{e}_{z_1}] \begin{bmatrix} \dot\xi \\ \dot\eta \\ \dot\zeta \end{bmatrix}$$

Third kind:

$$\boldsymbol{\omega} = [\mathbf{e}_{x_1}\ \mathbf{e}_{y_2}\ \mathbf{e}_{z_0}] \begin{bmatrix} \dot\xi \\ \dot\eta \\ \dot\zeta \end{bmatrix}$$

Fourth kind:

$$\boldsymbol{\omega} = [\mathbf{e}_{x_0}\ \mathbf{e}_{y_2}\ \mathbf{e}_{z_1}] \begin{bmatrix} \dot\xi \\ \dot\eta \\ \dot\zeta \end{bmatrix}$$

Fifth kind:

$$\boldsymbol{\omega} = [\mathbf{e}_{x_1}\ \mathbf{e}_{y_0}\ \mathbf{e}_{z_2}] \begin{bmatrix} \dot\xi \\ \dot\eta \\ \dot\zeta \end{bmatrix}$$

Sixth kind:

$$\boldsymbol{\omega} = [\mathbf{e}_{x_2}\ \mathbf{e}_{y_1}\ \mathbf{e}_{z_0}] \begin{bmatrix} \dot\xi \\ \dot\eta \\ \dot\zeta \end{bmatrix}.$$

First kind:

$$\begin{bmatrix} \omega_x \\ \omega_y \\ \omega_z \end{bmatrix} = \begin{bmatrix} \cos\eta\cos\zeta & \sin\zeta & 0 \\ -\cos\eta\sin\zeta & \cos\zeta & 0 \\ \sin\eta & 0 & 1 \end{bmatrix} \begin{bmatrix} \dot\xi \\ \dot\eta \\ \dot\zeta \end{bmatrix}$$

Second kind:

$$\begin{bmatrix} \omega_x \\ \omega_y \\ \omega_z \end{bmatrix} = \begin{bmatrix} 1 & \sin\zeta & 0 \\ 0 & \cos\xi\cos\zeta & \sin\xi \\ 0 & -\sin\xi\cos\zeta & \cos\xi \end{bmatrix} \begin{bmatrix} \dot\xi \\ \dot\eta \\ \dot\zeta \end{bmatrix}$$

Third kind:

$$\begin{bmatrix} \omega_x \\ \omega_y \\ \omega_z \end{bmatrix} = \begin{bmatrix} \cos\eta & 0 & -\cos\xi\sin\eta \\ 0 & 1 & \sin\xi \\ \sin\eta & 0 & \cos\xi\cos\eta \end{bmatrix} \begin{bmatrix} \dot{\xi} \\ \dot{\eta} \\ \dot{\zeta} \end{bmatrix}$$

Fourth kind:

$$\begin{bmatrix} \omega_x \\ \omega_y \\ \omega_z \end{bmatrix} = \begin{bmatrix} \cos\eta\cos\zeta & 0 & -\sin\eta \\ -\sin\zeta & 1 & 0 \\ \sin\eta\cos\zeta & 0 & \cos\eta \end{bmatrix} \begin{bmatrix} \dot{\xi} \\ \dot{\eta} \\ \dot{\zeta} \end{bmatrix}$$

Fifth kind:

$$\begin{bmatrix} \omega_x \\ \omega_y \\ \omega_z \end{bmatrix} = \begin{bmatrix} 1 & 0 & -\sin\xi \\ 0 & \cos\xi & \sin\xi\cos\eta \\ 0 & -\sin\xi & \cos\xi\cos\eta \end{bmatrix} \begin{bmatrix} \dot{\xi} \\ \dot{\eta} \\ \dot{\zeta} \end{bmatrix}$$

Sixth kind:

$$\begin{bmatrix} \omega_x \\ \omega_y \\ \omega_z \end{bmatrix} = \begin{bmatrix} \cos\zeta & \cos\xi\sin\zeta & 0 \\ -\sin\zeta & \cos\xi\cos\zeta & 0 \\ 0 & -\sin\xi & 1 \end{bmatrix} \begin{bmatrix} \dot{\xi} \\ \dot{\eta} \\ \dot{\zeta} \end{bmatrix}.$$

Yaw, Pitch, and Roll

Engineers describe the attitude of vehicles by angles of *yaw, pitch,* and *roll.* For ships, airplanes, railway cars, and automobiles *yaw* is the angle ζ measured about the space-fixed axis z_0 (Figure C.6) from vehicles mass center to the zenith, ***pitch*** is the angle η measured about the carried axis y_1 from vehicle mass center to the horizon, and ***roll*** is the angle ξ measured about the body-fixed axis x_2 from the vehicle mass center to the front (bow) of the vehicle. The system just described consists of Cardan angles of the fifth kind.

For spacecraft, the normal to the orbital plane can be looked upon as fixed in space. Engineers therefore prefer the sequence *pitch-yaw-roll* for space vehicles. *Pitch* of a spacecraft is the angle ζ about the space-fixed normal z_0 (Figure C.4) to the orbital plane, *yaw* is the angle ξ about the carried axis x_1 from the spacecraft mass center and lying within the orbital plane, and *roll* is the angle about the body-fixed axis y_2 from the spacecraft mass center to the spacecraft bow. The system thus consists of Cardan angles of the third kind.

In sports, the terms *twist* (yaw), *somersault* (pitch), and *tilt* (roll) are in common use, but for an athlete they typically are (non-integrable) angular velocity components of the ω_x, ω_y, ω_z type.

Problems

C.1. A gyro satellite with body-fixed $Cxyz$ coordinate axes has rotated through Cardan angles of the first kind

$$\xi = 90^\circ$$

$$\eta = 90^\circ$$

$$\zeta = 90^\circ.$$

A small instrument is located inside the satellite at

$$\mathbf{r} = [\mathbf{e}_x \ \mathbf{e}_y \ \mathbf{e}_z] \begin{bmatrix} 2 \\ 1 \\ 3 \end{bmatrix} \text{ m.}$$

(a) Express the position $\mathbf{r}$ of the instrument in space-fixed $CXYZ$ coordinates. (b) If, say, for analysis purposes, it turns out that Cardan angles of the second kind are more suitable, determine their values.

C.2. A sailing boat has a single vertical mast. (a) At a roll angle of 10°, a pitch angle of 8°, and a yaw angle of 5°, what is the angle between the mast and the local zenith? (b) At a roll angle of 90°, a pitch angle of 90°, and a yaw angle of 5°, what is the angle between mast and the local zenith?

C.3. A gyro with a body-fixed $Cxyz$ coordinate system has an attitude given by the Euler angles $\psi = 0^\circ$, $\nu = 15^\circ$, $\sigma = 30^\circ$, and at that very moment has Euler frequencies $\dot{\psi} = 2$ rad/s, $\dot{\nu} = 0$, $\dot{\sigma} = 2$ rad/s. (a) What is the attitude expressed in Cardan angles of the first kind? (b) What are the Cardan frequencies?

C.4. Cardan angles of the first kind and Euler angles are related by

$$\tan \xi_1 = \cos \psi \tan \nu$$

$$\sin \eta_1 = \sin \psi \sin \nu$$

$$\tan(\zeta_1 - \sigma) = \tan \psi \cos \nu.$$

Sketch (a) the appropriate spherical triangles and derive equations that relate Cardan angles of the second kind to Euler angles, and (b) Cardan angles of the third kind to Euler angles. The subscript 1 has been used to indicate that the Cardan angles are of the *first* kind.

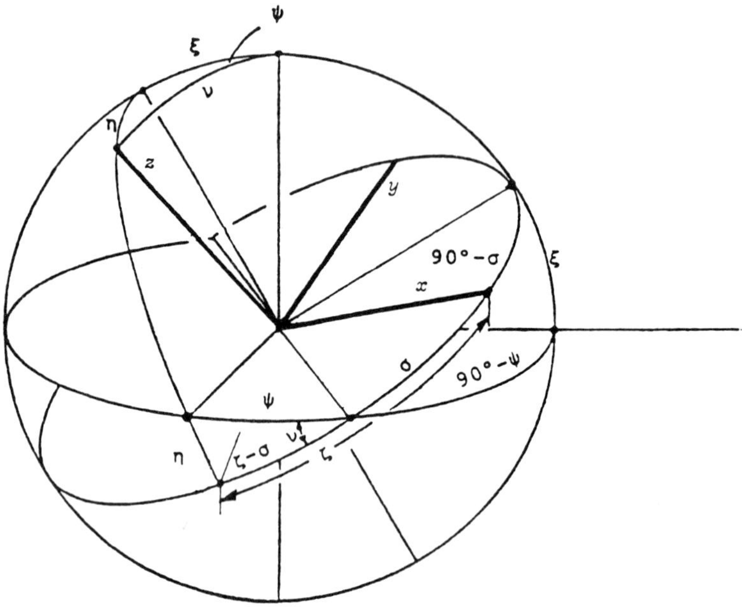

C.5. The figure shows the Hermes (†1976) satellite with roll, yaw, and pitch axes indicated. Sketch the (geosynchronous) orbit into the diagram. Which kind of Cardan angle system is used, if x is chosen as roll axis, z_1 as yaw, and y_0 as pitch axis?

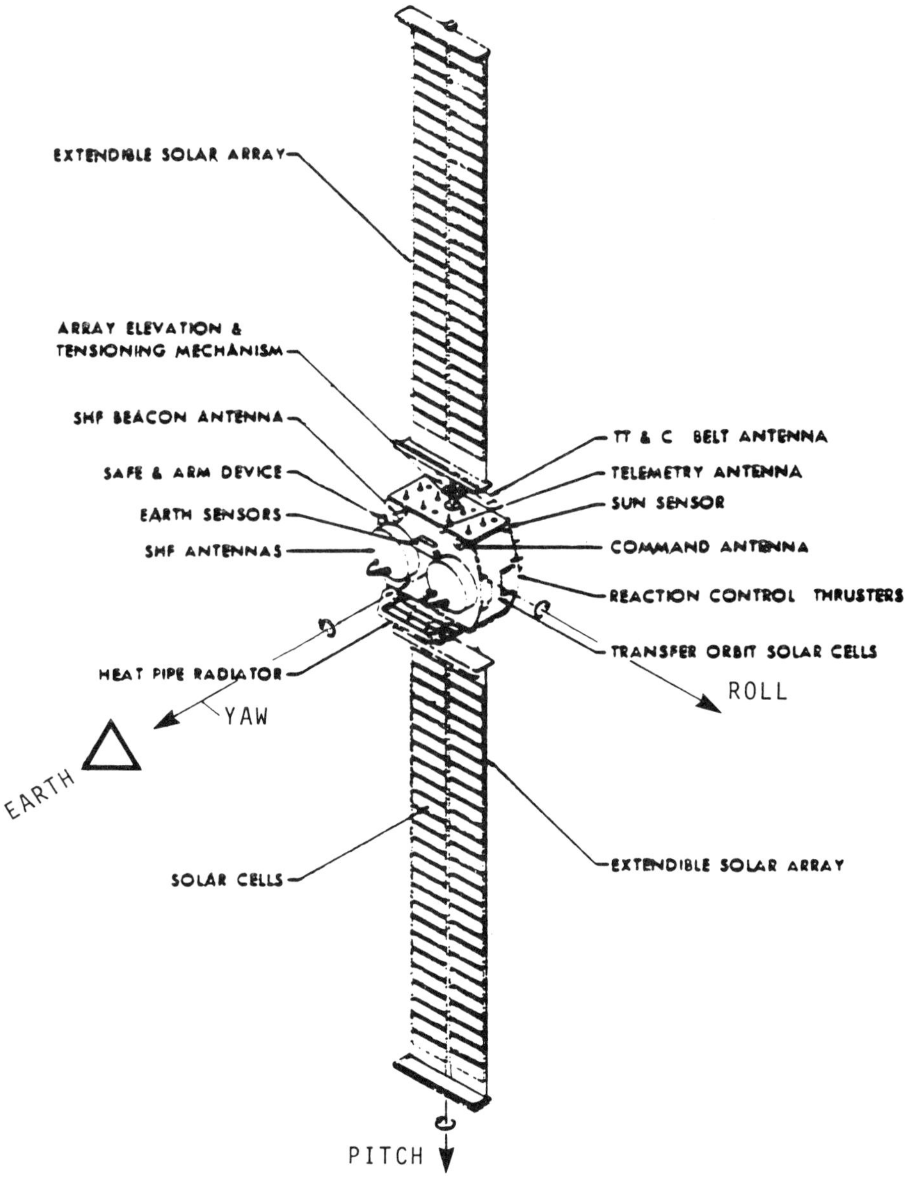

Appendix D

Engineering Data

Systems of Units

System	Basic Units				
	Length	*Mass*	*Force*	*Time*	
SI = physical MKS*	meter, m	kilogram[1], kg		second, s	Force: 1 newton = 1 N = 1 kg m/s^2
Engineering MKS	meter, m		kilopond[1,2], kp	second, s	Mass: 1 hyl = 1 kp s^2/m
Physical FPS	foot, ft	pound[3], lb_m		second, s	Force: 1 poundal = 1 $lb_m ft/s^2$
Engineering FPS**	foot, ft		pound[3], lb	second, s	Mass: 1 slug = 1 lb ft/s^2

*In use throughout world.
**In use in U.S.

Standard gravitational attraction constant
$g = 9.980665\ m/s^2 \approx 32.2\ ft/s^2$

[1]REMEMBER: Weight force W = 1 kp when mass M = 1 kg (on Earth surface).
[2]1 kp = 1 kgf = 1 kilogram-force
[3]REMEMBER: Weight force W = 1 lb when mass M = 1 lb_m (on Earth surface).

Notes: 1 = 0.3048 m/ft; 1 = 9.80665 N/kp; 1 = 0.453 592 37 kp/lb; 1 ≈ 4.448 N/lb; 1 ≈ 3.28 ft/m; 1 ≈ 0.102 kp/N; 1 ≈ 2.205 lb/kp; 1 ≈ 0.2248 lb/N; 1 ≈ 0.4536 kg/lb_m.

Larger and Smaller Units in the Metric System

Factor	Prefix	Symbol
Larger units		
10^{18}	Exa-	E
10^{15}	Peta-	P or PA
10^{12}	Tera-	T
10^{9}	Giga-	G
10^{6}	Mega-	M
10^{3}	Kilo-	k or K
10^{2}	Hecto-	h or H
10^{1}	Deka-	da or D or dk
Smaller units		
10^{-1}	deci-	d
10^{-2}	centi-	c
10^{-3}	milli-	m
10^{-6}	micro-	μ
10^{-9}	nano-	n
10^{-12}	pico-	p
10^{-15}	femto-	f
10^{-18}	atto-	a

Example: $r_{\text{Sun-Earth}} = 150$ Gm $= 150\ 000\ 000$ km $= 150(10^9)$m.

Principal Moments of Inertia

1. Thin rod: $A = (1/12)ml^2$; $B = (1/12)ml^2$; $C = 0$

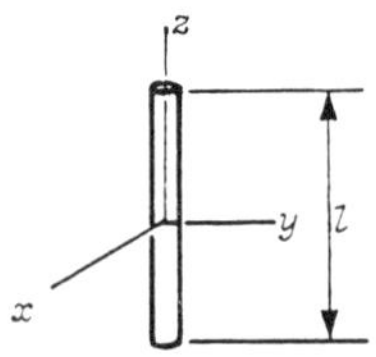

2. Thin disk: $A = \frac{1}{4}ma^2$; $B = \frac{1}{4}ma^2$; $C = \frac{1}{2}ma^2$

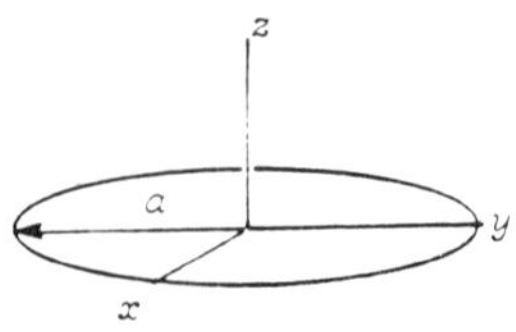

3. Cylinder: $A = (1/12)m(3a^2 + l^2)$; $B = (1/12)m(3a^2 + l^2)$;
   $C = \frac{1}{2}ma^2$

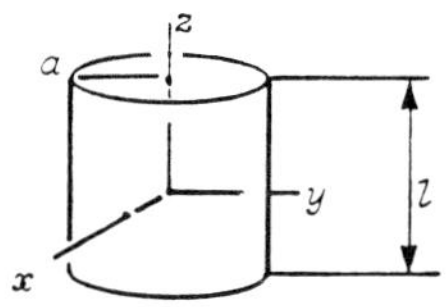

4. Thin plate: $A = (1/12)mb^2$; $B = (1/12)ma^2$;
   $C = (1/12)m(a^2 + b^2)$

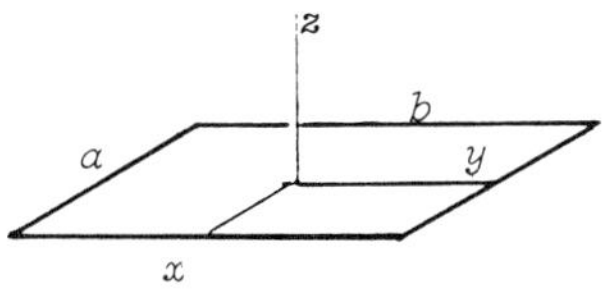

5. Rectangular parallelepiped: If $A > B > C$ then $a < b < c$
   $A = (1/12)m(b^2 + c^2)$; $B = (1/12)m(c^2 + a^2)$; $C = (1/12)m(a^2 + b^2)$

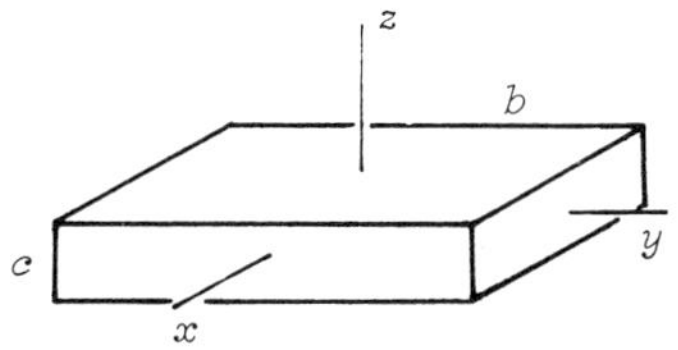

6. Sphere: surface $= 4\pi a^2$; volume $= \frac{4}{3}\pi a^3$
   $A = (2.5)ma^2$; $B = (2.5)ma^2$; $C = (2.5)ma^2$

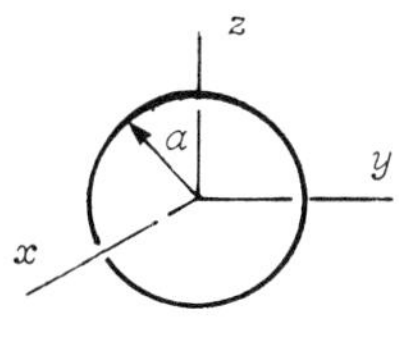

7. Spatial dumbbell: If $A > B > C$ then $a < b < c$
   $A = (1/12)m(b^2 + c^2)$; $B = (1/12)m(c^2 + a^2)$;
   $C = (1/12)m(a^2 + b^2)$
   m = total mass

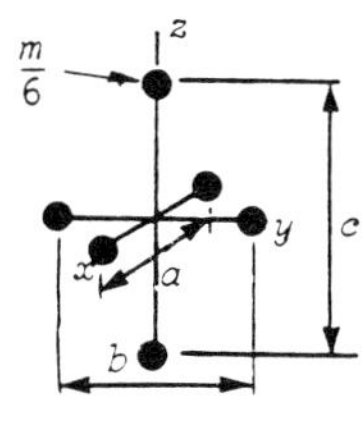

8. Spatial cross boom ("inertia skeleton"):
If $A > B > C$ then $a < b < c$
$A = (1/12)m(b^3 + c^3)/(a + b + c)$; $B = (1/12)m(c^3 + a^3)/(a + b + c)$;
$C = (1/12)m(a^3 + b^3)/(a + b + c)$; m = total mass

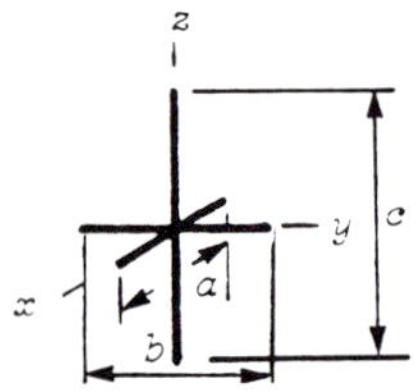

9. Ellipsoid: $V = (4/3)\pi abc$
$A = (1.5)m(b^2 + c^2)$; $B = (1.5)m(a^2 + c^2)$; $C = (1/5)m(a^2 + b^2)$

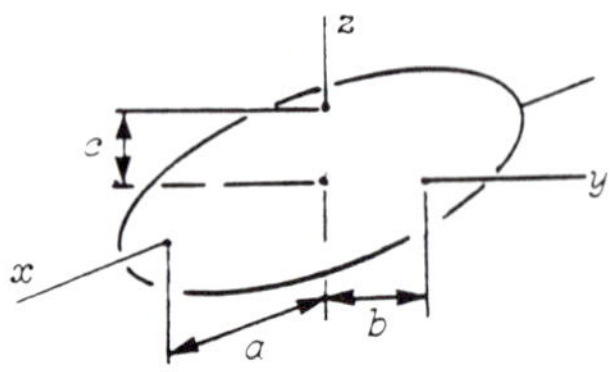

10. Thin-walled cylindrical shell: $A = B = (1/12)m(6a^2 + l^2)$;
$C = ma^2$

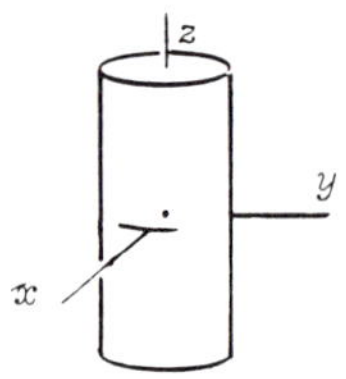

11. Thin-walled spherical shell: $V = 4\pi a^2 h$
$A = B = C = (2/3)ma^2$

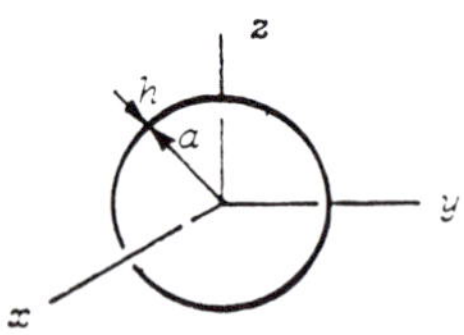

12. Simple dumbbell: $A = B = \frac{1}{4}mc^2$; $C = 0$

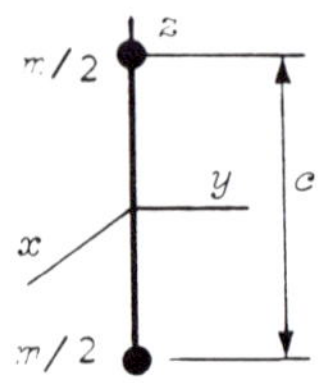

Material Properties

Material	Density ρ kg/m^3	Young's Modulus E MPa	Speed* of Sound c m/s	Tensile Strength σ_{ult} MPa
Air (0°C)	1.293	0.14	331	
Aluminum (2014–T6)	2 690	55 000	5 240	240
Beryllium copper (C17200)	8 290	120 000	4 410	1 500
Brass (naval)	8 450	100 000		420
Brick	1 800	17 300	3 600	240
Bronze(phosphor ASTM B 159)	8 930	100 000		700
Concrete	2 400	18 300	3 200	175**
Copper	9 810	84 900	3 580	210–450
Dacron (polyester)	1 380	13 800		1 120
Glass	2 590	60 000	5 560	90(900**)
Gold	19 300	59 100	2 030	140
Ice	900			13**
Iron (cast, gray, No. 60)	7 210	130 000	5 170	420
Kapton (polyimide film)	1 400	3 000	4.8	172
Kevlar 29 (aramid fibre)	1 440	12 000		2 760
Lead	11 370	13 200	1 250	14
Magnesium (AZ80A–T5)	1 800	43 000	1 310	380
Mercury	13 570		100	
Nylon (728)	1 140	5 520		985
Oil	90	830		
Rubber	920	2	54	15
Soil, wet	1 760			
Soil, dry	1 280			
Steel (Spring alloy SAE 4068)	7 830	200 000	5 850	1 720
Titanium	3 080	110 000	6 420	800
Water, fresh	1 000	2 000	1 400	
Water, salt	1 030	2 000	1 560	
Wood, oak	800	12 000	3 380	490**
Wood, pine	480	14 000	4 180	590**

* $c = \sqrt{\dfrac{1-\nu}{(1+\nu)(1-2\nu)} \dfrac{E}{\rho}}$.

**Compressive strength.

Series

$$\frac{1}{(1+x)^2} = 1 - 2x + 3x^2 - 4x^3 + 5x^4 - \dots$$

$$\frac{1}{(1-x)^2} = 1 + 2x + 3x^2 + 4x^3 + 5x^4 + \dots$$

$$\sin x = x - \frac{1}{3!}x^3 + \frac{1}{5!}x^5 - \frac{1}{7!}x^7 + \dots$$

$$\cos x = 1 - \frac{1}{2!}x^2 + \frac{1}{4!}x^4 - \frac{1}{6!}x^6 + \dots$$

$$\tan x = x + \frac{1}{3}x^3 + \frac{2}{15}x^5 + \frac{17}{315}x^7 + \dots$$

$$e^x = 1 + x + \frac{1}{2!}x^2 + \frac{1}{3!}x^3 + \frac{1}{4!}x^4 + \dots$$

$$\ln(1+x) = x - \frac{1}{2}x^2 + \frac{1}{3}x^3 - \frac{1}{4}x^4 + \frac{1}{5}x^5 - \dots$$

$$(1 \pm x)^n = 1 \pm nx + \frac{n(n-1)}{2!}x^2 \pm \frac{n(n-1)(n-2)}{3!}x^3 + \dots .$$

Expansions of $\rho/\rho_c = \sqrt{1 + 2(x/\rho_c)\cos\alpha + (x^2/\rho_c^2)}$

$$\frac{\rho}{\rho_c} = 1 + \frac{x}{\rho_c}\cos\alpha + \frac{1}{2}\frac{x^2}{\rho_c^2}(1 - \cos^2\alpha) + \dots .$$

$$\frac{\rho^2}{\rho_c^2} = 1 + 2\frac{x}{\rho_c}\cos\alpha + \frac{x^2}{\rho_c^2}$$

$$\frac{\rho^3}{\rho_c^3} = 1 + 3\frac{x}{\rho_c}\cos\alpha + \frac{3}{2}\frac{x^2}{\rho_c^2}(1 + \cos^2\alpha) + \dots .$$

$$\frac{\rho_c}{\rho} = 1 - \frac{x}{\rho_c}\cos\alpha + \frac{1}{2}\frac{x^2}{\rho_c^2}(3\cos^2\alpha - 1) - \dots .$$

$$\frac{\rho_c^2}{\rho^2} = 1 - 2\frac{x}{\rho_c}\cos\alpha + \frac{x^2}{\rho_c^2}(4\cos^2\alpha - 1) - \dots .$$

$$\frac{\rho_c^3}{\rho^3} = 1 - 3\frac{x}{\rho_c}\cos\alpha + \frac{3}{2}\frac{x^2}{\rho_c^2}(5\cos^2\alpha - 1) - \dots .$$

Appendix E

Nomenclature

Symbols

A	area, principal inertia moment about x-axis
A_1	principal inertia moment about x_1-axis
B	principal inertia moment about y-axis
B_1	principal inertia moment about y_1-axis
$\mathbf{C}$	centrifugal force
C	principal inertia moment about z-axis
C_1	principal inertia moment about z_1-axis
C'	reduced principal inertia moment about z-axis
C	mass center
D	dissipation energy
E	total energy
$\mathbf{F}$	force
G	universal gravitational constant
$[G]$	gyric matrix
G	gravity center
$\mathbf{H}$	angular momentum
$[I]$	inertia tensor
$\mathbf{K}$	Kepler force, internal angular momentum
K	complete elliptic integral of the first kind
$[K]$	stiffness matrix
K_x, K_y, K_z	interdependent constants, components of $\mathbf{K}$
M	mass, moment
$\mathbf{M}$	moment
O	origin
P	function
P_x, P_y, P_z	interdependent constants
Q	function, effective dissipation ratio
Q_x, Q_y, Q_z	interdependent constants
R	radius, nominal radius of master mass sphere
$[R]$	rotation matrix
S	satellite
T	kinetic energy
$[T]$	transition matrix
U	potential energy, potential, work done by external force or torque

V	elastic energy
W	weight force
W_μ	microweight force
X, Y, Z	coordinates

$\mathbf{a}$	acceleration
a	semi-minor axis of ellipsoid, distance, radius, reduced inertia moment ratio C'/C
b	semi-minor axis of ellipsoid, distance, radius
c	semi-axis of ellipsoid, damping coefficient
e	eccentric distance
$\mathbf{e}_x$	unit vector in x-direction
g	gravitational attraction constant on surface of master mass
g_μ	microgravitational attraction coefficient
i	$\sqrt{I/m}$ = gyration radius
k	modulus of Jacobi elliptic function, spring stiffness, Smelt parameter
l	length
m	mass
$\mathbf{r}$	position vector
r	radial distance
r_0	initial distance
t	time
u	Jacobi parameter
u, v, z	floating coordinates
$\mathbf{v}$	velocity
x	ratio $\omega_z^{\mathrm{R}}/\omega_z$
x, y, z	coordinates

α	direction angle with respect to x-axis, arcsin$k$ of Jacobi elliptic function, parameter elastic deformation coefficient
β	direction angle with respect to y-axis, internal energy dissipation coefficient
γ	direction angle with respect to z-axis
δ	deflection
δ_{ij}	Kronecker delta
$\boldsymbol{\epsilon}$	Euler vector
ϵ	Euler parameter
ζ	damping ratio
η	spin change allotment factor

θ	polar angle, rotation angle
μ	gravitational parameter, Euler angle
$\boldsymbol{\lambda}$	rotation axis unit vector
ν	nutation angle
ξ, η, ζ	coordinates, Cardan angles
$\boldsymbol{\rho}$	Rodrigues vector
ρ	density, radial distance, Rodrigues parameter
σ	spin angle
τ	Kepler period
ϕ	angle between two vectors
ψ	precession angle
$\boldsymbol{\Omega}$	angular velocity of coordinate system, orbital angular velocity on circular orbits
$\boldsymbol{\omega}$	angular velocity

Subscripts

C	center, mass center
G	gravity center
M	Moon
S	satellite
f	final
O	original, initial
r	radial
θ	transverse
μ	micro
ρ	about ρ

Superscripts

E	excess
P	of platform
R	of rotor

Appendix F

Answers to Selected Problems

Chapter 1

1.1. (a) $[I]_C = \begin{bmatrix} 12.5 & 0 & 0 \\ 0 & 12.5 & 0 \\ 0 & 0 & 25 \end{bmatrix} \mathrm{m^2kg}$

(b) $[I]_O = \begin{bmatrix} A_O & 0 & 0 \\ 0 & A_O & 0 \\ 0 & 0 & 25 \end{bmatrix} \mathrm{m^2kg}$ with $A_O = 12.5 + 50l^2$

(c) $\boldsymbol{\omega} = [\mathbf{e}_u \ \mathbf{e}_v \ \mathbf{e}_z] \begin{bmatrix} 0 \\ 0.01 \\ 3000 \end{bmatrix} \mathrm{rad/s}$

(d) $\boldsymbol{\Omega} = [\mathbf{e}_u \ \mathbf{e}_v \ \mathbf{e}_z] \begin{bmatrix} 0 \\ 0.01 \\ 0 \end{bmatrix} \mathrm{rad/s}$

(e) $\mathbf{H} = [\mathbf{e}_u \ \mathbf{e}_v \ \mathbf{e}_z] \begin{bmatrix} 0 \\ 0.125 \\ 75000 \end{bmatrix} \mathrm{Ws^2}$

(f) $\dot{\mathbf{H}} = [\mathbf{e}_u \ \mathbf{e}_v \ \mathbf{e}_z] \begin{bmatrix} 750 \\ 0 \\ 0 \end{bmatrix} \mathrm{Ws}$

(g) $\mathbf{M} = [\mathbf{e}_u \ \mathbf{e}_v \ \mathbf{e}_z] \begin{bmatrix} 750 \\ 0 \\ 0 \end{bmatrix} \mathrm{Nm}$

(h) 1.529 m

1.3. (a) $[T_{iJ}] = \begin{bmatrix} \cos\theta & \sin\theta & 0 \\ -\sin\theta & \cos\theta & 0 \\ 0 & 0 & 1 \end{bmatrix}$

1.8. $[I] = \frac{1}{12} m a^2 \begin{bmatrix} 2 & 0 & 0 \\ 0 & 1 & 0 \\ 0 & 0 & 3 \end{bmatrix}$

1.9. $c = \sqrt{2}a$

1.10. $2/a = \sqrt{3}$

1.11. $[I_{ij}] = \begin{bmatrix} \frac{7}{5}mR^2 & 0 & 0 \\ 0 & \frac{7}{5}mR^2 & 0 \\ 0 & 0 & \frac{2}{5}mR^2 \end{bmatrix}$

1.12. $\alpha = 22.5°$

1.13. $\lambda_1 = 2;\quad \lambda_2 = 4;\quad \lambda_3 = 4$

1.14. (a) $[I] = \begin{bmatrix} 0.1667 & 0 & 0 \\ 0 & 0.9167 & 0 \\ 0 & 0 & 0.9167 \end{bmatrix} \mathrm{m^2kg}$

(b) $\alpha = \beta = \gamma = 54.74°$

1.15. (a) $\frac{m}{12} \begin{bmatrix} 52 & -6 & -9 \\ -6 & 40 & -18 \\ -9 & -18 & 20 \end{bmatrix} \mathrm{m^2kg}$

(b) $\frac{m}{12} \begin{bmatrix} 54.68 & 0 & 0 \\ 0 & 50.49 & 0 \\ 0 & 0 & 6.826 \end{bmatrix} \mathrm{m^2kg}$

(c) New "1" axis (associated with $A = 54.68$)

$$\alpha = 15° \quad \beta = 103.3° \quad \gamma = 97°$$

New "2" axis (associated with $B = 50.49$)

$$\alpha = 82° \quad \beta = 33.8° \quad \gamma = 122.6°$$

New "3" axis (associated with $C = 6.826$)

$$\alpha = 76.44° \quad \beta = 60° \quad \gamma = 33.5°$$

1.16. $T = \frac{1}{2}(I_x\omega_x^2 + I_y\omega_y^2 + I_z\omega_z^2 + 2(I_{xy}\omega_x\omega_y + I_{yz}\omega_y\omega_z + I_{xz}\omega_x\omega_z))$

1.17. (b) $[\mathbf{e}_x\ \mathbf{e}_y\ \mathbf{e}_z] \begin{bmatrix} 18.046 \\ 4.877 \\ 3.950 \end{bmatrix}$

(c) $T = 12.68$ mJ

1.18. $\begin{bmatrix} 0 & \cos\psi & \sin\nu\sin\psi \\ 0 & \sin\psi & \sin\nu\cos\psi \\ 1 & 0 & \cos\nu \end{bmatrix}$

1.19. $\begin{bmatrix} \frac{\sin\sigma}{\sin\nu} & \frac{\cos\sigma}{\sin\nu} & 0 \\ \cos\sigma & -\sin\sigma & 0 \\ -\frac{\sin\sigma}{\tan\nu} & -\frac{\cos\sigma}{\tan\nu} & 1 \end{bmatrix}$

1.21. (a) $[I]_0 = \frac{m}{12}\begin{bmatrix} 8 & 0 & 0 \\ 0 & 5 & 0 \\ 0 & 0 & 5 \end{bmatrix} \mathrm{Ws}^3$

(b) $[I]_2 = \frac{m}{12}\begin{bmatrix} 40 & 0 & 0 \\ 0 & 37 & 0 \\ 0 & 0 & 5 \end{bmatrix} \mathrm{Ws}^3$

(c) $[I] = \frac{m}{12}\begin{bmatrix} 8(1+t) & 0 & 0 \\ 0 & 8(1+t) & 0 \\ 0 & 0 & 0 \end{bmatrix} \mathrm{Ws}^3$

1.22.

$$\begin{aligned} M_x &= \dot{A}\omega_x + A\dot{\omega}_x - (B - C)\omega_y\omega_z \\ M_y &= \dot{B}\omega_y + B\dot{\omega}_y - (C - A)\omega_z\omega_x \\ M_z &= \dot{C}\omega_z + C\dot{\omega}_z - (A - B)\omega_x\omega_y \end{aligned}$$

1.23.

$$A = \frac{m}{12}\left(b^2 + \frac{c^2}{\tau}t\right) \qquad \dot{A} = \frac{m}{12}\frac{c^2}{\tau}$$

$$B = \frac{m}{12}\left(\frac{c^2}{\tau}t + a^2\right) \qquad \dot{B} = \frac{m}{12}\frac{c^2}{\tau}$$

$$C = \frac{m}{12}(a^2 + b^2)$$

1.26. $k_1 = 0.1111 \quad k_2 = -0.6667 \quad k_3 = 0.6$

1.28. (a) $[I] = \begin{bmatrix} 4000 & 0 & 0 \\ 0 & 4000 & 0 \\ 0 & 0 & 0 \end{bmatrix} \mathrm{m^2kg}$

(b) $\mathbf{H}_0 = [\mathbf{e}_x\ \mathbf{e}_y\ \mathbf{e}_z]\begin{bmatrix} 8000 \\ 0 \\ 0 \end{bmatrix} \mathrm{Ws}^2$

(c) $T = 80$ J

(d) $\nu = 1$ m/s

(e) $[I] = \begin{bmatrix} 500 & 0 & 0 \\ 0 & 500 & 0 \\ 0 & 0 & 0 \end{bmatrix} \mathrm{m^2kg}$

(f) $\mathbf{H}_{C_1} = [\mathbf{e}_x\ \mathbf{e}_y\ \mathbf{e}_z]\begin{bmatrix} 100 \\ 0 \\ 0 \end{bmatrix} \mathrm{Ws}^2$

(g) $T_{Trans} = 30$ J

(h) $T_{Rotor} = 10$ J

(i) Zero

(j) $\mathbf{H} = [\mathbf{e}_x\ \mathbf{e}_y\ \mathbf{e}_z] \begin{bmatrix} 400 \\ 0 \\ 0 \end{bmatrix} \mathrm{Ws}^2$

(k) No loss of momentum

1.29. (a) $\mathbf{H} = [\mathbf{e}_x\ \mathbf{e}_y\ \mathbf{e}_z] \begin{bmatrix} 0 \\ 0 \\ 250 \end{bmatrix} \mathrm{Ws}^2$

(b) $T_1 = 625$ J

(c) $\boldsymbol{\omega}_2 = [\mathbf{e}_x\ \mathbf{e}_y\ \mathbf{e}_z] \begin{bmatrix} 0 \\ 0 \\ 2.5 \end{bmatrix}$ rad/s

(d) $T_2 = 312.5$ J

(e) $\boldsymbol{\omega}_3 = \boldsymbol{\omega}_1$

(f) $T_3 = 625$ J

Chapter 2

2.1. $\rho_\mathrm{G} = \rho_\mathrm{C} \sqrt[4]{1 + \dfrac{l^2}{4\rho_\mathrm{C}^2}}$

2.2. $[N]^{-1} = -\dfrac{1}{2} \begin{bmatrix} 3\cos^2\alpha - 2 & 3\cos\alpha\cos\beta & 3\cos\alpha\cos\gamma \\ 3\cos\alpha\cos\beta & 3\cos^2\beta - 2 & 3\cos\beta\cos\gamma \\ 3\cos\alpha\cos\gamma & 3\cos\beta\cos\gamma & 3\cos^2\gamma - 2 \end{bmatrix}$

2.3. (a) $[I] = \begin{bmatrix} 16\,000 & 0 & 0 \\ 0 & 15\,300 & 0 \\ 0 & 0 & 2\,500 \end{bmatrix} \mathrm{m}^2\mathrm{kg}$

(b) $\mathbf{r}_\mathrm{G} = [\mathbf{e}_x\ \mathbf{e}_y\ \mathbf{e}_z] \begin{bmatrix} 0.976 \\ 0.831 \\ -1.808 \end{bmatrix} \mu\mathrm{m}$

2.6. (a) $[I] = \begin{bmatrix} 1042 & 0 & 0 \\ 0 & 6667 & 0 \\ 0 & 0 & 6375 \end{bmatrix} \mathrm{m}^2\mathrm{kg}$

(b) $\mathbf{r}_\mathrm{G} = [\mathbf{e}_x\ \mathbf{e}_y\ \mathbf{e}_z] \begin{bmatrix} -0.0073 \\ 0.0033 \\ 0.0028 \end{bmatrix}$ mm

2.7. $$[I] = \begin{bmatrix} 2\,500 & 0 & 0 \\ 0 & 16\,000 & 0 \\ 0 & 0 & 15\,300 \end{bmatrix} \mathrm{m^2kg}$$

$$\mathbf{r}_G = [\mathbf{e}_x\ \mathbf{e}_y\ \mathbf{e}_z] \begin{bmatrix} -1.4 \\ 0 \\ 2.56 \end{bmatrix} \mu\mathrm{m}$$

2.8. (a) $\rho_\mathrm{C} = \dfrac{9\sqrt{3}}{10} a = 1.559\, a$

(b) $\rho_\mathrm{G} = \sqrt{\dfrac{15}{4\sqrt{3}}} a = 1.471\, a$

(c) $K = \dfrac{4\sqrt{3}}{15} \dfrac{\mu\mathrm{m}}{a^2} = 0.462 \dfrac{\mu\mathrm{m}}{a^2}$

2.9. (a) $X_\mathrm{C} = r \dfrac{\sin\alpha}{\alpha}$

(b) $X_\mathrm{G} = r \sqrt{\dfrac{\alpha}{\sin\alpha}}$

(c) $K = \dfrac{\sin\alpha}{\alpha} \dfrac{\mu\mathrm{m}}{r^2}$

Table I

α	r_C	r_G	K
0°	r	r	$\dfrac{\mu\mathrm{m}}{r^2}$
$90^\circ = \dfrac{1}{2}\pi$	$0.637r$	$1.253r$	$0.637\dfrac{\mu\mathrm{m}}{r^2}$
$150^\circ = \dfrac{5}{6}\pi$	$0.191r$	$2.288r$	$0.191\dfrac{\mu\mathrm{m}}{r^2}$
$180^\circ = \pi$	0	0	0
$210^\circ = \dfrac{7}{6}\pi$	$-0.136r$	$-2.707r$	$-0.136\dfrac{\mu\mathrm{m}}{r^2}$

2.10. (a) $\rho_\mathrm{C} = \dfrac{3}{4} a$

(b) $\rho_\mathrm{G} = \sqrt{2} a$

(c) $K = \dfrac{\mu\mathrm{m}}{2a^2}$

2.11. (a) $\rho_\mathrm{C} = \dfrac{a}{2}$

(b) $\rho_\mathrm{G} = a$

(c) $K = \dfrac{\mu\mathrm{m}}{a^2}$

(d) Unstable, because $\rho_\mathrm{G} > \rho_\mathrm{C} = a/2$

2.13. (a) $C = mr\Omega^2$ through mass center

(b) $C = mr\Omega^2$ through mass center

(c) $\rho_\mathrm{G} = 1.409882\, r$

2.16. (a) $K \approx \dfrac{\mu\mathrm{m}}{\rho_\mathrm{C}^2}\left(1 - \dfrac{3}{2}\dfrac{x^2}{\rho_\mathrm{C}^2}\right)$

(b) $\rho_\mathrm{G} = \rho_\mathrm{C}\left(1 + \dfrac{3}{4}\dfrac{x^2}{\rho_\mathrm{C}^2}\right)$

2.17. (a) $K \approx \dfrac{\mu\mathrm{m}}{\rho_\mathrm{C}^2}\left(1 - \dfrac{1}{2}\dfrac{a^2}{\rho_\mathrm{C}^2}\right)$

(b) $\rho_\mathrm{G} = \rho_\mathrm{C}\left(1 + \dfrac{1}{4}\dfrac{a^2}{\rho_\mathrm{C}^2}\right)$

2.18. $K \approx \dfrac{2\mu\mathrm{m}}{a^2 2}\left[2 + \sqrt{a^2 + \left(r_\mathrm{C} - \dfrac{l}{2}\right)^2} - \sqrt{a^2 + \left(r_\mathrm{C} + \dfrac{l}{2}\right)^2}\right]$

2.19. (a) $\tau = 2\pi\sqrt{\dfrac{r^2}{\mu}}\left(1 + \dfrac{1}{16}\dfrac{l^2}{r^2}\right)$

(b) $H = m\sqrt{\mu r}\left(1 + \dfrac{1}{48}\dfrac{l^2}{r^2}\right)$

(c) $U = -\dfrac{\mu\mathrm{m}}{r}\left(1 - \dfrac{1}{24}\dfrac{l^2}{r^2}\right)$

(d) $T = \dfrac{1}{2}\dfrac{\mu\mathrm{m}}{r}\left(1 - \dfrac{1}{24}\dfrac{l^2}{r^2}\right)$

(e) $E = -\dfrac{\mu\mathrm{m}}{2r}\left(1 - \dfrac{1}{24}\dfrac{l^2}{r^2}\right)$

(f) $\tau = 2\pi\sqrt{\dfrac{r_2^3}{\mu}}\left(1 - \dfrac{1}{8}\dfrac{l^2}{r_2^2}\right)$

(g) $H_2 = m\sqrt{\mu r_2}\left(1 + \dfrac{5}{24}\dfrac{l}{r_2^2}\right)$

(h) $U = -\frac{\mu m}{r}\left(1 + \frac{11}{24}\frac{l^2}{r^2}\right)$

(i) $T = \frac{1}{2}\frac{\mu m}{r}\left(1 - \frac{17}{24}\frac{l^2}{r^2}\right)$

(j) $E = -\frac{\mu m}{2r}\left(1 + \frac{5}{24}\frac{l^2}{r^2}\right)$

(k) $\Delta E = -\frac{1}{8}\frac{l^2}{r^2}\frac{\mu m}{r}$

(l) $2.3\,\mu$W

(m) $\Delta r = 53.571$ mm

2.20. $U = -72.473\,[1 - 1.122 \times 10^{-12}]$ J

$$\mathbf{M} = [\mathbf{e}_x\ \mathbf{e}_y\ \mathbf{e}_z]\begin{bmatrix} 0.114 \\ -0.649 \\ 0.804 \end{bmatrix} 10^{-9}\,\mathrm{Nm}$$

2.21. (a) $[I] = \begin{bmatrix} \frac{2}{3}ma^2 & 0 & 0 \\ 0 & \frac{5}{48}ma^2 & 0 \\ 0 & 0 & \frac{19}{48}ma^2 \end{bmatrix}$

(b) $x_G = 89$ mm

(c) unstable because $x_G = +ve$

2.22. $M = 0.0306$ Nm

2.23. (a) $K = \frac{\mu m}{r^2}\frac{1}{(1 - \frac{l^2}{4r^2})}$

(b) $r_G = r\left(1 - \frac{l^2}{4r^2}\right)$

(c) $\tau = 2\pi\sqrt{\frac{r^3}{\mu}}\left(1 - \frac{l^2}{4r^2}\right)$

(d) Tension $= \frac{\mu m}{r^2}\left(\frac{3}{2}\frac{l}{r}\right)$

2.24. (a) $r_C = R + \frac{l}{2}$

(b) $r_G = r_C\sqrt{1 - \frac{l^2}{4r_C^2}}$

(c) $N = \frac{\mu m}{r_C^2} - \frac{l^2}{4} - m\omega^2 r_C$

(d) $l = 143\ 877$ km

(e) $\sigma = \rho y \left[\left(l' - \frac{y}{2} \right) \omega^2 - \frac{\mu}{l'(l'-y)} \right] \qquad x = l - y$

(f) 35 793 km from the surface of the Earth

(g) $\sigma_{max} = 69\,832$ MPa (exceeds σ_{ult})

2.26. (a) $U = -\frac{GMm}{\rho}\left(1 + \frac{C-A}{2M\rho^2}(1 - 3\sin^2\Delta) + \frac{c-a}{2m\rho^2}(1 - 3\sin^2\delta)\right)$

(b) $K = -\frac{GMm}{\rho^2}\left(1 + \frac{3}{2}\frac{C-A}{M\rho^2} + \frac{3}{2}\frac{c-a}{m\rho^2}\right)$

(c) $K = -\frac{GMm}{\rho^2}\left(1 + \frac{3}{2}\frac{C-A}{M\rho^2} - 3\frac{c-a}{m\rho^2}\right)$

(d) $K = -\frac{GMm}{\rho^2}\left(1 - 3\frac{C-A}{M\rho^2} - 3\frac{c-a}{m\rho^2}\right)$

(e) $K = -\frac{GMm}{\rho^2}\left(1 - \frac{3}{4}\frac{C-A}{M\rho^2} + \frac{3}{2}\frac{c-a}{m\rho^2}\right) \qquad M_y = \frac{3}{2}\frac{Gm}{\rho^3}(A - C)$

(f) $K = -\frac{GMm}{\rho^2}\left(1 + \frac{3}{2}\frac{C-A}{M\rho^2} - \frac{3}{4}\frac{c-a}{m\rho^2}\right) \qquad M_y = \frac{3}{2}\frac{GM}{\rho^3}(a - c)$

2.27. (a) $\rho_G = \rho_C\left(1 - \frac{3}{16}\frac{a^2}{\rho_C^2}\right)$

(b) $\rho_G = \rho_C\left(1 + \frac{3}{8}\frac{a^2}{\rho_C^2}\right)$

2.28. (a) $\Omega = \Omega_0\left(1 - \frac{1}{16}\frac{l^2}{r^2}\right)$

(b) $\Omega_0 = \sqrt{\frac{\mu}{r_3}}$

(c) $U_1 = -\frac{\mu \mathrm{m}}{r}\left(1 - \frac{1}{24}\frac{l^2}{r^2}\right)$

(d) $T_1 = \frac{1}{2} m \Omega_0^2 r^2 \left(1 - \frac{1}{24}\frac{l^2}{r^2}\right)$

(e) $T_1 + U_1 = -\frac{\mu \mathrm{m}}{2r}$

2.29. $F = 145$ mN

2.30. $W_\mu = 53.312(10^{-6})$N

Chapter 3

3.1. $\tau = 56.08$ min

3.2. $\zeta = 3.982^{\circ}$

3.3. $\omega = 2\Omega$

Chapter 4

4.3. (a) $A^* < 1/3$ is insufficient

(b) Other conditions: C^* must not be intermediate inertia moment; $B^* > A^*$ otherwise satellite would be turned by gravity gradient; $C^* > A^*$ otherwise satellite would be turned by gravity gradient

4.4.

$$\lambda_1 = +\sqrt{-\frac{p}{2} + \sqrt{\frac{p^2}{4} - q}}$$

$$\lambda_2 = -\sqrt{-\frac{p}{2} + \sqrt{\frac{p^2}{4} - q}}$$

$$\lambda_3 = +\sqrt{-\frac{p}{2} - \sqrt{\frac{p^2}{4} - q}}$$

$$\lambda_4 = -\sqrt{-\frac{p}{2} + \sqrt{\frac{p^2}{4} - q}}$$

are all always imaginary if

$$p > 0$$

$$q > 0$$

$$p^2 - 4q > 0$$

4.5. $b = 9$ m; $c = 11$ m

4.6. See Appendix C

4.7. Fifth kind

4.8. $$[T]_{3\leftarrow 0} = \begin{bmatrix} \cos\eta\cos\zeta & \sin\xi\sin\eta\cos\zeta + \cos\xi\sin\zeta & -\sin\xi\sin\eta\cos\zeta + \sin \\ -\cos\eta\sin\zeta & \sin\xi\sin\eta\sin\zeta + \cos\xi\sin\zeta & \cos\xi\sin\eta\sin\zeta + \sin\xi \\ \sin\eta & -\sin\xi\cos\eta & \cos\xi\cos\eta \end{bmatrix}$$

4.9. $\cos(x_0, y_3) = \sin\xi\sin\eta\cos\zeta + \cos\xi\sin\zeta$

4.10. Second kind

4.11. $\dfrac{\omega_0}{\Omega} = 364.438$; $\dfrac{C}{A} = 1.003$

Chapter 5

5.2. (a) $H^2 = 23125\,\mathrm{kg^2m^4/s^2}$
$2TA = 68325\,\mathrm{kg^2m^4/s^2}$
$2TB = 45550\,\mathrm{kg^2m^4/s^2}$
$2TC = 22775\,\mathrm{kg^2m^4/s^2}$

(b) $\omega_x = 5 = 7.6376\,\mathrm{cn}\,u_0 \approx 7.6376\,\mathrm{cn}\,u_0$
$\omega_y = 10 = -13.2288\,\mathrm{sn}\,u_0 \approx -13.2288\,\mathrm{sn}\,u_0$
$\omega_z = 150 = 150.333\,\mathrm{dn}\,u_0 \approx 150.333$

$$u_0 = -0.857\,068\,627$$

$$t_0 = 0.009\,874\,656\,\mathrm{s}$$

(c) $t - t_0 = 0.01 - t_0 = 0.000\,125\,344\,\mathrm{s}$
$u = 0.010\,879\,204$
$\omega_x = 7.637\,148$ rad/s
$\omega_y = -0.143\,915\,974$ rad/s
$\omega_z = 150.333$ rad/s

(d) $\alpha_0 = 84.399°$ $\quad \alpha = 81.334°$
$\beta_0 = 82.442°$ $\quad \beta = 89.894°$
$\gamma_0 = 9.462°$ $\quad \gamma = 8.666°$

5.3. (a) $t = 0$

$$\omega_x = 0.00\,74\,\mathrm{rad/s} \approx +0$$

$$\omega_y = 0.8\,\mathrm{rad/s}$$

$$\omega_z = 0.001\,92\,\mathrm{rad/s} \approx +0$$

(b) $t = 20\mathrm{s}$

$$\omega_x = 0.378\,\mathrm{rad/s}$$

$$\omega_y = 0$$

$$\omega_z = 0.98\,\mathrm{rad/s}$$

(c) $t = 30\mathrm{s}$

$$\omega_x = -0.799\,\mathrm{rad/s}$$

$$\omega_y = -0.799\,\mathrm{rad/s}$$

$$\omega_z = 0.06\,\mathrm{rad/s}$$

(d) $t = 40\mathrm{s}$

$$\omega_x = 0.00074\,\mathrm{rad/s}$$

$$\omega_y = -0.8\,\mathrm{rad/s}$$

$$\omega_z = 0.001\,92\,\mathrm{rad/s}$$

5.6. $\omega_x = 0.555$ rad/s when $\dot{\omega}_x = 0$
$\omega_x = 0.141$ rad/s when $\dot{\omega}_x = 0$
$\dot{\omega}_y = 0.215$ rad/s^2 when $\omega_y = 0$
$\omega_y = 0.803$ rad/s when $\dot{\omega}_y = 0$
$\dot{\omega}_z = 0.044$ rad/s^2 when $\omega_z = 0$
$\omega_z = 0.659$ rad/s when $\dot{\omega}_z = 0$

5.7. (a) $\mathbf{v} = [\mathbf{e}_x \ \mathbf{e}_y \ \mathbf{e}_z] \begin{bmatrix} 20 \\ 0 \\ 0 \end{bmatrix}$ m/s

(b) $\mathbf{a} = [\mathbf{e}_x \ \mathbf{e}_j \ \mathbf{e}_z] \begin{bmatrix} 79 \\ 187 \\ -125 \end{bmatrix}$ m/s^2

5.8. $a = 8.889$ rad/s ; $b = 5$ rad/s; $c = 3.2$ rad/s

5.9. $a = 1.333$ rad/s; $b = 1$ rad/s; $c = 0.8$ rad/s

5.10. (a) $\mathbf{H} = [\mathbf{e}_x \ \mathbf{e}_y \ \mathbf{e}_z] \begin{bmatrix} 180 \\ 640 \\ 1500 \end{bmatrix}$ Ws^2

(b) $H = 1641\,\text{Ws}^2$

(c) $T = 29\,800$ J

5.11. (a) $H = 25\,\text{Ws}^2$

(b) $T = 127.50$ J

5.12. (a) $A = 18\,\text{Ws}^3$; $B = 32\ \text{Ws}^3$; $C = 50\,\text{Ws}^3$

(b) $\mathbf{H}_1 = [\mathbf{e}_x \ \mathbf{e}_y \ \mathbf{e}_z] \begin{bmatrix} 180 \\ 640 \\ 1500 \end{bmatrix}$ Ws^2

$H = 1640.7\,\text{Ws}^2$

(c) $T_1 = 29\,800\,\text{Ws}$

(d) $\Delta T = 2880\,\text{Ws}$

5.14. (b) $r = 2.582\,\text{m}$
$l = 3.162\,\text{m}$

(c) $r = 3.651\,\text{m}$
$l = 3.162\,\text{m}$

(d) $a = b = 105.2\,\text{m}$
$c = 66.28\,\text{m}$

(e) $a = b = 16.67\,\text{m}$
$c = 8.33\,\text{m}$

Chapter 6

6.1. (a) $$[I] = \begin{bmatrix} 61.252 & 0 & 0 \\ 0 & 61.252 & 0 \\ 0 & 0 & 122.5 \end{bmatrix} \text{m}^2\text{kg}$$

(b) $$\mathbf{H} = [\mathbf{e}_u \ \mathbf{e}_v \ \mathbf{e}_z] \begin{bmatrix} 0 \\ 1837 \\ 4901 \end{bmatrix} \text{Ws}^2$$

(c) $T = 125.579\,40$ kJ

(d) 40 rad/s about z-axis

(e) 85.45 rad/s about **H**-axis

6.2. (a) $$[I] = \begin{bmatrix} 164\,800 & 0 & 0 \\ 0 & 164\,800 & 0 \\ 0 & 0 & 9\,600 \end{bmatrix} \text{m}^2\text{kg}$$

(b) $\dot{\sigma} = 0.014\,107\,8$ rad/s

(c) $\omega = 0.015\,056\,4$ rad/s

(d) $$\mathbf{H} = [\mathbf{e}_u \ \mathbf{e}_v \ \mathbf{e}_z] \begin{bmatrix} 0 \\ 249.078\,72 \\ 143.811\,84 \end{bmatrix} \text{Ws}^2$$

(e) $T = 1.265\,408\,2$ J

(f) $$\mathbf{v} = [\mathbf{e}_u \ \mathbf{e}_v \ \mathbf{e}_z] \begin{bmatrix} 0.015\,114 \\ 0 \\ 0 \end{bmatrix} \text{m/s}$$

(g) $\dot{\psi}' = 0.003\,548$ rad/s

(h) $\dot{\sigma}' = 0.011\,770\,4$ rad/s

(i) $$\mathbf{H}_1 = [\mathbf{e}_u \ \mathbf{e}_v \ \mathbf{e}_z] \begin{bmatrix} 0 \\ 33.855\,36 \\ 71.905\,92 \end{bmatrix} \text{Ws}^2$$

6.3. (b) $$\mathbf{H} = [\mathbf{e}_x \ \mathbf{e}_y \ \mathbf{e}_z] \begin{bmatrix} 6000 \\ 8000 \\ 48000 \end{bmatrix} \text{Ws}^2$$

(c) 12 rad/s

(d) 5 rad/s

(e) 24 rad/s

6.4. (a) $$[I] = \begin{bmatrix} 45 & 0 & 0 \\ 0 & 45 & 0 \\ 0 & 0 & 90 \end{bmatrix} \text{m}^2\text{kg}$$

(b) $\mathbf{H} = [\mathbf{e}_u \ \mathbf{e}_v \ \mathbf{e}_z]\begin{bmatrix} 0 \\ 900 \\ 900 \end{bmatrix} \mathrm{Ws}^2$

(c) $\nu = 45°$

(d) $\gamma = 63.435°$

(e) $\dot{\psi} = 28.28$ rad/s

(f) $\dot{\sigma} = -10$ rad/s

(g) $\omega_z' = 10$ rad/s

(h) $\mathbf{v} = [\mathbf{e}_x \ \mathbf{e}_y \ \mathbf{e}_z]\begin{bmatrix} 0 \\ 15 \\ -30 \sin 10t \end{bmatrix} \mathrm{m/s}$

6.5. $\dot{\psi} = 50.99$ rad/s $\qquad \nu = 78.690$ rad/s
$\dot{\sigma} = -4.99$ rad/s

6.6. Force $= \{\mathbf{e}\}^T \begin{bmatrix} 0 \\ 33\,671 \\ -19\,440 \end{bmatrix}$ N

6.7. (a) $[I] = \begin{bmatrix} 20\,000 & 0 & 0 \\ 0 & 20\,000 & 0 \\ 0 & 0 & 40\,000 \end{bmatrix} \mathrm{m^2kg}$

(b) $\begin{bmatrix} H_x \\ H_y \\ H_z \end{bmatrix} = \begin{bmatrix} 0 \\ 16 \\ 40 \end{bmatrix} \mathrm{Ws}^2$

(c) $\omega_z' = 10 \times 10^{-4}$ rad/s

(d) $\dot{\sigma} = -10 \times 10^{-4}$ rad/s

(e) $\nu = 24.22°$

(f) $\dot{\psi} = 21.54 \times 10^{-4}$ rad/s

6.8. (a) $\mathbf{H} = [\mathbf{e}_u \ \mathbf{e}_v \ \mathbf{e}_z]\begin{bmatrix} 0 \\ 3\,125 \\ 4\,500 \end{bmatrix} \mathrm{Ws}^2$

(b) 2.77778 rad/s

(c) 1.7123 rad/s

6.9. (a) $\dot{\sigma} = -\omega_z' = 8$ rad/s
$\sigma = 114.592$ rev

(b) $\dot{\psi} = 6.403$ rad/s
$\psi = 91.7$ rev

(c) $\nu = 51.34°$

(d) prolate $C/A = 1/3$.

6.10. $\mathbf{a} = [\mathbf{e}_u\ \mathbf{e}_v\ \mathbf{e}_z] \begin{bmatrix} -u\omega^2 \\ -v\omega_z^2 + \frac{C}{A} z\omega_v\omega_z \\ -z\omega_v^2 + \left(2 - \frac{C}{A}\right) v\omega_v\omega_z \end{bmatrix}$

6.11. $\mathbf{f} = [\mathbf{e}_u\ \mathbf{e}_v\ \mathbf{e}_z] \begin{bmatrix} 0 \\ 1535.30 \\ -95.66 \end{bmatrix} \mathrm{N}$

$f = 1538.3$ N

6.12. $a = -8.1\,\mathrm{m/s^2}$ in u-direction

Chapter 7

7.1. $[I] = \begin{bmatrix} 24.0175 & 0.268 & -0.2845 \\ 0.268 & 15.0226 & 0.0356 \\ -0.2845 & 0.0356 & 10.011 \end{bmatrix} \mathrm{Ws^3}$

7.2. $[I] = \begin{bmatrix} 24.0256 & -0.002 & -0.2872 \\ -0.002 & 24.0226 & 0.1256 \\ -0.2872 & 0.1256 & 10.012 \end{bmatrix} \mathrm{Ws^3}$

7.3. (a) $[I] = \begin{bmatrix} 40 & 0 & 0 \\ 0 & 40.000\,905 & 0 \\ 0 & 0 & 19.999\,095 \end{bmatrix} \mathrm{Ws^3}$

(b) $\xi = 0.286°$ $\eta = 0.258°$

(c) for $\lambda_1 = 40$:
$\alpha = 42°$
$\beta = 48°$
$\sigma = 90°$
for $\lambda_2 = 40.000\,905$:
$\alpha = 132°$
$\beta = 42°$
$\sigma = 89.615°$
for $\lambda_3 = 19.999\,095$:
$\alpha = 89.742°$
$\beta = 90.286°$
$\sigma = 0.385°$

7.4. (a) $[I'] = \begin{bmatrix} 40 & 0 & 0 \\ 0 & 40.000\,405 & 0 \\ 0 & 0 & 19.999\,595 \end{bmatrix} \mathrm{Ws^3}$

(b) $\xi = 0°$ $\eta = 0.258°$

(c) for $\lambda_1 = 40$:
$\alpha = 90°$
$\beta = 0°$
$\sigma = 90°$
for $\lambda_2 = 40.000\,405$:
$\alpha = 179.742°$
$\beta = 90°$
$\sigma = 89.742°$
for $\lambda_3 = 19.999\,595$:
$\alpha = 89.742°$
$\beta = 90°$
$\sigma = 0.258°$

7.5. (a) $[I] = \begin{bmatrix} 84 & 0 & 0 \\ 0 & 84 & 0 \\ 0 & 0 & 120 \end{bmatrix} \text{Ws}^3$

(b) $[I] = \begin{bmatrix} 84.006 & 0 & 0 \\ 0 & 84 & 0.12 \\ 0 & 0.12 & 120.006 \end{bmatrix} \text{Ws}^3$

(c) $\xi = -0.19°$

(d) $\dfrac{\Delta A}{A} = 0.000\,007 \qquad \dfrac{\Delta B}{A} = 0 \qquad \dfrac{\Delta C}{C} = 0.000\,005$

(e) $J_{vz} = 0$
$J_{zv} = 0.12 \text{ Ws}^3$

(f) $Y_{vz} = 0.17 \text{ Ws}^3$

7.6. (a) $[I] = \begin{bmatrix} 15\,714 & 0 & 0 \\ 0 & 15\,714 & 0 \\ 0 & 0 & 720 \end{bmatrix} \text{Ws}^3$

(b) $[I] = \begin{bmatrix} 15714.0006 & 0 & 0 \\ 0 & 15\,714 & 0.12 \\ 0 & 0.12 & 720.0006 \end{bmatrix} \text{Ws}^3$

(c) $\xi = 0.000\,458°$

(d) $\dfrac{\Delta A}{A} = 0.000\,000\,038 \qquad \dfrac{\Delta B}{A} = 0 \qquad \dfrac{\Delta C}{C} = 0.000\,000\,833$

(e) $J_{vz} = 0$
$J_{zv} = 0.12 \text{ Ws}^3$

(f) $Y_{vz} = 0.045\,819 \text{ Ws}^3$

7.7. (a) $[I] = \begin{bmatrix} 15\,690.14 & 0 & 0 \\ 0 & 15\,690.14 & 0 \\ 0 & 0 & 740.28 \end{bmatrix} \text{Ws}^3$

(b) $[I] = \begin{bmatrix} 15\,690.1406 & 0 & 0 \\ 0 & 15\,690.1406 & 0.078 \\ 0 & 0.078 & 740.28 \end{bmatrix} \text{Ws}^3$

(c) $\xi = 0.000\,2990$

(d) $\dfrac{\Delta A}{A} = 3.824\,(10^{-8}) \qquad \dfrac{\Delta B}{A} = 3.824\,(10^{-8}) \qquad \dfrac{\Delta C}{C} = 0$

(e) $J_{vz} = 0.078 \text{ Ws}^3$
$J_{zv} = 0$

(f) $Y_{vz} = 0.152 \text{ Ws}^3$

7.8. (a) $[I] = \begin{bmatrix} 15\,714.14 & 0 & 0 \\ 0 & 15\,714.14 & 0 \\ 0 & 0 & 740.28 \end{bmatrix} \text{Ws}^3$

(b) $[I] = \begin{bmatrix} 15\,714.142\,136 & 0 & 0 \\ 0 & 15\,714.141\,536 & 0.2448 \\ 0 & 0.2448 & 740.2806 \end{bmatrix} \text{Ws}^3$

(c) $\xi = 0.000\,9370$

(d) $\dfrac{\Delta A}{A} = 1.359\,(10^{-7}) \qquad \dfrac{\Delta B}{A} = 0.977\,(10^{-7}) \qquad \dfrac{\Delta C}{C} = 8.105\,(10^{-7})$

(e) $J_{vz} = 0.1248 \text{ Ws}^3$
$J_{zv} = 0.12$

(f) $Y_{vz} = 0.249 \text{ Ws}^3$

7.9. (a) $[I] = \begin{bmatrix} 15\,714.14 & 0 & 0 \\ 0 & 15\,714.14 & 0 \\ 0 & 0 & 740.28 \end{bmatrix} \text{Ws}^3$

(b) $[I] = \begin{bmatrix} 15\,714.142\,358 & 0 & -0.0588 \\ 0 & 15\,714.141\,752 & 0.2448 \\ -0.0588 & 0.2448 & 740.286\,006 \end{bmatrix} \text{Ws}^3$

(c) $\xi = 0.000\,937° \qquad \eta = 0.000\,225°$

(d) $\dfrac{\Delta A}{A} = 1.500\,(10^{-7}) \qquad \dfrac{\Delta B}{A} = 1.115\,(10^{-7}) \qquad \dfrac{\Delta C}{C} = 81.131\,(10^{-7})$

(e) $J_{vz} = 0.1248 \text{ Ws}^3$
$J_{zv} = 0.12 \text{ Ws}^3$
$J_{uz} = -0.0468 \text{ Ws}^3$
$J_{zu} = -0.012 \text{ Ws}^3$

(f) $Y_{vz} = 0.249 \text{ Ws}^3$
$Y_{uz} = -2.092 \text{ Ws}^3$

7.10. (a) $\boldsymbol{\omega} = [\mathbf{e}_u \ \mathbf{e}_v \ \mathbf{e}_z] \begin{bmatrix} 0 \\ 2 \\ 5 \end{bmatrix} \text{rad/s}$

(b) $\mathbf{H} = [\mathbf{e}_u\ \mathbf{e}_v\ \mathbf{e}_z] \begin{bmatrix} 0 \\ 4.764 \\ 23.818 \end{bmatrix} \mathrm{Ws}^2$

(c) $\boldsymbol{\Omega} = [\mathbf{e}_u\ \mathbf{e}_v\ \mathbf{e}_z] \begin{bmatrix} 0 \\ 2 \\ 0 \end{bmatrix}$ rad/s

(d) $\mathbf{M} = [\mathbf{e}_u\ \mathbf{e}_v\ \mathbf{e}_z] \begin{bmatrix} 47.636 \\ 0 \\ 0 \end{bmatrix}$ Nm

(e) $\dot{U} = 0$

(f) $\dot{U} = 0.1$ W

7.11. (a) $F = 2m\,\omega_v\omega_z l \sin\omega t \qquad \omega = \omega_z$

(b) $m\ddot{e} + c\dot{e} + ke = F_0 \sin\omega t \qquad F_0 = 2m\omega_v\omega l$

(c) $\tan\phi = \dfrac{2\zeta_r}{1-r^2}$ with $r = \dfrac{\omega}{\omega_n} \qquad \zeta = \dfrac{c}{2\sqrt{km}}$

(d) $I_{vz} = \dfrac{1-r^2}{(1-r^2)+(2\zeta r)^2}\dfrac{4m^2l^2}{k}\omega_v\omega_z$

(e) $I_{uz} = \dfrac{2\zeta r}{(1-r^2)^2+(2\zeta r)^2}\dfrac{4m^2l^2}{k}\omega_v\omega_z$

Chapter 8

8.1. (a) $T_0 = 123.370055$ J

(b) $\omega_1 = 0.049087$ rad/s

(c) $T_1 = 19.276539$ J

(d) $\Delta T = 104.093517$ J (lost)

8.2. (a) $[I] = \begin{bmatrix} 4000 & 0 & 0 \\ 0 & 4000 & 0 \\ 0 & 0 & 800 \end{bmatrix} \mathrm{m}^2\mathrm{kg}$

(b) $\omega_v = 5$ rad/s

(c) $\nu_0 = 64.359°$

(d) $T_0 = 107\,600$ J

(e) $T_f = 61\,577$ J

(f) $\Delta t = 1.46$ years

8.5. (a) 108 Nms

(b) 46.477 J

(c) -0.340 rad/s

(d) 1.079 rad/s

(e) 0.551 rad/s

(f) 29.729 J

(g) -0.528 rad/s

(h) 1.079 rad/s

8.6. (a) $H_0 = 47.50\ \mathrm{Ws}^2$

(b) $T = 27.807$ J

(c) $\dot{\sigma} = -0.409$ rad/s

(d) $\dot{\psi} = 1.484$ rad/s

(e) $\omega_1 = 0.95$ rad/s

(f) $T_1 = 22.563$ J

(g) $T_0 - T_1 = 5.244$ J

(h) $\dot{\sigma}_1 = -0.534$ rad/s

(i) $\dot{\psi}_1 = 1.484$ rad/s

8.8. (a) $[I] = \begin{bmatrix} 1300 & 0 & 0 \\ 0 & 1300 & 0 \\ 0 & 0 & 2500 \end{bmatrix} \mathrm{Ws}^3$

(b) $\omega = 13$ rad/s $\qquad \omega_v = 5$ rad/s

(c) $\mathbf{H} = [\mathbf{e}_u\ \mathbf{e}_v\ \mathbf{e}_z] \begin{bmatrix} 0 \\ 6500 \\ 30\,000 \end{bmatrix} \mathrm{Ws}^2$

(d) $\nu = 12.225°$

(e) $T_0 = 196\,250$ J

(f) $T_f = 188\,450$ J

(g) 90.28 days

8.9. (a) $\omega'_z = 27.143$ rad/s
$\omega_v = 32.991$ rad/s
$\omega = 44.614$ rad/s
$\dot{\sigma} = -27.143$ rad/s
$\dot{\psi} = 65.982$ rad/s
$H_v = 3464.1\ \mathrm{Ws}^2$
$H_z = 6000\ \mathrm{Ws}^2$
$H = 6928.2\ \mathrm{Ws}^2$
$T = 147\,141$ J

(b) $\omega_z = 34.641$ rad/s
$\omega_z' = 31.341$ rad/s
$\omega_v = 0$
$\omega = 34.641$ rad/s
$\dot{\sigma} = -31.341$ rad/s
$\dot{\psi} = 65.982$ rad/s
$H_v = 0$
$H = H_z = 6928.2\,\text{Ws}^2$
$T = 120\,000$ J

8.11. $\mu = 19.47°$

8.12. $\mu = -41.81°$

8.13. (a) $\omega_n = 100$ rad/s

(b) $\zeta = 0.5$

(c) $$[I] = \begin{bmatrix} 1859.12 & 0 & 0 \\ 0 & 1859.12 & 0 \\ 0 & 0 & 118.24 \end{bmatrix} \text{Ws}^3$$

(d) $\omega_v = 3.227$ rad/s
$\omega_z = 87.891$ rad/s
$\dot{\psi} = 6.455$ rad/s
$\dot{\sigma} = 82.301$ rad/s

(e) $F = 878.736$ N
$\delta = 0.099\,406$ m

(f) $J_{vz} = 0.116\ \text{Ws}^3$
$J_{zv} = 0$
$J_{uz} = -0.296\ \text{Ws}^3$
$J_{zu} = 0$
$\phi = 68.59°$

(g) $V = 99$ J

(h) $\Delta\nu = 0.007645°$

(i) $\nu = 0.047$ rad/s

(j) $\dot{D} = 13\,386$ W

(k) $T_{rigid} = 466\,380$ J
$T_{def} = 466\,281$ J

8.14. (a) $\omega_n = 82.3$ rad/s

(b) $\zeta = 0.5$

(c) $$[I] = \begin{bmatrix} 1859.12 & 0 & 0 \\ 0 & 1859.12 & 0 \\ 0 & 0 & 118.24 \end{bmatrix} \text{Ws}^3$$

(d) $\omega_v = 3.227$ rad/s
$\omega_z = 87.891$ rad/s
$\dot{\sigma} = 82.301$ rad/s
$\dot{\psi} = 6.455$ rad/s

(e) $F = 878.736$ N

(f) $J_{vz} = J_{zv} = 0$
$J_{uz} = -0.4152$ Ws^3
$J_{zu} = 0$
$\phi = 90°$

(g) $V = 114$ J

(h) $\Delta\nu = 0.013\,282°$

(i) $\dot{\nu} = 0.019\,629$ rad/s

(j) $\dot{D} = 9692$ W

(k) $T_{rigid} = 466\,380$ J
$T_{def} = 466\,266$ J

8.15. (a) $\omega_n = 60$ rad/s

(b) $J = 0.5$

(c) $$[I] = \begin{bmatrix} 1859.12 & 0 & 0 \\ 0 & 1859.12 & 0 \\ 0 & 0 & 118.24 \end{bmatrix} \text{Ws}^3$$

(d) $\omega_v = 3.227$ rad/s
$\omega_z = 87.891$ rad/s
$\dot{\sigma} = 82.301$ rad/s
$\dot{\psi} = 6.455$ rad/s

(e) $F = 878.736$ N

(f) $J_{vz} = -0.1982$ Ws^3
$J_{zv} = 0$
$J_{uz} = -0.4301$ Ws^3
$J_{zu} = 0$
$\phi = 114.7°$

(g) $V = 79$ J

(h) $\Delta\nu = 0.0156°$

(i) $\dot{\nu} = 0.02$ rad/s

(j) $\dot{D} = 9950$ W

8.17. (a) $\omega_n = 100$ rad/s

(b) $\zeta = 0.01$

(c) $$[I] = \begin{bmatrix} 1886 & 0 & 0 \\ 0 & 1886 & 0 \\ 0 & 0 & 108 \end{bmatrix} \text{m}^2\text{kg}$$

(d) $\omega_v = 3.181$ rad/s
$\omega_z = 96.225$ rad/s
$\dot{\sigma} = 90.715$ rad/s
$\dot{\psi} = 6.363$ rad/s

(e) $F = 70.112$ N
$\delta = 0.007\,010$ m

(f) $\phi = 0.009\,425°$
$J_{zv} = 0.056$ Ws^3
$J_{vz} = 0$
$J_{uz} = 0$
$J_{zu} = -0.000\,009\,225$ Ws^3

(g) $V = 70$ J

(h) $\Delta\nu = 0.007\,277°$

(i) $\dot{\nu} = 26.95\,(10^{-9})$ rad/s $= 0.133°$/day

(j) $\dot{D} = 0.014\,716$ W

(k) $T_{rigid} = 509\,544$ J
$T_{def} = 509\,474$ J

8.18.

$$\omega_n = 1.2 \text{ rad/s}; \quad \omega_z = 0.6 \text{ rad/s}; \quad r_0 = 0.5$$

for stability

$$C > A + 2ml^2 \frac{r_0^2}{1 - r_0^2}; \quad 1800 \not> 1750 + 80$$

not satisfied, unstable!

Chapter 9

9.1. (a) 31.66 m
(b) $F = 31\,142$ N

9.2. (a) $l = 21.287$ m
(b) $t = 10.164$ s

9.3. 549.3 m

9.4. (a) $H_x = K_x + I_x\omega_x + I_{xy}\omega_y + I_{xz}\omega_z$
$H_y = K_y + I_{xy}\omega_x + I_y\omega_y + I_{yz}\omega_z$
$H_z = K_z + I_{xz}\omega_x + I_{yz}\omega_y + I_z\omega_z$

(b) $M_x = \dot{H}_x + \omega_y H_z - \omega_z H_y$
$M_y = \dot{H}_x + \omega_y H_z - \omega_z H_y$
$M_y = \dot{H}_y + \omega_z H_x - \omega_x H_z$
$M_z = \dot{H}_z + \omega_x H_y - \omega_y H_y$

9.5. (a) $\omega = \dfrac{C\omega_0}{C + \frac{2}{3}\frac{m}{l}x^2}$

(b) $\omega = \dfrac{C\omega_0}{C + \frac{2}{3}ml^2}$

9.7. (a) $M = m(x\dot{\omega} + 2\dot{x}\omega - e\omega^2)(x - b)$

(b) $e = \dfrac{\dot{x}}{\omega_0}\dfrac{2(C + me^2)}{C + 2m(e^2 + b^2)}$ = constant if $\dot{x}$ = const.

Chapter 10

10.1. (a) $H_{0_v} = 10\,000$ m kg/s
$H_{0_z} = 17\,320$ m kg/s

(b) $\omega_{0_v} = 10$ rad/s
$\omega_{0_z} = 8.66$ rad/s

(c) $|\omega_0| = 13.23$ rad/s

(d) $\omega'_{0_z} = 8.66$ rad/s

(h) $H_{f_v} = 10\,000$ m kg/s
$H_{fz} = 27\,472$ m kg/s

(i) $|\mathbf{H}| = 29\,235$ Ws^2

(j) $\omega_{f_v} = 10$ rad/s

(k) $\omega_f = 17$ rad/s

(l) $\omega'_{fz} = 13.736$ rad/s

(m) $\omega'_{fz} = 11.20$ rad/s

(o) 0.94 Nm

10.2. (a) $\omega_{fv} = \sqrt{\omega_{x0}^2 + 2\omega_{x0}\dfrac{M}{A\omega'_{z0}} + 4\dfrac{M^2}{A^2\omega'_{z0}}}$

(b) Nothing has changed

10.3. $\omega_{z0} = 0.4$ rad/s
$\omega'_{z0} = 0.2$ rad/s
$R = 1$ rad/s
$\omega_x = 0.809$ rad/s
$\omega_y = 0.412$ rad/s
$\omega_z = 0.4$ rad/s

Chapter 11

11.1. (a) $\mathbf{H} = [\mathbf{e}_u\ \mathbf{e}_v\ \mathbf{e}_z]\begin{bmatrix} 0 \\ 59.787 \\ 50.161 \end{bmatrix}$ Ws^2

(b) $\mathbf{H}^{\mathrm{P}} \doteq [\mathbf{e}_u\ \mathbf{e}_v\ \mathbf{e}_z]\begin{bmatrix} 0 \\ 56.775 \\ 1.461 \end{bmatrix}$ Ws^2

(c) $\mathbf{H}^{\mathrm{R}} = [\mathbf{e}_u\ \mathbf{e}_v\ \mathbf{e}_z]\begin{bmatrix} 0 \\ 3.028 \\ 48.70 \end{bmatrix}$ Ws^2

(d) $\boldsymbol{\omega} = [\mathbf{e}_u\ \mathbf{e}_v\ \mathbf{e}_z]\begin{bmatrix} 0 \\ 0.757 \\ 0.487 \end{bmatrix}$ rad/s

(e) $\boldsymbol{\omega}^{\mathrm{R}} = [\mathbf{e}_u\ \mathbf{e}_v\ \mathbf{e}_z]\begin{bmatrix} 0 \\ 0.757 \\ 48.70 \end{bmatrix}$ rad/s

(f) $T = 1208.8$ J

(g) $T^{\mathrm{P}} = 21.8$ J

(h) $T^{\mathrm{R}} = 1187$ J

(i) $\dot{\sigma} = -0.148$ rad/s

(j) $\dot{\sigma}^{\mathrm{R}} = 48.065$ rad/s

(k) $\dot{\psi} = 0.988$ rad/s

(l) $\dot{\psi}^{\mathrm{R}} = 0.988$ rad/s

11.2. (a) $\omega_z' = 0.148$ rad/s

(b) 0.635 rad/s

(c) $F = 87.358$ N

11.3. (a) $\mathbf{f}^{\mathrm{P}} = [\mathbf{e}_u\ \mathbf{e}_v\ \mathbf{e}_z]\begin{bmatrix} 50.405 \\ -15.9 \\ 0.81 \end{bmatrix}$ N

(b) $\mathbf{f}^{\mathrm{P}} = [\mathbf{e}_u\ \mathbf{e}_v\ \mathbf{e}_z]\begin{bmatrix} 0 \\ 34.1 \\ -0.24 \end{bmatrix}$ N

Chapter 12

12.1. (a) $\Delta\nu = -10 = -0.017453$ rad
$\Delta\omega_v = -0.000\,377$ rad

(b) $\Delta\omega_z = 0.002\,181$ rad

(c) $\Delta\omega_z^{\mathrm{R}} = 0$

(d) $\Delta T^{\mathrm{R}} = -0.047$ mJ

(e) $\Delta T^{\mathrm{P}} = -3.518$ mJ

(f) $\Delta D = 3.565$ mJ

(g) $\dot{\nu} = 0.101\,(10^{-6})$ rad/s

(h) $Y_{uz}^{\mathrm{P}} = 0.019$ Ws^3

(i) $M_u^{\mathrm{R}} = 0.261$ Nm
$M_v^{\mathrm{R}} = 0.000\,005$ Nm

(j) $\beta^{\mathrm{P}} = 17\,701$ Ws^6

(k) $T = 44.695$ J

(l) $\Delta\omega_v = -0.000\,377$ rad/s
$\Delta\omega_z = 0.002\,181$ rad/s
$\Delta\omega_z^{\mathrm{R}} = 0$
$\Delta T^{\mathrm{R}} = -0.047$ mJ
$\Delta T^{\mathrm{P}} = 0.027$ mJ
$\Delta D = 0.020$ mJ
$\dot{\nu} = -0.101\,(10^{-6})$ rad/s

12.4. (a) $\omega_v = 3.68$ rad/s

(b) stable, since $A < C' < Ax$ and $\eta < 0$

(c) -18.52 W

(d) 18.36 W

(e) -9.86 W

(f) 10.02 W

(g) -0.16 W means the energy flow is out of the rotor, into the platform

12.5. (a) $0.828\ \omega$

(b) $0.171572\ C\omega^2$

(c) 32.8%

12.6. (a) $0.166\ \omega$

(b) 46.6%

12.7. (a) $C\omega$

(b) $\frac{1}{2}C\omega^2 + \frac{1}{2}C\omega^2(1-\cos\nu)^2 + \frac{C^2\omega^2}{4A}\sin^2\nu$

(c) $\boldsymbol{\omega}$

(d) 0

(e) $\frac{1}{2}C\omega^2$

(f) $(1-\cos\nu)^2 + \frac{C}{2A}\sin^2\nu$

(g) 26.8%

(h) 8%

(i) 1.8%

Appendix B

B.1. (a)
$$\begin{matrix} 0.985 & 0.173 & 0.012 \\ -0.174 & 0.982 & 0.069 \\ 0 & -0.070 & 0.998 \end{matrix}$$

(b) $\epsilon_1 = 0.035 \quad \epsilon_2 = -0.003$
$\epsilon_3 = 0.087 \quad \epsilon_4 = 0.0996$

(c) $\lambda = [\mathbf{e}_x\ \mathbf{e}_y\ \mathbf{e}_z]\begin{bmatrix} 0.371 \\ -0.032 \\ 0.925 \end{bmatrix}$

(d) $\theta = 10.8°$

Appendix C

C.1. (a) $X = 3$ m
$Y = -1$ m
$Z = 2$ m

(b) $\eta = -90°$
$\zeta = 0°$
$\xi = 180°$

C.2. (a) $\gamma = 12.78°$

(b) $\gamma = 90°$

C.3. (a) $\xi = 15°$
$\eta = 0°$
$\zeta = 30°$

(b) $\dot{\xi} = 0$

$\dot{\eta} = -0.517$ rad/s

$\dot{\zeta} = 3.932$ rad/s

C.4. (a) $\tan \eta_2 = \dfrac{\sin \nu \sin \sigma}{\sin \psi \cos \nu \sin \sigma - \cos \psi \cos \sigma}$

$\sin \zeta_2 = \sin \psi \cos \sigma + \cos \psi \cos \nu \sin \sigma$

$\tan \xi_2 = \dfrac{\sin \nu}{\cos \nu \cos \sigma - \tan \psi \sin \sigma}$

(b) $\tan \zeta_3 = \dfrac{\tan \sigma + \tan \psi \cos \nu}{\cos \nu - \tan \psi \tan \sigma}$

$\sin \xi_3 = \sin \nu \cos \sigma$

$\tan \eta_3 = -\tan \nu \sin \sigma$

C.5. Second kind

Author Index

Annett, R. 293, 305
Arnold, R.N. 1, 27

Bauer, H.F. 65
Beletskii, V.V. 85
Bianchi, G. 306
Blanton, J.N. 132
Buckens, F. 175

Canavin, J.R. 169, 198
Cardano, G. 87, 329
Cherchas, D.B. 155, 160
Chesser, H.C. 27, 178, 198
Cochran, J.E. 305

DeBra, D.B. 24, 27
deCoriolis, G. 153
Delp, R.H. 24, 27
Doetsch, K.H. 156, 160
Dong, W.N. 230, 233
Duffing, G. 112

Elias, J.H. 86
Emde, F. 132
Euler, L. 1, 13, 14, 15, 311, 321

Faulkner, M.G. 306

Giacaglia, G.E.O. 85
Goldeich, P. 85
Görtler, H. 101
Gossain, D.M. 155, 160
Graham, J.D. 155, 160
Grammel, R. 256, 262
Greenwood, D.T. 27

Hagedorn, P. 265, 287
Halfman, L. 65
Hamilton, W.R. 230
Hetényi, M. 65
Hughes, P.C. 1, 27, 100, 155, 160
Huygens, C. 8

Iorillo, A.J. 305
Ishlinskij, A.Y. 1, 27

Jacobi, C.G.J. 106
Jacobson, D. 132
Jahnke, E. 108, 132
Jarmolow, K. 256, 262
Johnson, T.V. 86
Junkins, J.L. 115, 132

Kane, T.R. 1, 27, 94, 100, 273, 287
Kaplan, M.H. 1, 27, 287
Kelvin, W.T. 264
Kepler, J. 40, 46
Kim, Y.I. 28
Kohler, P. 85
Kolbe, O. 65

Landon, V.D. 207, 233, 293, 305
Leech, C.M. 28
Leipholz, H.H.E. 230, 233
Levinson, D.A. 27
Li, Y.S. 299, 306
Liapunov, A.M. 230
Likins, P.W. 27, 132, 169, 198, 269, 287, 293, 306
Ling, F.W. 189, 199, 291, 306
Liu, Y.Z. 1, 27
Lohmeier, P. 129, 132
Lösch, F. 132

MacCullagh, J. 118, 129, 224
MacNaughton, J.D. 155, 160, 253
Magnus, K. 1, 22, 27, 98, 100, 127, 132, 184, 198, 287
Marsh, E.L. 100
Maunder, L. 1, 27
McCord, T.B. 86
Meirovitch, L. 230, 234
Misra, A.K. 86
Modi, V.J. 86

Newton, I. 1

Okhotsimsky, D.E. 101

Peale, S.J. 85
Poinsot, L. 115
Poisson, S.D. 327

Rasmussen, H. 27
Rimrott, F.P.J. 27, 83, 86, 123, 132, 178, 189, 198, 234, 253, 299, 301, 302, 306
Robe, T.R. 1, 27
Roberson, R.E. 273, 287, 306
Rodrigues, O. 328

Sarychev, V.A. 101
Schiehlen, W. 65, 287, 306
Schlack, A.L. Jr. 230, 233
Shackcloth, W.J. 305
Sherman, B.C. 155, 160
Smelt, R. 24
Steiner, J. 8
Stewart, B. 293, 305

Tabarrok, B. 1, 28
Thomson, W.T. 1, 28, 95, 101
Tisserand, F. 175, 177, 198
Tonkin, S.W. 305
Torzhevsky, A.P. 101
Truesdell, C.A. 1, 28
Tyc, G. 96, 101

Vigneron, F. 287
Vincenti, W.G. 65

Wagner-Bartak, C.G. 156, 160
Wilson, W.G. 100
Wittenburg, J. 1, 28, 87, 101, 287

Xu, D.M. 86

Yu, Y. 189, 198, 234, 306

Zhang, W. 189, 198, 199, 291, 306
Zlatoustov, V.A. 99, 101

Subject Index

Angular momentum 1
Angular momentum ellipsoid 117
Angular momentum law 1
Angular velocity 18
Apparent libration 83
Approach 156
Attitude 40, 87
Attitude diagram 130, 150, 152
Attitude drift 205, 209
Attitude matrix 324
Attitude stability 83, 95, 98, 122, 205, 268
Auxiliary frequency 140, 270
Auxiliary inertia moments 170
Auxiliary inertia products 170
Axisymmetric gyros 139
Axisymmetric gyrostats 270

Bearing friction 273
Body cone 146
Boom extension 244
Booms 59

Capture 156
Cardan angles 87, 329
Carrier 264
Center of gravity 40
Center of mass 40
Central force field 87, 281
Centrifugal moment 5
Changing inertia moments 154
Characteristic equation 10, 93
Circular orbit 91
Collinearity theorems 131
Complete elliptic integral 108
Configuration change 249
Constant configuration 170
Coordinate transformation 17
Coriolis force 153, 187

Damping ratio 183, 189
Deformable 124
Deformable gyros 166
Degeneracy 19
Despin 240
Deviation moment 5
Direct precession 146
Dissipation energy 132
Dissipation ratio 302
Dissipative gyros 198
Dissipative gyrostats 290
Dissipative platform 292, 299
Dissipative rotor 297, 299
Dissipative solid bodies 198
Diurnal libration 85
Drift rate 205
Dual spinners 265
Duffing equation 112

Eccentric booms 244
Effective dissipation ratio 302
Eigenvalue 9
Elastic deformation 182, 187, 225
Elastic deformation coefficient 182, 187
Elastic energy 132
Elliptic functions 106
Elliptical orbit 98
End effector 156
Energy dissipation rate 132, 186, 205, 297, 299, 302
Epicycloidal motion 106
Euler angles 15, 311
Euler equations 13
Euler frequencies 15
Euler parameters 321
Euler's law 1
Excess angular velocity 267

Floating coordinates 140, 169, 273
Frame rate of change 4, 5, 14
Frequency ratio 183, 189

Grapple fixture 156
Gravitational attraction constant 40
Gravitational force 40, 52
Gravity center 40
Gravity gradient 59, 62
Gravity gradient stabilization 83
Guide rate of change 4
Gyric matrix 95
Gyro 1, 265
Gyroscope 1, 265
Gyroscopic stability 95, 98
Gyrostats 265

Huygens–Steiner parallel axis
theorem 8

Inertia ellipsoid 117
Inertia moment 5
Inertia product 5
Inertia tensor 5
Inertial force field 153, 278
Integrability 20
Interdependent constants 112
Internal angular momentum 249
Internal energy dissipation
coefficient 183, 192, 208, 211
Internal torque 273, 293, 297, 298,
299, 300, 303

Jacobi elliptic functions 106

Kepler force 40, 52, 87
Kepler period 40
Kinematic equations 327
Kinetic energy 11
Kinetic energy ellipsoid 115

Libration 76
Line of nodes 15, 16, 19
Locked rotation 41, 55, 76, 95, 264
Long run (attitude stability) 124

MacCullagh ellipsoid 118
Magnus shape triangle 22, 97, 123
Mass center 40, 95
Maximum inertia moment 123
Microgravitational attraction
coefficient 64
Microgravity 63
Microweight force 64
Minimum inertia moment 123
Modified inertia product 173, 301
Moment of inertia 5
Momentum wheel 264
Moon 83

Natural frequency 183
Near-rigid solid 167
Node line 15, 16, 19
Nutation 15

Oblate 117
Ogives 210, 212
Orbital period 40, 55

Parallel axes theorem 8
Parametric coefficients 100
Partial inertia products 173
Pericycloidal motion 108
Period 40
Phase plane 113
Pitch 338
Platform 264
Poinsot ellipsoid 115
Poisson's kinematic equations 327
Polar inertia moment 5
Potential 54
Practical stability 123
Precession 15
Principal axes 10, 14, 123
Principal inertia moments 7, 8, 9,
344
Product of inertia 5
Prolate 117

Quasi-locked rotation 98

Reduced inertia moment 267
Relative rate 5, 13
Remote manipulator system 156
Retrograde precession 146
Rigid 122
Rigid platform 297
Rigid rotor 292
Rigidification 131
Rigidization 156
RMS 156
Rod satellite 41
Rodrigues parameters 328
Rodrigues vector 328
Roll 338
Rotation matrix 320
Rotation theorem 324
Rotor 264

Satellite 40
Secular attitude drift 205, 292, 297, 299
Self-excited gyros 256
Shape petal 27
Shape triangle 22
Short run 123, 268
Shuttle 156
Singularity 19
Smelt parameters 24
Somersault 338
Space cone 146
Spacecraft 59
Spin 15, 183, 189
Spin change allotment factor 299
Stability 95, 98, 122, 124, 205, 268
Static stability 95, 98
Steiner parallel axes theorem 8
STEM tubing 83
Stiffness matrix 95

TEE tubing 83
Tethered bodies 85
Theta functions 135
Tilt 338
Tisserand-type frame 178
Torque 1, 53, 273, 293, 297, 298, 299, 300, 303
Torquefree gyros 105
Transition matrix 89, 318
True libration 83
Twist 338

Wheel 264

Yaw 338
Yo-yo masses 240